History of Computing

The History of Computing series publishes high-quality books which address the history of computing, with an emphasis on the 'externalist' view of this history, more accessible to a wider audience. The series examines content and history from four main quadrants: the history of relevant technologies, the history of the core science, the history of relevant business and economic developments, and the history of computing as it pertains to social history and societal developments.

Titles can span a variety of product types, including but not exclusively, themed volumes, biographies, 'profile' books (with brief biographies of a number of key people), expansions of workshop proceedings, general readers, scholarly expositions, titles used as ancillary textbooks, revivals and new editions of previous worthy titles.

These books will appeal, varyingly, to academics and students in computer science, history, mathematics, business and technology studies. Some titles will also directly appeal to professionals and practitioners of different backgrounds.

More information about this series at http://www.springer.com/series/8442

William Aspray

Women and Underrepresented Minorities in Computing

A Historical and Social Study

 Springer

William Aspray
Department of Information Science
University of Colorado Boulder
Boulder, CO, USA

ISSN 2190-6831   ISSN 2190-684X   (electronic)
History of Computing
ISBN 978-3-319-79681-9   ISBN 978-3-319-24811-0   (eBook)
DOI 10.1007/978-3-319-24811-0

Printed on acid-free paper

This Springer imprint is published by Springer Nature
The registered company is Springer International Publishing AG Switzerland

Preface

This book is the second of two books written by this author – and published in the *Springer History of Computing* series – on the history of underrepresentation in the computing disciplines in the United States. These two books can be read entirely independently of one another. The first book, entitled *Participation in Computing*, discusses the role of the National Science Foundation in this effort. This second book presents a more general account of women, African Americans, American Indians, and Hispanics in the computing disciplines in the United States over time.

The writing of these two books was stimulated by a recent grant program of the Alfred P. Sloan Foundation to better understand the underlying causes for the persistent underrepresentation of women in the computing disciplines. African Americans, American Indians, and Hispanics have also been persistently underrepresented in the computing disciplines despite numerous efforts to correct this problem; and thus, this book considers all of the main demographic groups that are underrepresented in computing in the United States.

The author of this book is principally a historian of science and technology, and this work is primarily a history – with particular attention to the formal education that prepares individuals from these underrepresented demographic groups for computing careers. However, one way in which this book differs from the existing historical literature is that it pays considerable attention to the social science literature, which it employs to bolster historical understanding.

Organizational Structure of the Book

This book has two main sections, which can be read entirely separately from one another. The first section presents a digest of relevant literatures, while the second section contains historical case studies.

There are many types of scholars interested in underrepresentation in computing. These include computer scientists, social scientists studying science and technology, education faculty and psychologists studying the learning of science and technology, race and gender scholars, education historians, policy scholars, and historians of computing. The relevant literature on this topic extends across all of these academic literatures. While any interested scholar is probably already familiar with the relevant literature from her own academic discipline, she may have less familiarity with the literatures from the other relevant academic disciplines.

The first section of the book digests these various relevant literatures on behalf of the interested reader. A computer scientist interested in this topic might be willing to read ten pages on the history of women in American science, but he is unlikely to read all of Margaret Rossiter's comprehensive three-volume study of this topic. This first section of the book provides a historical overview, drawing from some of the leading historical and social science sources concerning: college education for women, science and engineering education for women, and higher education as well as STEM (science, technology, engineering, and mathematics) education and careers for African Americans, Hispanics, and American Indians. The value added by this section thus comes from the digesting of a large set of disparate literatures, not from new historical scholarship.

The second section of the book serves a different function. It provides new historical case studies – first about organizations interested in broadening participation in the STEM disciplines generally and in computing in particular and then about college and university departments of computer science and engineering that have had success in attracting, retaining, and advancing women in STEM and computing careers.

Many women and underrepresented minorities who have sought to enter the computing field have received support from various nonprofit organizations. From the 1950s to the 1970s, these organizations were focused on the STEM disciplines broadly, not specifically on computing. While these early organizations continued to operate in the 1980s and beyond, they were joined in the 1980s by organizations specifically focused on computing. Both the early and later organizations are often focused on a single demographic group: women (SWE, AWIS, WEPAN, MentorNet, ABI, CRA-W, ACM-W, NCWIT), African Americans (NSBE, NACME, GEM, BDPA, ADMI, CDC, CMD-IT), Hispanics (SACNAS, MAES, SHPE), or American Indians (AIHEC, AISES).[1] This section of the book provides profiles of each of the organizations listed above – brief ones for the early, STEM-focused organizations, longer ones for the computing-focused organization. Two of these STEM-focused organizations, MentorNet and WEPAN, receive more detailed treatment because of their significant work in the computing community.

These nonprofits are not the only organizations that support the development of human resources for the computing field in the United States. In the author's companion book, *Participation in Computing*, he discusses the Broadening Participation in Computing alliances created with funding from the National Science Foundation,

---

[1] This statement is a simplification of the target audiences. For example, SACNAS has been interested in American Indians since its founding, although its founders and its target audience were primarily Hispanic. MentorNet began in 1997 focused solely on women, but in 2003 expanded its charter to include diversity more broadly.

such as the Expanding Computing Education Pathways and STARS alliances.[2] The companion book also shows the important role of professional organizations, most notably the ACM, through its Computer Science Teachers Association and its Educational Policy Committee. The National Science Foundation itself is also an important player because of the research and implementation it funds relating to broadening participation in computing.

The second section of this book concludes with a long, single chapter that profiles colleges and universities that have been successful in opening up computer science or engineering to female students. The chapter opens with a discussion of several social science studies that discuss what characteristics make for a department able to attract, retain, and advance female students. Then five case studies of successful departments are given: University of California Berkeley/Mills College, Carnegie Mellon University, Olin College, Smith College, and Harvey Mudd College. Several other colleges and universities are discussed briefly after these five detailed case studies.

Caveats and Acknowledgments

In order to write this and the companion volume so quickly, certain shortcuts were taken. No trips were made to archives to find source materials. There has been an extensive, if not exhaustive, search of the published literature for source materials. More than 900 sources have been consulted in writing these two books. This book relies not only on published books and articles but also on websites, project reports, white papers by nonprofit organizations, existing oral histories, and other sources. The project also involved the recording of a number of new oral histories, and these interviews provide the largest value added to this work. They are housed at the Charles Babbage Institute at the University of Minnesota–Twin Cities and will eventually be made available to other scholars.

Some 25 computer scientists, historians, and social scientists have kindly volunteered their time to advise on this project. Their names and affiliations appear in Table 1. They have devoted many hours providing guidance, opening doors, and critiquing draft chapters. Thanks also to the two doctoral students in the University of Texas at Austin School of Information, Steve McLaughlin and Rachel Simons, who provided research assistance, and to another doctoral student, Melissa Ocepek, who helped to render the references into a form suitable to the publisher. Everyone interviewed for this book as well as all the principal investigators in the Sloan Foundation program that supported this project, in addition to the Project Advisory Group members listed in Table 1, were given a chance to comment on a complete first draft of the manuscript. Their comments led to many improvements in the text;

---

[2] The one organization covered in both that book and in this one is the National Center for Women & IT (NCWIT).

Table 1 Project Advisory
Group

Rick Adrion (U. Massachusetts)
Atsushi Akera (Rensselaer P.I.)
Lecia Barker (U. Colorado Boulder and NCWIT)
Bruce Barnow (George Washington U.)
Paul Ceruzzi (National Air and Space Museum)
Jan Cuny (NSF)
Nathan Ensmenger (Indiana U.)
Mary Frank Fox (Georgia Tech)
Peter Freeman (Georgia Tech)
Juan Gilbert (U. Florida)
Jonathan Grudin (Microsoft)
Thomas Haigh (U. Wisconsin–Milwaukee)
Evelynn Hammonds (Harvard U.)
Peter Harsha (Computing Research Association)
Mary Jane Irwin (Penn State U.)
Martin Kenney (U. California, Davis)
Ed Lazowska (U. Washington)
Ephraim McLean (Georgia State U.)
Thomas Misa (Charles Babbage Institute)
Andrew Russell (Stevens I.T.)
Lucy Sanders (National Center for Women & IT)
Robert Schnabel (ACM)
Bruce Seely (Michigan Tech U.)
Eugene Spafford (Purdue U.)
Moshe Vardi (Rice U.)
Roli Varma (U. New Mexico)
Stuart Zweben (Ohio State U.)

all factual errors and unreasonable interpretations are the sole responsibility of the author.

This study was enabled in part by a grant from the Alfred P. Sloan Foundation, which helped the author to buy out of his teaching for a year and pay for transcription of interviews. The author is also grateful for support from the School of Information at the University of Texas at Austin, which relieved him of some administrative responsibilities for a year and paid for a part-time research assistant for a semester, and for a grant from the Institute of Museum and Library Services, which supported a doctoral student for a semester to assist with the research. Most of the research and writing was conducted while the author was the Bill and Lewis Suit Professor of Information Technologies at Texas, but the last stages of the work were carried out at his new academic home, the Department of Information Science at the University of Colorado Boulder.

Boulder, CO, USA   William Asfor

Contents

Chapter 1
Introduction

Abstract This chapter provides an introduction to the two main sections of this book. The first section of the book provides a summary of the historical and social science literature about the history of higher education – and science, technology, and computing education in particular – for women, African Americans, Hispanics, and American Indians in the United States. The second section provides case studies of organizations interested in broadening participation in the science and technology disciplines in general and in computing in particular; as well as case studies about college and university departments of computer science and engineering that have had success in attracting, retaining, and advancing women in engineering and computing careers. The chapter discusses overarching themes that run through the book: exogenous forces (war, civil rights, reverse discrimination, and IT workforce needs); the conceptualization of the underrepresentation problem in terms of a pipeline instead of a pathway; solutions that involve fixing people contrasted with those that involve fixing the system; the role of nonprofit organizations and individual change agents in broadening participation in computing; and the issues surrounding intersectionality, i.e. cases in which someone belongs to two or more underrepresented groups such as being both female and African American.

We felt like anomalies.... The women felt the difference most keenly during breaks, when they couldn't join in the inside jokes and casual conversations into which their male colleagues seemed to fall so easily. In a profession so dependent on teamwork and learning new technology, being part of the community is not just a matter of feeling comfortable. It's essential to being competitive. (Katherine Jarmul, co-founder of PyLadies, as quoted in Shah 2012)

This book addresses the history of underrepresentation of women and certain groups of minorities in the computing field in the United States.[1] The focus is primarily on the formal education of information workers rather than on the workforce

---

[1] This book employs the contemporary policy language when it speaks of "underrepresented minorities" instead of the term "race", which is more commonly used by historians. Race is, of course, a social construct. At one time in American history, Jews, the Irish, and Eastern Europeans were segregated from Whites as separate racial groups, but today they are all considered as Whites. African Americans, however, have historically been racially segregated throughout American history and continue to be segregated today. Not all minorities are underrepresented in the computing

© Springer International Publishing Switzerland 2016

W. Aspray, *Women and Underrepresented Minorities in Computing*, History of Computing, DOI 10.1007/978-3-319-24811-0_1

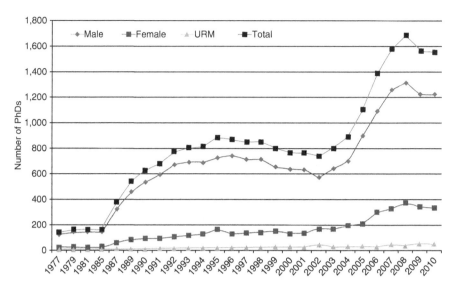

Fig. 1.1 Computer science doctoral degrees granted (Data Source: IPEDS Completion by Race, from https://webcaspar.nsf.gov). This figure is published as Figure 1 in http://cra.org/crn/2013/05/expanding_the_pipeline_diversity_drives_innovation/]. (Source: McKinley and Camp 2013)

itself. As Fig. 1.1 and Table 1.1 show, women, African Americans, Hispanics, and American Indians have been consistently underrepresented in the computing field throughout the entire era of modern computing, i.e. since 1945.[2] Figure 1.1 shows underrepresentation of women and minorities in receiving doctoral degrees in computing from 1977 to 2010. Table 1.1 shows similar underrepresentation at the bachelor's level for 2014.

There are various reasons why this underrepresentation is important. It a *social equity* issue that these high-paying, fulfilling, socially transformative jobs are less available to individuals from other demographic groups than they are to many White and Asian men. Shortages of skilled professionals occur every few years in the computing fields, and many scholars and policymakers believe both that these shortages have been harmful to American competitiveness and that larger participation of these underrepresented groups would go a long way towards meeting this *skilled workforce need*. Other scholars and policymakers (Page 2008; Barker et al. 2014)

---

field. For example, Asians from India, China, and Korea are over-represented in computing in the United States. Other Asians, such as the Hmong and Vietnamese, are underrepresented.

[2] The numbers about representation of women in traditional scientific and computing occupations may tell only part of the story. For example, the treatment of the ENIAC women at the time – and in many historical treatments since then – has treated these women mostly as assistants in a way that undervalued their scientific contributions. Similarly, occupational classifications over time may not count the large numbers of women in the data processing industry as information workers. See, for example, Light (1999), Grier (2005) and Misa (2010).

Table 1.1 Bachelor's production in the computing fields (2014) – percentages by demographic group

	Computer science	Computer engineering	Information systems/science/ technology	US population
Women	14.1	11.7	20.3	51
African Americans	3.2	3.3	8.2	13
American Indians	0.4	1.0	0.3	2
Hispanics	6.8	8.4	10.7	17

Sources: 2014 CRA Taulbee Survey; US Census
Note: The Taulbee Survey only includes data about Ph.D.-granting institutions, so it does not include data on most of the for-profit universities, which have higher percentages of minority student enrollment than the Taulbee schools

have pointed out that the work products of *diverse work teams* are more innovative and more likely to meet the needs of a wide range of customers than those created by a White male monoculture of technology developers.[3]

There is a small body of historical literature about women in computing in the United States. In addition to a number of articles and books focused on narrowly defined pieces of this history, there are three general books on this history: monographs by Janet Abbate (2012) and Nathan Ensmenger (2012), and an edited volume by Thomas Misa (2010). The literature on the history of underrepresented minorities and computing is much thinner; and in fact there are not yet any book-length general histories of this topic.

The social science literature on underrepresentation in computing is perhaps two orders of magnitude larger than the historical literature mentioned above. Like the historical literature, this social science literature is stronger and more numerous on women than on underrepresented minorities. Unfortunately, none of the major historical studies in this area have been informed in any substantial way by this social science literature, even though it has promise to offer insights into underlying causes.

This book is intended primarily as an historical study, even though it pays considerably more attention to the social science literature than do the other historical works on underrepresentation and computing mentioned above. This book is not a definitive history of underrepresentation and computing in the United States, but it provides useful background material drawn from the historical and social studies of computing, education (especially technical education), and race and gender that should help prepare some future scholar to write a more definitive history.

---

[3] This paragraph is copied almost verbatim from the introduction to the author's companion book, *Participation in Computing: The National Science Foundation's Expansionary Programs.*

1.1   Overarching Themes

We will close this introduction with a brief discussion of some of the themes that run throughout this book and its companion volume – and, indeed, are likely to be present in any historical study of broadening participation in computing in the United States.

1.1.1   Four Exogenous Forces

In this section, we briefly identify four exogenous forces that have shaped efforts to broaden participation in computing in the United States since 1945. The first is the return after the Second World War of male veterans who displaced a number of women from science and engineering jobs. There were many fewer women working in jobs during the war as scientists or engineers than there were women working in manual manufacturing positions, but those who were displaced from scientific occupations were often unhappy about this change. One outcome of this dissatisfaction was the creation of the Society of Women Engineers in 1950. It was the first of a number of organizations formed from the 1950s to the 1970s with the purpose of helping to broaden participation in the STEM disciplines. This story is told in passing in Chaps. 2 and 6, and will not be discussed further here.

The second exogenous force shaping broadening participation in computing in the United States were the civil rights and women's rights movements. For example, the women's rights movement led directly to the establishment in 1980 of the Committee on Equal Opportunities in Science and Technology with the purpose of advising the Director of the National Science Foundation on matters of broadening participation across the STEM disciplines.[4] The role of the women's rights movement in these broadening participation activities is told in passing in Chaps. 2 and 6 and in another publication by this author (Aspray 2016), and so will not be discussed further here.

The impact of the civil rights movement is less well known, so we will take some extra space to discuss one important episode related to this particular exogenous force. The second half of the 1960s was punctuated by race riots in Rochester, NY, Harlem, and Philadelphia in 1964; in the Watts section of Los Angeles in 1965; in Cleveland and Omaha in 1966; in Newark and Plainfield, New Jersey as well as in Detroit and Minneapolis in 1967; in Chicago, Washington DC, Baltimore, and again in Cleveland in 1968; and again in Omaha in 1969. In response to the wanton destruction and lack of opportunities for minorities, in 1973 the National Academy of Engineering, together with the Commission on Education, convened a Symposium

---

[4] There were other important pieces of federal legislation as well, e.g. the Civil Rights Act of 1964, which was used as a tool to open university admission to African Americans and other racial minorities, and the Title IX Education Amendments of 1972, which had the effect of enabling much higher female admission in higher education programs in the STEM and medical fields.

Table 1.2 Major recommendations from the 1974 Sloan Report

(1) Reaching the parity number of 18 % in minority participation in engineering;
(2) Forming a new national organization to raise and distribute financial aid to cover 5 years of financial aid for minority college engineering students;
(3) Having industry channel its funding through this new organization and having foundations assist this organization, especially during its first 5 years;
(4) Having the National Academy of Engineering coordinate efforts of many organizations and many programs to increase minority participation in engineering;
(5) Supporting the six historically Black universities that operated engineering programs (Howard, North Carolina A&T, Prairie View A&M, Southern, Tennessee State, and Tuskegee) with special funding so that they could double their enrollments over 5 years;
(6) Identifying colleges with large concentrations of Chicanos, Puerto Ricans, and American Indians to provide them with additional funds so that they would become strong providers of minority engineers;
(7) Improving articulation between 2-year and 4-year colleges concerning engineering education;
(8) Enhanceing counseling and cooperative engineering programs with industry at these colleges;
(9) Improving Ph.D. production of minority students in engineering so as to have adequate minority representation among faculty and academic administrators;
(10) Collaborating with the educational programs of the armed services to increase the number of minority veterans who enter into the engineering professions;
(11) Increasing the number of elementary and high school teachers in bilingual and bicultural schools teaching science and mathematics;
(12) Initiating school-year and summer programs for minority students to increase their interest in science and mathematics; and
(13) Encouraging the U.S. Department of Education to establish programs that support these goals

Source: *Minorities in Engineering: A Blueprint for Action* (Sloan 1974)

on Increasing Minority Participation in Engineering. The more than 250 attendees included government officials, industrial representatives, minority leaders, and students. The major outcome of this meeting was a call for "equitable participation" of minorities within a decade.

To understand the dimensions of the problem and operationalize the call for action at the symposium, the Alfred P. Sloan Foundation sponsored a 7-month study carried out in late 1973 and early 1974 at Stanford University. The study provides today's historian with a good snapshot of the situation in the early 1970s. For the contemporary reader of the 1970s, the study intended to identify numerical targets for minority representation in engineering; ascertain the feasibility, steps, and costs to achieve these goals; and identify organizations that could carry out various aspects of the work. The report appeared in 1974 under the title *Minorities in Engineering: A Blueprint for Action* (Sloan 1974).

The report focused on four underrepresented minority groups in engineering: Black, Chicano, Puerto Rican, and American Indian (using the terms that appeared in the report). These four minorities represented 14.4 % of the population in 1970 but only 2.8 % of the engineers. Of these four minorities, Blacks had the poorest representation, with 11.1 % of the U.S. population but only 1.2 % of the engineering profession. Table 1.2 lists the report's 13 major recommendations.

The report found that Black students take longer to complete a high school degree and have a higher dropout rate than White students. While Black enrollment in college had at least doubled in every decade of the twentieth century, and while there had been a particularly significant growth in Black freshmen enrollment in college between 1970 and 1973, there was still a significant gap between college entry for Blacks and Whites. A greater percentage of Blacks dropped out of college than Whites. The community colleges were a particularly important educational feeder for Blacks, with more than 40 % of full-time Black students enrolled in community colleges; but there were concerns that a two-tiered higher educational system would emerge with the community colleges becoming "dumping grounds" for minorities, where career opportunities would be limited. Until 1960, approximately two-thirds of Blacks enrolled in college were enrolled at historically Black colleges and universities, but there was concern about the financial soundness of these institutions.

The report found that Chicanos were less likely than Blacks to be successful at the elementary school, high school, and college levels. Unlike Blacks, who were widely dispersed across the nation, Chicanos were primarily concentrated in five states in the American southwest. For every 100 Chicano children entering first grade, only 60 graduated from high school, only 22 entered college, and only 6 graduated from college. This indicated that programs intended to increase Chicanos in engineering would have to focus on precollege as well as college education. The study found that precollege Chicano students are held back by language, both reading skills in English and the low number of Chicano and bilingual teachers. Family expectations that females are less likely than males to continue their education was also cited as a factor.[5]

Puerto Ricans were concentrated mostly in the states of New York and New Jersey. The report found that high school and college completion rates for Puerto Ricans were lower than for Chicanos. English illiteracy was higher for Puerto Ricans than Chicanos. Puerto Ricans were channeled more than White or other minorities into vocational rather than college preparatory public education programs.

The report concurred with a Presidential message to Congress that American Indians are "the most deprived and most isolated minority group" in the nation. (Sloan 1974, p. 53) Many American Indians received their public school education through special Bureau of Indian Affairs-run schools or Church-run mission schools. American Indians experienced elevated high-school dropout rates, and of those who did graduate from high school, they were half as likely as White high-school graduates to matriculate in college. Once matriculated in college, American Indian students had attrition rates not much different from other underrepresented minorities. The report argued that close ties to family and tribe, as well as differences between American Indians and Whites regarding cultural values about individual

---

[5] Another barrier for Hispanic women in 1970 was that most engineering schools did not admit women (of any race) or only admitted a very few, typically representing well under 10 % of the student population.

versus collaborative action, placed stress on American Indian students. There were even more acute problems in finding bilingual and bicultural teachers for American Indian children than there were for Spanish-speaking children.

Responding to this situation described in the 1974 Sloan report, a number of new organizations were formed to support the advancement of ethnic and racial minorities in the science and engineering disciplines. These included the American Indian Higher Education Consortium (AIHEC) in 1972, Society for Advancement of Chicanos and Native Americans in Science (SACNAS) in 1973, the National Action Council for Minorities in Engineering (NACME) and the Society of Hispanic Professional Engineers (SHPE) in 1974, the National Society of Black Engineers (NSBE) in 1975, and the American Indian Science and Engineering Society (AISES) in 1977. Background material on this exogenous force is covered in passing in Chaps. 3, 4, and 5, and the stories of these support organizations are told in Chap. 7.[6]

The third important exogenous force shaping the efforts to broaden participation in computing is the reverse discrimination environment that developed in the United States in the 1990s and that is still in force to some degree today. It was a conservative reaction to civil rights legislation passed between the 1960s and the 1980s; and it tried to undo preferential policies that had been given on the basis of race or gender through affirmative action programs, e.g. relating to employment or admission to higher education institutions. This issue is discussed in connection with access to higher education for African Americans and Hispanics in Chaps. 3 and 4 and in connection to computer science re-entry programs for women in Chap. 10. This author has also discussed the impact the reverse discrimination environment of the 1990s had on the programs of the National Science Foundation to broaden participation in the STEM disciplines in Aspray (2016).

The fourth exogenous force that we mention here as shaping efforts to broaden participation in computing is IT workforce needs. The demand of employers for IT workers has inexorably increased in the United States over the past 75 years as information and communication technologies have become embedded in ever more aspects of work. However, the demand has not increased at a steady or predictable rate; instead, IT workforce demand is characterized by cycles of demand and glut, of uneven duration and ferocity. A good example is the late 1990s and early 2000s, when there was an insatiable demand for workers to resolve Y2K problems of legacy computer systems and fuel the dot-com boom – only to be followed in 2002 with a dot-com bust that led to numerous IT worker layoffs. One might think that a demand for IT workers would be a good situation for women and minorities who were underrepresented in the workforce. However, employers generally preferred to fill open IT positions with foreign workers rather than with Americans from underrepresented groups. This was the story of the fierce battles around H-1B visas

---

[6] We make no claim to have been exhaustive in the list of STEM broadening participation organization. We list in the main body of the text only those that we profile later in the book. For example, the National Society of Black Physicists was created in 1977 and is not discussed here. There may well have been additional organizations of this type.

around 2000 and over outsourcing of IT work 5 years later. A second way in which this exogenous force played out is that higher education enrollments in computer science increased rapidly when industry increased demand for IT workers. The higher education system is not very flexible in its ability to meet rapidly changing enrollments. In the face of too many students, many computer science departments introduced weed-out courses that had the impact of weeding out many students with weak high school education or low self-confidence. These courses had the inadvertent effect of disproportionately weeding out women and underrepresented minorities. They sometimes also introduced higher entrance standards that also often affected women and underrepresented minorities in a disproportionate manner. This exogenous force is discussed in passing in Chaps. 8 and 9.

1.1.2 Pipeline Versus Pathway

Many analyses of the issues related to broadening participation in computing use the metaphor of a pipeline: if you do not take the right preparatory courses in middle school and high school, you cannot be admitted into an undergraduate major in one of the computing disciplines; if you do not have receive an undergraduate degree in a computing discipline, you cannot be admitted into a graduate program in a computing discipline; if you do not receive a graduate degree in a computing discipline, you cannot obtain a position in a high-level computing occupation, such as professor or senior researcher in industry. The iconic embodiment of this argument is a paper that appeared in *Communications of the ACM* in 1997, written by the computer science professor Tracy Camp and entitled "The Incredible Shrinking Pipeline." (Camp 1997) Camp's paper described a pipeline that leads from high school to a good computing career, but one that is leaky in many places, losing people from the pipeline at each of these transitions; so that only a few – too few – individuals were able to achieve the desired job at the end of the pipeline.

Camp is by no means the only person to use this metaphor; indeed it is commonly found in discussions of broadening participation in computing and also in efforts to craft solutions: for example, how do we make the pipeline less leaky so that more people can make the transition from high school to the undergraduate computer science major? The metaphor perhaps makes the most sense in the case of the computing occupation that is of primary interest to Camp, becoming a professor of computer science in a research-intensive university. It is not a perfect metaphor even in this occupational setting, however, because there are people who take a different pathway to their computing professorship, for example through the reentry program discussed in Chap. 9 that operated at Berkeley (to enable people who majored in a non-computing subject as an undergraduate to prepare for graduate student in computer science) or by training in a traditional way for the professoriate in a different, computing-intensive field, such as physics or economics, and then being hired into a computer science faculty position.

In point of fact, there are many different computing occupations; and these occupations have widely varying educational requirements.[7] It is hard to count the numbers accurately, but perhaps only a quarter of the people who hold positions in computing occupations have any formal education in computing. (See Freeman and Aspray 1999.) In 2004, the Committee on Equal Opportunity in Science and Engineering, the esteemed advisory body created by the U.S. Congress in 1980 to advise the Director of the National Science Foundation on issues of broadening participation in the various science and engineering disciplines, specifically renounced the pipeline metaphor and advocated replacing it with a pathways metaphor that emphasizes the many different possible pathways to a computing career.[8]

The 2004 CEOSE report to Congress included a review of NSF's broadening participation efforts. Analyzing NSF programs since 1980, the Executive Summary of this report highlighted the need for a changing framework for understanding the nature of the process by which one prepared for and entered a STEM occupation. This is where they discussed moving from the pipeline metaphor to the pathway metaphor[9]:

> Early efforts to broaden participation focused primarily on encouraging individuals from underrepresented segments of the population to enter STEM disciplines. This "pipeline" metaphor is a way of looking at the persistence of women, minorities, and persons with disabilities in STEM statistically. It emphasizes attracting students into the STEM "pipeline" when they are young, and spotlights the points at which "leaks" occur, differentially draining away individuals from underrepresented groups. Today, many efforts to make science and engineering more inclusive are paying attention instead to the multiplicity of "pathways" by which persons from underrepresented groups can enter and progress through STEM careers. Creating viable pathways requires addressing the tough issues related to what invites children to learn science (attraction), what causes young people to choose to keep learning mathematics and science (retention), and what then leads students to graduate (persistence) and continue into STEM careers (attachment). (CEOSE 2004)

This emphasis on pathways rather than pipeline is probably even more appropriate today, given the many informal pathways to a computing education through online courses, computing boot camps, and hackathons.[10]

---

[7] See, for example, the analysis in Freeman and Aspray (1999) or the federal occupational categories related to computing at http://www.bls.gov/ooh/computer-and-information-technology/home.htm

[8] Two years earlier, Carol Muller (the founder of MentorNet) and Susan Metz (a co-founder of WEPAN) had written an editorial calling for an abandonment of the pipeline metaphor and renewed focus on multiple entry points into STEM careers. (Muller and Metz 2002)

[9] Evelynn Hammonds has also pointed out another problem of the pipeline metaphor: it led policy-makers for a number of years to the mistaken belief that "the factors leading to the production of white male scientists were the same ones that lead to the production of women and minority scientists." (Personal communication to the author, 18 March 2016)

[10] It is an unexplored question which computing occupations are open to people who take these informal pathways, and which kinds of employers are willing to hire people with an informal computing education; but it is clear that this informal education opens up some computing occupations with some employers.

Several scholars studying broadening participation in computing have criticized the pipeline metaphor. Jolene Jesse, a program officer at NSF who studies issues of women in science, is one. (Jesse 2006) An earlier study in which she participated, jointly conducted by the American Association for the Advancement of Science and the Committee on Professionals in Science and Technology, examined nontraditional pathways into the computing workforce.

A 'nontraditional pathway' is defined as the path taken by a nontraditional student, i.e., someone who: delays enrollment at least three years after graduating from high school or earning a GED; attends college mostly part-time; takes longer than six years to complete a degree; is employed full-time during most of their studies; or has dependents while attending college. (MIT Press, Scholarship Online, http://mitpress.universitypressscholarship. com/view/10.7551/mitpress/9780262033459.001.0001/upso-9780262033459-chapter-8)

These pathways are completely missed by the pipeline metaphor.

Mark Guzdial, a computer scientist at Georgia Tech who is a leading scholar in computer science education, writes in his computer education blog:

By using the "leaky pipeline" metaphor, we stigmatize and discount the achievements of people (women, in particular in this article) who take their technical knowledge and apply it in non-computing domains. Sure, we want more women in computing, but we ought not to blame the women who leave for the low numbers. ... [N]ew research of which I am the coauthor shows this pervasive leaky pipeline metaphor is wrong for nearly all postsecondary pathways in science and engineering. It also devalues students who want to use their technical training to make important societal contributions elsewhere. (Guzdial 2015)

Lecia Barker, a professor in Department of Information Science at the University of Colorado Boulder and a senior research scientist at NCWIT, has noted that the pipeline metaphor can be counterproductive when trying to broaden participation in computing because it accepts as lost those individuals who leaked out of the pipeline at an earlier stage rather than finding a way to prepare them for a computing career. (Private communication to author, November 2015)

Historian of computing Thomas Haigh has also criticized the pipeline metaphor, but his criticism is that the pipeline metaphor focuses too much on the formal educational system instead of on the workplace. Haigh (2010) writes:

History broadens our perspectives. The literature on women in computing is dominated by discussion of computer science education. Fixing computer science is equated with fixing computing. This is justified by the metaphor of the pipeline carrying women from specialist education into IT work. Yet we saw that the gender dynamics of data processing were well formed by the 1960s, before undergraduate computer science education was an appreciable factor. Gender dynamics were shaped instead by the specific historical legacy of data processing work and the broader gender politics of corporate society. So to understand gender segmentation in the workforce, we must study the workplace as well as the classroom.

Some of the more recent social science research on underrepresentation is beginning to adopt a pathway metaphor. See, for example, Fox and Kline (2016), which discusses women faculty in computing from a pathways perspective.

1.1.3 Fixing the People Versus Fixing the System

Many of the interventions employed in broadening participation in computing principally address what is sometimes called "fixing the women". These interventions include, for example, women's support groups, special summer-before-college bridging programs for minority students, special sections of introductory computer science for people with limited computing experience, special financial aid programs for minorities, special mentoring opportunities for underrepresented groups, and conferences such as the Grace Hopper and Tapia celebrations, which target women and minority populations, respectively. These interventions can help to rectify shortcomings an individual has experienced, such as coming from a poor family or having a poor public school education – and some of these activities, such as the Grace Hopper conference – have fiercely loyal supporters.

However, there is also a negative side to these interventions: they can stigmatize underrepresented groups and reinforce stereotypes and unconscious bias about who should be doing computer science. Moreover, some social scientists believe that the track record of such interventions show that they are not effective at making the wholesale change of broadening participation in computing sought at the national level. Some organizations such as the National Center for Women & IT, and some individual researchers such as Mary Frank Fox and Gerhard Sonnert, believe that substantial change comes only through fixing the system rather than fixing the individuals. Systemic change is difficult, and it is unsettling to those who have existed in the established system. It is not surprising that, for many years, technology companies were more engaged in recruiting new women and minority hires rather than changing the practices and environments within their organizations so as to improve retention and advancement of women and underrepresented minorities. Chapter 10 discusses some of the issues associated with broadening participation in computer science departments; and Chaps. 8 and 9 discuss this issue in passing as various organizations, including NCWIT, are discussed.

1.1.4 Nonprofit Organizations and Individual Change Agents

Much of this book (Chaps. 6, 7, 8 and 9) focuses on nonprofit organizations. Many of these organizations had their origins in the concerns of a few scientists, engineers, and computer scientists who wanted the STEM disciplines in general or the computing discipline in particular to be equally open to all people, regardless of race or gender. These organizations can achieve what the concerned individuals – who often want to devote their main efforts to their scientific or engineering career – cannot do alone. Organizations have staff members who can handle the management of programs and events, provide effective communication to multiple stakeholders, and conduct business with safeguards that enable these organizations to act professionally and ethically. All of these organizations have strong advisory committees populated by members of the scientific professional community – many of whom volunteer large amounts of time and have strong identification with the organization. As nonprofit organizations, the most common funding model is project grants from either government (especially the National Science Foundation) or corporate entities. The most successful of these broadening-participation organizations focused on computing at this time are the CRA Committee on Women in Computing Research (CRA-W) and the National Center for Women & IT (NCWIT).

Equally important in this story of broadening participation in computing are individuals who serve as change agents – people who have served to transform their organizations or their profession through their individual actions to make them more effective at carrying out this broadening participation mission. Some of them are known for their roles in specific organizations, e.g. Jan Cuny's work in creating the Broadening Participation in Computing and the CS10K programs at the National Science Foundation; Anita Borg's work in creating Systers, the Grace Hopper conference, and what is now called the Anita Borg Institute; Lucy Sanders in creating NCWIT; and Richard Tapia in his inspirational work at Rice University. Others, such as the ethnographer Jane Margolis who studied both Carnegie Mellon University and the Los Angeles School District, have had a reach that has had more importance for its national dissemination of ideas than its contributions to a single institution.[11]

1.1.5 Intersectionality

This book is organized with chapters specifically about women or about a particular racial group (African Americans, Hispanics, and American Indians). The world is not nearly so tidy a place as this chapter organization suggests. There are many variations across these individual racial categories, e.g. tribal differences among American Indians, or to take another example the geographical, racial, language, and cultural differences between Chicanos living in the western and southwestern United States and Puerto Ricans living in the eastern United States. Many individuals are of more than one race. In addition, there are many other groups that are underrepresented that are not discussed in detail in this book, e.g. as organized by age, various sexual orientations, or various disabilities. People who are underrepresented in computing by belonging to more than one of these underrepresented groups often have multiple and simultaneous issues at play. This is an important topic, but it is not addressed except briefly in passing several times in this book.[12]

---

[11] There are, of course, many more individual change agents than I can mention here. It has been a highlight of this author's career to get to know many of these people and watch them in action.

[12] Some examples of recent literature on intersectionality and STEM include Bruning et al. (2012), Herrera et al. (2013), Ko et al. (2013), Charleston et al. (2014) and O'Brien et al. (2015).

References

Abbate, Janet. 2012. *Recoding gender: Women's changing participation in computing*. Cambridge, MA: MIT Press.

Aspray, William. 2016. *Participation in computing: The national science foundation's expansionary programs*. Cham: Springer.

Barker, Lecia, Cynthia Mancha, and Catherine Ashcraft. 2014. *What is the impact of gender diversity on technology business performance?* NCWIT Research summary. http://www.ncwit.org/sites/default/files/resources/impactgenderdiversitytechbusinessperformance_print.pdf. Accessed 9 Sept 2015.

Bruning, Monica, Jill Bystydzienski, and Margaret Eisenhart. 2012. *Intersectionality as a framework for understanding diverse young women's interest in engineering*. Getting to the Heart of it All: Connecting Gender Research, WIE Programs, Faculty & Corporate Partners, WEPAN National Conference. June 25–27. Columbus, OH. 14.

Camp, Tracy. 1997. The incredible shrinking pipeline. *Communications of the ACM* 40(10): 103–110.

Charleston, LaVar J., Ryan P. Adserias, Nicole M. Lang, and Jerlando F.L. Jackson. 2014. Intersectionality and STEM: The role of race and gender in the academic pursuits of African American women in STEM. *Journal of Progressive Policy and Practice* 2(3): 273–293.

Ensmenger, Nathan. 2012. *The computer boys take over: Computers, programmers, and the politics of technical expertise*. Cambridge, MA: MIT Press.

Fox, Mary Frank, and Kathryn Kline. 2016. Women faculty in computing: A key case of women in science. Chapter 3. In *Pathways, potholes, and the persistence of women in science: Reconsidering the pipeline*, ed. Enobong Hannah Branch. Lexington Books. Lanham, Maryland

Freeman, Peter, and William Aspray. 1999. *The supply of information technology workers in the United States*. Washington, DC: Computing Research Association.

Grier, David Alan. 2005. *When computers were human*. Princeton: Princeton University Press.

Guzdial, Mark. 2015. End the 'leaky pipeline' metaphor when discussing women in science: Technical knowledge can be used in many domains. *Computing Education Blog*, April 27. https://computinged.wordpress.com/2015/04/27/end-the-leaky-pipeline-metaphor-when-discussing-women-in-science-technical-knowledge-can-be-used-in-many-domains/. Accessed 31 Dec 2015.

Haigh, Thomas. 2010. Masculinity and the machine man: Gender in the history of data processing. In *Gender codes: Why women are leaving computing*, ed. Thomas J. Misa, 51–72. Hoboken: IEEE Computer Society Press.

Jesse, Jolene. 2006. The poverty of the pipeline metaphor: The AAAS/CPST study of nontraditional pathways into IT/CS education and the workforce. In *Women and information technology: Research on underrepresentation*, ed. Cohoon Joanne and Aspray William, 239–278. Cambridge, MA: MIT Press.

Ko, Lily T., Rachel R. Kachchaf, Maria Ong, and Apriel K. Hodari. 2013. Narratives of the double bind: Intersectionality in life stories of women of color in physics, astrophysics and astronomy. *Physics Education Research Conference, AIP Conference Proceedings* 1513: 222–225.

Light, Jennifer. 1999. When computers were women. *Technology and Culture* 40(3): 455–483.

Misa, Thomas. 2010. *Gender codes: Why women are leaving computing*. Hoboken: IEEE Computer Society.

Muller, Carol B., and Susan Staffin Metz. 2002. Burying the pipeline and opening avenues to engineering. *ASEE Prism* 12(4). http://www.prism-magazine.org/dec02/lastword.cfm. Accessed 26 Feb 2016.

National Science Foundation Committee on Equal Opportunities in Science and Engineering (CEOSE). 2004. *Broadening participation in America's science and engineering workforce*. The 1994–2003 decennial & 2004 biennial reports to Congress. http://www.nsf.gov/od/iia/activities/ceose/reports/ceose2004report.pdf. Accessed 26 Nov 2014.

O'Brien, Laurie T., Alison Blodorn, Glenn Adams, Donna M. Garcia, and Elliott Hammer. 2015. Ethnic variation in gender-STEM stereotypes and STEM participation: An intersectional approach. *Cultural Diversity and Ethnic Minority Psychology* 21(2): 169–180.

Page, Scott E. 2008. *The difference*. Princeton: Princeton University Press.

Shah, Angilee. 2012. Geek chicks, PyLadies, a gang of female computer programmers. *LA Weekly*, February 16. http://www.laweekly.com/publicspectacle/2012/02/16/geek-chicks-pyladies-a-gang-of-female-computer-programmers?showFullText=true. Accessed 27 Oct 2015.

Sloan Foundation. 1974. *Minorities in engineering: A blueprint for action. Report.* New York: Sloan Foundation.

Part I
Digest of Relevant Literatures

Chapter 2
Opening STEM Careers to Women

Abstract This chapter examines the history of higher education for women, as well as the history of careers for women in science and engineering, in the United States. The first section discusses women's matriculation in college generally from 1900 to the present day. The next section presents a statistical overview of women in science and engineering from the early twentieth century to the present. The third section provides a qualitative analysis of the history of women in science since 1820. The final section provides a qualitative analysis of the history of women in engineering since 1918.

This chapter examines the history of higher education for women, as well as the history of careers for women in science and engineering, in the United States. The chapter relies heavily on a few sources, principally Margaret Rossiter's three-volume history of women in science and Amy Sue Bix's book on the history of engineering education for women. (Rossiter 1982, 1995, 2012; Bix 2013) Those readers who are familiar with the history of higher education for women or the history of science and engineering careers for women in America might not find much new in this chapter. However, many of the computer scientists, education specialists, and other social scientists who study women and the STEM disciplines, including computing, are not familiar with this material. The selection, slant, and augmentation of these two major authors have been designed to appeal to these readers who are carrying out research or interventions related to women and STEM (including computing).

2.1 College Matriculation of Women – A Brief History

In this section, we discuss women's matriculation in college generally. For many of the higher-level professional positions in the STEM and computing disciplines, a baccalaureate degree is necessary, typically with a significant amount of course instruction in the STEM or computing disciplines. A study by three Harvard

© Springer International Publishing Switzerland 2016
W. Aspray, *Women and Underrepresented Minorities in Computing*,
History of Computing, DOI 10.1007/978-3-319-24811-0_2

economists (Goldin et al. 2006) provides a useful place to start this account, by trac-ing the history of the college matriculation of women in the United States.[1]

The numbers of women and men attending college were at approximate parity in the period from 1900 to 1930. However, women's enrollment was suppressed some-what in the 1930s in part because of an increase in the number of so-called "mar-riage bar rules", which barred married women from working as teachers. These rules made less valuable the completion of a degree in education – one of the most common majors available to and chosen by female students. The number of men attending college began to increase at a faster rate in the 1930s and grew especially rapidly after the Second World War ended in 1945 (on account of the GI Bill) and during the Viet Nam War of the late 1960s and early 1970s (as a means of exemption from military service). For example, in the year 1947 there were 2.3 men attending college for every woman. Beginning in the 1970s, the trend began to reverse. Women began to attend college at a higher rate than they previously had, and gender parity in college attendance was once again achieved in 1980. Since that time, there have been more women than men enrolled in college. As of 2003, there were 1.3 female undergraduates for every male undergraduate.

Until the early 1970s, most female students selected (or in some cases channeled towards) college majors in traditional female-intensive disciplines such as teaching, English, and literature and regarded college primarily as a way to prepare for a tra-ditionally female-intensive occupation such as teaching or social work, or as a means to find a husband.[2] This pattern changed radically in the 1970s and 1980s, and during this time the mean age of a woman's marriage rose significantly and contraceptive technology meant greater choice in when to have children. During these two decades, the women's labor force increased substantially and high school girls began to take more math and science courses in preparation for college.

The authors report gender difference in academic performance and subject area selection over time by reporting on national surveys conducted in the years 1957, 1972, and 1990. In all three of these surveys, high-school girls had significantly higher grades than boys, but in the first two of these periods, boys took more math and science courses in high school. No separate records on twelfth-grade math and reading achievement scores are available for 1957. In 1972, however, these break-outs are available: boys did better than girls in math, while girls did better than boys in reading. By 1992, girls had narrowed the lead of boys in math scores and had widened their lead in reading scores – and had a composite score that was higher than that of the boys. As for high school courses taken, in 1957 boys took far more courses in math and science than girls – and the differences were especially strong in advanced math, chemistry, and physics. Between 1972 and 1982 there was a strong increase in the number of girls taking math and science courses, and by 1992

---

[1] Other books on the history of women in higher education in America include Newcomer (1959), Graham (1978), Soloman (1985), Gordon (1992), Nash (2005) and Lucas (2006).

[2] I am following Goldin et al. (2006) here on the common intent of using college to find a husband. I do not know, for example, if it was any more common for women to seek a husband in college than for a man to seek a wife – not to mention homosexual or other relationships.

there was rough parity between boys and girls in the math and science courses taken (while girls still maintained a large lead in taking foreign language courses).

In the early postwar years, the socioeconomic background and parental education level mattered. In 1957, families in which the parents were more highly educated or of higher socio-economic status were gender-neutral about sending their children to college; however, in families with less-educated parents or lower socio-economic status, male children were much more likely than female children to be sent to college. By 1992, however, this pattern had completely changed, and at every socio-economic level parents sent more female children to college than male children. Interestingly, the gap between female and male children sent to college was greatest for those in the lowest half of the family socio-economic status distribution.

2.2   A Statistical Overview of Women in Science

We next present some statistical information about women in science. This material is drawn mostly from Margaret Rossiter's authoritative, three-volume history of women in science (1982, 1995, 2012).[3] Much of her data is compiled from the publication *American Men of Science*. Culling data from the 1906, 1910, and 1921 editions, Rossiter (1982, Table 1.1) found that only 439 women had received science baccalaureate degrees prior to 1920. The largest numbers of science degrees women received were awarded in botany (80), zoology (80), and psychology (67), with fewer than 50 degrees given in mathematics and every other scientific discipline. Engineering was not even listed. The top four producers of women scientists (i.e. where these scientists received their undergraduate degree) were all women's colleges: Wellesley (36), Vassar (34), Smith (29), and Mount Holyoke (26). The other schools that graduated 10 or more female students on their way to science careers were several additional women's colleges (Bryn Mawr, Barnard, and Goucher) and five coeducational universities (Cornell, Chicago, Michigan, Pennsylvania, and Nebraska). The leading messages from this data are the overall low numbers of female science graduates, the emphasis on the biological sciences, and the importance of the women's colleges.[4]

---

[3] Also see Rossiter's earlier research on women scientists in America, e.g. Rossiter (1974).

[4] Rossiter (1982, Table 2.1) also considered doctoral education of women. By the year 1900, 228 women had received doctoral degrees in any field in the United States. 56 had received science degrees, with the largest numbers in chemistry (13), math (9), and psychology (9). Bryn Mawr and Yale were tied with the largest number of doctoral degrees awarded to women in scientific fields, at four apiece. By the 1938 edition of *American Men of Science* the number of doctorates awarded overall and the number to females had increased significantly. The data is not reported in a way to make precise numerical comparisons, but Rossiter (Rossiter 1982, p. 152) infers from the literature that the largest producers of female scientific doctorates at the time were Chicago (strong in botany, mathematics, medical sciences, biochemistry, and nutrition) and Columbia (strong in nutri-

Comparing the 1938 publication with the 1921 version of *American Men of Science*, Rossiter (1982, Table 6.2) found that the number of female scientists had more than quadrupled over this period. The only scientific fields with more than 200 female practitioners in 1938 were zoology (281), psychology (277), and botany (256). Engineering shows up for the first time in the 1938 edition, with a total of eight female engineers. There is little change in the undergraduate institutions from which these female scientists receive their degrees. The same women's colleges are prominent, but among coeducational universities, Wisconsin and UC Berkeley climb into the top ten. (Rossiter 1982, Table 6.5)

Just after the Second World War, in 1946–1947, Rossiter (1995, Table 2.1) finds that women comprise 2.7 % of the science and engineering workforce, with only 0.3 % of engineers being women but 20.1 % of mathematicians being women. Over the period from 1954 to 1970, the percentage of women overall in the science and engineering workforce increased from 6.67 to 9.37 % of the workforce.

Considering advanced higher education, Rossiter (1982, Table 4.4) finds that, between 1947 and 1961, only 0.3 % of the engineering doctorates are awarded to women. The 5.5 % of the mathematics and statistics doctorates awarded to women is a surprisingly low number, given the active participation of women in the mathematical sciences workforce during the war.

Between 1970 and 2000, the total number of female scientists and engineers in the United States in all fields grew from about 80,000 to just under 200,000. (Rossiter 2012, Figure 3.1) The vast majority were scientists, not engineers. The number of baccalaureate degrees in engineering awarded to women increased from a few hundred (less than 1 % of all these degrees) in 1970 to approximately 12,000 (approximately 20 %) in 2000. (Rossiter 2012, Figures 3.2, 3.3) In computer science, the number of baccalaureates offered to women increased from just a few hundred in 1970 – with a steep rise between 1980 and 1987 – to approximately 14,000; and then they began to decline, stabilizing in the mid-1990s at around 8000. (Rossiter 2012, Figure 3.11) This pattern of decline in the late 1980s has perplexed many scholars and is the subject of continuing discussion.

2.3   Science Education for Women – A Brief History

Now that we have shared a few of the statistics from Rossiter's three-volume historical survey of American women in science, we selectively summarize from her narrative. Women participated in science in the United States in small numbers, beginning in the 1820s. Throughout the nineteenth century, gains were made in secondary and college education for women, thus preparing more women to pursue scientific careers. However, many of these women were channeled into the "nurturing sciences", especially domestic science and the medical sciences. In the late

---

tion, zoology, anthropology, and psychology) – followed distantly by Cornell, Johns Hopkins, and Yale.

nineteenth century, a new barrier to women's participation in science arose: the requirement of a doctoral degree to pursue a scientific career in the academy. This was a barrier because of the lag of 20–30 years before most doctoral programs began to admit women. As professional scientific societies were formed in the second half of the nineteenth century, women were often excluded from being members or were restricted to lower categories of membership. The number of women employed in scientific positions in the federal government increased between the two World Wars, but the hiring patterns showed gendered occupational differentiation, with most women working in jobs involving home economics, botany, microbiology, statistics, or clinical psychology.[5] Individual women scientists during the years before the Second World War often experienced isolation. As of 1940, women scientists, even if they held doctoral degrees, experienced significant barriers to scientific career advancement.

Although the 25 years following the end of the Second World War comprised a period of rapid scientific development, it was a low period for women, who often departed early from their scientific education or careers or were marginalized or underutilized in their scientific careers. Women experienced low pay compared to men in these scientific jobs and were often forced out if they got married; and there was little legal protection for their jobs or compensation. According to Rossiter, claims about the advancement of women in science during the Second World War are often exaggerated. Her data shows only a marginal increase in the participation of women in science created by the Second World War. Approximately 100 women scientists found temporary appointments in government and industrial war projects. Some women scientists who had been unemployed during the 1930s found temporary college teaching positions to replace men who were off at war. The total size of the scientific workforce almost doubled during the war years, but the percentage of scientists who were female grew only from 4.0 % in 1941 to 4.1 % in 1945. Rossiter argues that public opinion about women's proper role in society, even among some of the women scientists themselves, dampened growth in the number of women scientists.

The situation for women scientists deteriorated after the war was over. Overwhelming demand for a college education, paid for by the GI Bill, led to the rapid increase in college enrollment of male veterans. Partly to meet this demand, colleges set quotas or other restrictions on women's enrollment, beginning in 1946. The restrictions were especially widespread for graduate education.

If women had difficulties in studying to be scientists, there were also barriers to women scientists becoming faculty members in the 1950s and 1960s. Many of the places where women had traditionally taught – women's colleges, teachers colleges,

---

[5] Mathematics was one route into computer science, especially in the 1950s through the 1970s. Green and LaDuke (1989) describe the production of women receiving doctorates in the mathematical sciences prior to the 1940s. There is a surprisingly large literature on women mathematics through history. For an extensive bibliography, see the webpage entitled "Biographies of Women Mathematicians" created by Larry Riddle and colleagues at Agnes Scott College, https://www.agnesscott.edu/lriddle/WOMEN/women.htm (accessed 21 March 2016).

and home economics departments – were closing their doors to women faculty, and the newly created or expanded coeducational institutions did not hire women in large numbers. Some of the middling-quality teachers colleges tried to build institutional prestige in the first decades after the war by taking actions that were often antithetical to the interest of women scientists: getting rid of the "old girls" on their faculties, raising salaries and reducing teaching loads (to attract highly qualified male applicants), hiring more PhDs, and rebranding themselves as universities. The reinstatement of anti-nepotism laws after the war (focused primarily on husband and wife appointments), especially in the public universities, led to married women being forced to leave their current faculty positions or not being considered for faculty positions. Even single women were affected by these laws because university administrators worried that these women would eventually marry. At Smith College, historically one of the most important and most feminist-minded of the women's colleges, the percentage of women on the faculty dropped from 60 to 49 % between AY47-48 and AY 56–57. When women were hired as faculty members, it was often at a lower rank or to visiting professor or research assistant positions; typically at low pay; and in a number of cases the women faculty were not given full access to university facilities, e.g. not being permitted to use the faculty club where intellectual and power networks were built on campus. Few women received appointments as department chair or dean in the 1950s and 1960s, except occasionally at women's, junior, or teaching colleges.

In the early 1950s there was some interest in increasing opportunities for women in science to make sure the United States had enough experts to staff its national defense needs, but this interest subsided after the Korean War ended in 1953. Some limited interest appeared again in training and hiring women scientists in 1957, after the launch of Sputnik, on the basis of a need for more teachers in higher education for national defense purposes. In the late 1950s and early 1960s, while the government mildly encouraged young women to pursue science and engineering careers, graduate schools and employers did not show the same enthusiasm. Although the number of women entering graduate school increased in the 1950s and 1960s, women wanting to attend graduate school were at a disadvantage compared to men; they were severely limited in the set of schools that would admit them (often the most prestigious schools were not open to them), the fields open to them (in the sciences, women were often channeled to psychology and home economics), and the range of professors who would agree to supervise their studies.[6]

Career opportunities for women scientists were somewhat better in other nonprofit organizations during the 1950s and 1960s than in the universities. A significant number of women scientists found employment in federal labs, although leadership positions were seldom open to them. While opportunities for women scientists in industry grew during the 1950s and 1960s, most of the jobs available were in feminized occupations such as routine testing, home economics, and chemical librarianship – although there were also jobs for women as computer programmers. Women scientists were often underutilized in their work in industry.

---

[6] For more information on this topic, see Bix (2002).

In the 1950s, the American Association of University Women (AAUW) protested against the marginalization of women in the academy, but it had limited influence at the time due to complaints from the political Left that some AAUW chapters did not accept African American women, and from the political Right that AAUW members were Communist sympathizers. Betty Friedan, who had an undergraduate degree in psychology and a year of graduate study at UC Berkeley, was an inspirational figure in the 1960s for a number of the women (and a few men), agitating for a better situation for women scientists. In the late 1960s, increasing numbers of, and more strident, voices began to be heard.[7] 1972 was a watershed year for American women scientists with the passage of the Equal Employment Opportunity Act, the Educational Adjustments Act (including, most famously Title IX), and the Equal Rights Amendment – the last of these passed by Congress but not yet ratified by the states. Unfortunately, it became apparent at about the same time that the universities had produced a glut of science doctorates in the United States, reducing the opportunities for women newly entering the scientific workforce.

In the 1970s there was a growing infrastructure for women in science in the United States. A number of professional organizations became active, including the Association for Women in Science (discussed in Chap. 6), Association for Women in Mathematics, Association for Women in Psychology, and Sociologists for Women in Society. There were also women's committees within existing scientific societies such as the American Chemical Society and the American Physical Society, which were created or became more active in the 1970s. Also created in this decade was the low budget, but effective Office of Opportunities in Science, co-located with the American Association for the Advancement of Science. This Office produced reports and registries of women scientists. In 1975 the National Science Foundation created a Women in Science program, which by 1981 had supported 51 re-entry programs for women.

The politically well-connected chemist Lilly Hornig from the Higher Education Resource Service (HERS), a placement service, was highly critical of NSF's initial efforts concerning women in science.[8] She argued that the NSF program supporting women was buried in the education directorate and not associated with the more powerful research directorates; and that this program mainly sponsored remedial programs for women at the bachelor's and master's levels and did not address the problems found in the leading research universities. Her criticisms reached the ears of Senator Edward Kennedy, and he persuaded the NSF to convene a conference in 1977 of 60 young female scientists in the hopes of identifying new steps the federal government might take in improving the situation for women scientists. As part of

---

[7] One of the more strident voices was that of psychology professor Naomi Weisstein, who wrote a satirical piece entitled "How Can a Small Girl Like You Teach a Class Full of Big Men, and Other Things the Chairman Said" (Weisstein 1974, most readily found as a 1977 reprint) as well as "Woman as Nigger, or How Psychology Constructs the Female." (Weisstein 1969). This latter piece was reprinted in the *Congressional Record* as part of hearings on discrimination of women in higher education.

[8] For an excellent but dated (ending in the mid-1990s) account of women scientists and engineers at American research universities, see Part II of Hornig (2003).

this same story, the White House's Office of Science and Technology Policy provided funding to the National Research Council's Committee on the Education and Employment of Women in Science and Engineering, which was chaired by Hornig, to write two reports: *Climbing the Academic Ladder: Doctoral Women Scientists in Academe* (National Research Council 1979) and *Women Scientists in Industry and Government: How Much Progress in the 1970s?* (National Research Council 1980) These reports spelled out how poorly women scientists were faring in both the academic and industrial realms.[9]

During the 1970s, women scientists enhanced their political acumen and political power. The principal organization through which this happened was the Federation of Organizations for Professional Women, which represented more than 100 women's groups interested in Washington science policy. The organization was active from 1972 until the early 1990s. Perhaps the most important figure involved with this organization was the political scientist Janet Welsh Brown from the Office of Opportunities in Science. One of the largest political goals of women scientists in the 1970s was ratification of the Equal Rights Amendment. It had been introduced into every Congress since 1923 and was passed by both the House and Senate in 1972, but it needed to be ratified by 38 states. Getting approval from the final three states needed for ratification turned out to be a major political challenge. The National Organization of Women boycotted states that had not yet ratified the Amendment. The American Association for the Advancement of Science followed suit in 1979, when it broke a hotel contract in Chicago in 1979 and moved its annual meeting to Houston in support of the NOW boycott. Nevertheless, the Amendment was never ratified by the required 38 states.

Instead, the major political success for women scientists of that era was the passage in 1980 of the Science and Technology Equal Opportunity Act. The bill had failed to achieve passage in 1978, and the 1980 version was a weakened version. It created the Committee on Equal Opportunities in Science and Technology (CEOST) to advise the NSF director and provided modest funds for visiting professorships for women and some other programs.[10] (For more information on CEOST, later renamed CEOSE, see Aspray (2016).)

Political action was supplemented by legal action during the 1970s. While there was some modest voluntary improvement in the employment practices concerning women faculty at some universities, change in the 1970s came primarily through a series of lawsuits against universities under the Equal Employment Opportunity Act, which generally had lax enforcement until there was pressure of legal action. The three most significant cases of the 1970s were *Johnson v. University of Pittsburgh* and *Lamphere v. Brown University*, which both involved female assistant

---

[9] There was far from total sympathy for the plight of women scientists. One highly public example was the book published in 1979 by sociologist Jonathan Cole, entitled *Fair Science: Women in the Scientific Community*. (Cole 1979) Cole argued on the basis of a citation analysis that women deserved their second-tier status in science because of the quality of their work.

[10] For more information about the Women in Science and Technology Equal Opportunity Act, see Puaca (2014), Sheffield (2005) and National Research Council (2007).

professors being denied tenure as the universities practiced what came to be known as the "revolving door policy," and *Rajender v. University of Minnesota* in which Shymala Rajender was underemployed (denied a tenure-track appointment) on the basis of sex. These and other court cases moved universities to change their personnel policies. The 1970s also witnessed several lawsuits over equal pay for women academics.

Government support for women academics deteriorated in the 1980s during the Reagan Administration, when its fiscal conservatism limited funds that might have been used for improving the situation and because it relaxed enforcement of the equal opportunity laws and, in particular, gave a narrow interpretation of the Title IX provisions of the Educational Amendment Act as applying mainly to athletics. For all of these reasons, and for the fact that tenure-track positions turn over slowly, increases in the number of women in tenure-track positions increased slowly in the 1970s and 1980s. Change was particularly slow in the fields of chemistry, math, physics, economics, and engineering.

Another factor that affected women in academic positions in science and engineering were changing patterns in higher educational institutions. Between 1968 and 1985 approximately 40 major men's colleges and universities, including for example Bowdoin and Princeton, began admitting women. Since many of these schools offered strong programs in the STEM disciplines, this coeducation movement created a new opportunity for women students to obtain a STEM undergraduate education. However, in most of these schools there were only a few women majoring in the STEM disciplines, perhaps because of an environment that favored men over women, such as professors calling more frequently on men than women in class.[11]

Historically Black colleges and universities had been under duress during the late 1960s and early 1970s, when many African American students elected to attend majority institutions. But in the period 1978–1994, with strong federal support, the HBCUs began to grow again. Typical undergraduate enrollment at an HBCU was more than 50 % female, and about half of all African American women who earned science and engineering doctorates attended HBCUs as undergraduates. Spelman and Bennett Colleges, in particular, were important feeders into this system.

The number of women's colleges in the United States dropped from approximately 200 in 1970 to only 90 in the 1990s. However, the top women's colleges made concerted efforts to strengthen themselves in order to survive in these years – with stronger student recruitment, better salaries to attract more qualified faculty members, and improvements in the physical plant including science facilities. Although the women's colleges enrolled only about 2 % of college students, a disproportionately large number of these students attended graduate school and entered science careers.

---

[11] For more on why undergraduate women remain in or depart from science majors, see Seymour and Hewitt (1997).

Between 1970 and 2000, the number of women completing doctoral degrees in science and engineering rose from 1648 to 9396.[12] There seem to be many reasons for this increase: more women actively seeking professional careers, some improvement in the academic climate for women at least at some universities (although there was great variation by both institution and discipline), more equitable distribution of fellowship funding, a slightly larger number of women faculty members to serve as role models, increasing sizes of female cohorts in some schools and some disciplines (e.g. in the psychological and biological sciences, not so much in engineering) that helped reduce student feelings of isolation, creation of women in science programs on campus, and establishment of local campus chapters of national organizations to support women such as SWE and AWIS – as well as, of course, equal opportunity laws. Despite all of these changes, many graduate programs continued to have chilly climates or outright discrimination, and most scientific and engineering departments remained overwhelmingly male.

2.4   Engineering Education for Women – A Brief History

We now turn to the history of engineering education for women, relying heavily on the writings of Amy Sue Bix (especially 2013, but also 2000, 2002, 2004). The first engineering departments were formed in the United States in the 1810s and 1820s – at West Point, Norwich University, and Rensselaer Polytechnic Institute. Civil, mining, and mechanical engineering developed first among the engineering disciplines. The first electrical engineering departments were formed in the 1880s at MIT and Cornell. By the end of the nineteenth century, each of the major engineering fields had formed its own professional organization in the United States.

The first efforts to organize women in the engineering disciplines came in the 1910s.[13] The earliest identified group is a local group of women engineers, the T-Square Society, at the University of Michigan. It was formed in 1914 by the thirteen female students then enrolled in engineering and architecture, with the intention of providing a meeting place for these students to socialize and discuss topics of mutual interest.[14] According to a survey taken in 1919, there were 43 women engineers at the University of Michigan at that time – possibly the largest concentration of women engineers on any American campus. This survey of women engineering students and graduates was conducted by women engineers at the University of Colorado at Boulder, who in 1918 had founded what they called the American

---

[12] For more on the graduate education of women in science and engineering, see Sonnert and Holton (1995a, b) and Long (2001).

[13] The leading scholar of the history of women in engineering in the United States is Amy Sue Bix. She has consolidated a number of her earlier studies into a book (Bix 2013). For those who want to read an article-length version of the material in her book, see Bix (2004). The account in this section relies heavily on Bix's writings.

[14] See Bix (2013) for additional information about the T-Square Society.

Society of Women Engineers and Architects. They polled twenty universities and identified approximately 200 women enrolled in the study of engineering across the nation. The American Society of Women Engineers was never able to build a national membership, but two of the women from the University of Colorado involved with the survey, Hilda (Counts) Edgecomb and Elsie Eaves, helped to create the Society of Women Engineers after the Second World War.[15]

The timing of the survey is perhaps not coincidental. During the First World War, in which the United States participated during 1917 and 1918, there was "a sudden influx of women into such unusual occupations as bank clerks, ticket sellers, elevator operator, chauffeur, street car conductor, railroad trackwalker, section hand, locomotive wiper and oiler, locomotive dispatcher, block operator, draw bridge attendant, and employment in machine shops, steel mills, powder and ammunition factories, airplane works, boot blacking and farming" (*Seattle Union Record* (1918) as quoted in Kim (2003)). Women worked during the war in munitions factories, and there are occasional references to women holding engineering jobs. The war work done by women, as nurses on the war front and in offices and factories on the home front, was recognized by President Woodrow Wilson when he addressed the U.S. Senate in September 1918 to urge it to join the House in approving the 19th Amendment, which gave women suffrage. (Gavin 1997; Goldstein 2001) However, these women were treated as replacements for the men away at war and were expected to surrender those jobs to returning servicemen. Opportunities for women in engineering were uncertain in 1919, when the University of Colorado women sent out their survey.[16]

Until after the Second World War, the number of women who studied or practiced in any of the engineering fields remained small.[17] During the Second World War there was a shortage of engineers to work in the factories building guns, tanks, aircraft, and other devices for the war. A number of companies, such as General Electric and the Curtiss-Wright aircraft company, hired women with basic math and science skills and trained them to work as engineers. For example, Curtiss-Wright provided intensive 10-month training to its Cadettes on college campuses (Cornell, Iowa State, Minnesota, Penn State, Rensselaer, and Texas) to prepare them to move

---

[15] More detail about the University of Colorado activities can be found in Bix (2013).

[16] The situation for women working during the First World War was similar in Britain and the United States. For an insightful and detailed account of the British situation, see the keynote lecture by Patricia Fara of the University of Cambridge (Fara 2014). Human computers had been used in astronomy research since 1750, when Alexis Claude Clairaut had used Newton's laws of motion to calculate the return of Halley's comet in 1759. Most of the "computers" as these human calculators were called were men until the Harvard University astronomer Edward Charles Pickering employed a team of women to do his computations. It is possible that the First World War gave women a chance to serve as computers, taking up jobs that had been by men who had gone off to war. For information about human computers, see Grier (2005).

[17] Another organization, the American Society of Women Engineers and Architects, was founded in 1920 as a support network for practicing female engineers. It also encouraged and advised young women who wanted to be engineers. Membership was always small, and the organization had become largely inactive by the entry of the United States into the Second World War in 1941.

into the company's research and production facilities. But just as in the First World War, when the war ended most of the jobs reverted to the returning veterans.

A number of women who had practiced as engineers, plus a few supportive men, agitated for opening up the engineering professions to women in the postwar years. Female students on the campuses of Iowa State, Syracuse, and Cornell created women's groups or honorary societies to support women wanting to study engineering. In 1952 Georgia Tech opened admission to women, despite considerable opposition from students, faculty, and alumni.[18] The most important response perhaps was the creation of the Society of Women Engineers (discussed in Chap. 6).

During the 1950s and 1960s, although admission to some of the leading engineering schools in the United States was open to women, the numbers of female engineering students remained small and public sentiment was skeptical about whether women could be good engineers. In the 1950s, female engineering enrollment in the United States grew from approximately 700 to 1700 – thus growing in this time period from 0.35 to 0.60 % of all engineering students.[19]

Change started to occur at a faster pace in the mid-1960s. A major symposium on American Women in Science and Engineering was held at MIT in 1964.[20] That same year, Congress passed the Civil Rights Act barring discrimination on the basis of sex (as well as race, color, religion, or national origin), and the National Organization for Women pressured the Equal Employment Opportunity Commission to enforce the law. While major companies complied with the law and actively recruited women engineers, both the engineering educational system and the engineering workplace remained overwhelmingly male. As of 2000, university engineering departments had few female students and few female faculty members, especially at the higher ranks; while engineers in the workforce remained predominantly male, and female engineers were not promoted as rapidly or given advances in compensation as rapidly as males.[21]

---

[18] For more information about admission of women engineering students at Georgia Tech, as well as at Cal Tech, MIT, and RPI, see Bix (2000).

[19] These statistics are summarized from McGreaham (1963) as reprinted in Bix (2013).

[20] In fact, events had almost gone the other way at MIT: "in 1960, a fundamental change of course had occurred. An MIT faculty committee majority report recently had recommended that MIT cease admitting women, which would have placed MIT on the wrong side of history and counter the changes occurring at other technical universities where women were being admitted to engineering programs for the first time. Happily it was a minority report by Kenneth Wadleigh which won support by President James Killian and his Chancellor Julius Stratton, and the decision was made not only to continue to admit women, bit to actively work to improve the environment and resources available for women students." (http://1964.alumclass.mit.edu/s/1314/2015/club-class-main.aspx?sid=1314&gid=55&pgid=11879)

[21] On the choice by women of an engineering education or career, see Frehill 1997.

References

Aspray, William. 2016. *Participation in computing: The National Science Foundation's expansionary programs*. London: Springer.

Bix, Amy Sue. 2000. 'Engineeresses' invade campus: Four decades of debate over technical coeducation. *Technology and Society Magazine* 19(1): 20–26.

Bix, Amy Sue. 2002. Equipped for life: Gendered technical training and consumerism in home economics, 1920–1980. *Technology and Culture* 43(4): 728–754.

Bix, Amy. 2004. From 'engineeresses' to 'good engineers': A history of women's U.S. Engineering education. *NWSA Journal* 16(1): 27–49.

Bix, Amy Sue. 2013. *Girls coming to tech! A history of American engineering education for women*. Cambridge, MA: MIT Press.

Cole, Jonathan R. 1979. *Fair science: Women in the scientific community*. New York: Free Press.

Fara, Patricia. 2014. A lab of one's own? Women and science in World War One. *Revealing Lives, Royal Society, May*. London. Conference Presentation. http://womeninscience.net/?page_id=675#keynote2. Accessed 22 Sept 2014.

Frehill, Lisa. 1997. Education and occupational sex segregation: The decision to major in engineering. *Sociological Quarterly* 38: 225–249.

Gavin, Lettie. 1997. *American women in World War I – They also served*. Niwot: University Press of Colorado.

Goldin, Claudia, Lawrence F. Katz, and Ilyana Kuziemko. 2006. *The homecoming of American college women: The reversal of the college gender gap*. Working paper 12139, National Bureau of Economic Research, March. http://www.nber.org/papers/w12139. Accessed 25 Apr 2015.

Goldstein, Joshua S. 2001. *War and gender: How gender shapes the war system and vice versa*. Cambridge: Cambridge University Press.

Gordon, Lynn D. 1992. *Gender and higher education in the progressive era*. New Haven: Yale University Press.

Graham, Patricia Albjerg. 1978. Expansion and exclusion: A history of women in American higher education. *Signs* 3(4): 759–773.

Green, Judy, and Jeanne LaDuke. 1989. Women in American mathematics: A century of contributions. In *A century of mathematics in America, part II*, ed. Peter L. Duren, 379–398. Providence: American Mathematical Society.

Grier, David Alan. 2005. *When computers were human*. Princeton: Princeton University Press.

Hornig, Lilli S. (ed.). 2003. *Equal rites, unequal outcomes: Women in American research universities*. New York: Kluwer.

Kim, Tae H. 2003. Seattle general strike: Where women worked during World War I. The Great Depression in Washington State Project. http://depts.washington.edu/labhist/strike/kim.shtml. Accessed 26 Feb 2016.

Long, J. Scott. 2001. *From scarcity to visibility: Gender differences in the careers of doctoral scientists and engineers*. Washington, DC: National Academies Press.

Lucas, Christopher J. 2006. *American higher education: A history*, 2nd ed. New York: Palgrave Macmillan.

McGreaham, Ann. 1963. The opposite sex in engineering. *Purdue Engineer* May: 20–24.

Nash, Margaret A. 2005. *Women's education in the United States, 1780–1840*. New York: Palgrave Macmillan.

National Research Council. 1979. *Climbing the academic ladder: Doctoral women scientists in academe*. Committee on the Education and Employment of Women in Science and Engineering, Commission on Human Resources. Washington, DC: National Academies Press.

National Research Council. 1980. *Women scientists in industry and government: How much progress in the 1970s?* Committee on the Education and Employment of Women in Science and Engineering, Commission on Human Resources. Washington, DC: National Academies Press.

National Research Council. 2007. *Beyond bias and barriers: Fulfilling the potential of women in academic science and engineering*. Committee on Maximizing the Potential of Women in Academic Science and Engineering. Washington, DC: National Academies Press.

Newcomer, Mabel. 1959. *A century of higher education for American women*. New York: Wiley.

Puaca, Laura Micheletti. 2014. *Searching for scientific womanpower: Technocratic feminism and the politics of national security, 1940–1980*. Chapel Hill: The University of North Carolina Press.

Rossiter, Margaret. 1974. Women scientists in America before 1920: Career patterns of over five hundred women scientists of the period reveal that, while discrimination was widespread, many women were working hard to overcome it. *American Scientist* 62(3): 312–323.

Rossiter, Margaret. 1982. *Women scientists in America: Struggles and strategies to 1940*. Baltimore: Johns Hopkins University Press.

Rossiter, Margaret. 1995. *Women scientists in America: Before affirmative action, 1940–1972*. Baltimore: Johns Hopkins University Press.

Rossiter, Margaret. 2012. *Women scientists in America: Forging a new world since 1972*. Baltimore: Johns Hopkins University Press.

Seattle Union Record. 1918. *Protecting the working mothers*. April 24

Seymour, Elaine, and N.M. Hewitt. 1997. *Talking about leaving: Why undergraduates leave the sciences*. Boulder: Westview Press.

Sheffield, Suzanne. 2005. *Women and science: Social impact and interaction*. New Brunswick: Rutgers University Press.

Soloman, Barbara Miller. 1985. *In the company of educated women: A history of women in higher education in America*. New Haven: Yale University Press.

Sonnert, Gerhard, and Gerald Holton. 1995a. *Gender differences in science careers*. New Brunswick: Rutgers University Press.

Sonnert, Gerhard, and Gerald Holton. 1995b. *Who succeeds in science? The gender dimension*. New Brunswick: Rutgers University Press.

Weisstein, Naomi. 1969. Woman as nigger: Psychology constructs the female. *Psychology Today* 3: 20, 22, 58.

Weisstein, Naomi. 1977. How can a small girl like you teach a class full of big men, and other things the chairman said. Reprint of a 1974 note. In *Working it out: Twenty-three women writers, artists, scientists, and scholars talk about Their lives an work*, ed. Sara Ruddick and Pamela Daniels, 241–250, New York: Pantheon.

Chapter 3
Opening STEM Careers to African Americans

Abstract This chapter provides an overview of higher education for African Americans. It includes material about the history of African American education and the important role of federal legislation and court decisions in a sphere (education) that is largely left to the states. It provides an overview of science and technology education for African Americans. It discusses the important role of Historically Black Colleges and Universities. It also includes information about two organizations – the United Negro College Fund and the Thurgood Marshall College Fund – that are important funders of African American education generally.

This chapter provides an overview of higher education for African Americans. It includes material about the history of African American education and the important role of federal legislation and court decisions in a sphere (education) that is largely left to the states. It provides an overview of STEM education for African Americans. It discusses the important role of Historically Black Colleges and Universities. It also includes information about two organizations – the United Negro College Fund and the Thurgood Marshall College Fund – that are important funders of African American education generally. Some people who are mostly concerned with increasing the participation of African Americans in computing today might question whether we needed to go back as far in time in telling this history as we do, but the purpose is to better understand the social and political contexts in which African Americans have pursued higher education and STEM careers in the United States over time.

© Springer International Publishing Switzerland 2016   31
W. Aspray, *Women and Underrepresented Minorities in Computing*,
History of Computing, DOI 10.1007/978-3-319-24811-0_3

3.1 African Americans and Higher Education – A Brief History

African Americans were the largest minority group in the United States throughout the twentieth century.[1] With the rapid growth in the Hispanic population in the last decades of the century, however, as of the 2000 U.S. Census the Hispanics became the largest and the African-Americans the second largest minority group.

The first African Americans arrived in America early in the nation's history, in 1619, as indentured servants in the Jamestown, VA colony; and of course, until 1865 a large percentage of African Americans were bound in slavery.[2] Little attention was given to the education of African Americans until the time of the American Revolution. In 1774 the Abolitionist Society opened a school for Blacks in Philadelphia. Between 1790 and 1810, several schools were opened in the Carolinas for free Blacks, but a conservative White backlash led South Carolina to pass a law in 1834 that outlawed any instruction to Blacks, even to those who were free.[3]

From the 1820s until the American Civil War in the 1860s, only a few African Americans received a higher education. Amherst, Bowdoin, and Dartmouth Colleges admitted African-American students in the 1820s, and Harvard did so in the 1840s. Oberlin College, founded in the 1830s in Ohio, had a strong Abolitionist orientation; and one third of its students were African Americans before the Civil War. Berea College, founded in Kentucky in 1855, also admitted African-American students.[4] Several Black colleges were opened before the Civil War, including Lincoln College in Pennsylvania in 1854 and Wilberforce University organized by the African Methodist Episcopal Church in Ohio in 1856.

---

[1] The literature on African Americans and higher education covers a number of topics. Richards et al. (2013) and Thurgood Marshall College Fund (2013) provide information about scholarship organizations. Cowan and Maguire (1995), Jackson (2001), Willie et al. (1991), Bechtel (1989), Roebuck and Murty (1993) and Brown and Davis (2001) discuss the history of African-American higher education. Bauman (1998) covers educational attainment. Grier-Reed et al. (2011) covers transition to college. Blau (1999) covers community college. Freeman (1999), Brooks and Starks (2011), Elam (1989), Fleming (1984), Hale (2004), Harper and Newman (2010), Roebuck and Murty (1993) and Roscoe (1989) provide general information. For a bibliography of the early literature, see Fitch and Johnson (1988).

[2] It is hard to get good data, but as of 1860, there were approximately 4.5 million African Americans in the United States. Of the approximately four million African Americans in the South, approximately a quarter million were free. Of the half million African Americans in the North, practically all of them were free as of 1860. Abolitionist laws were enacted in Northern states beginning in the late eighteenth century, and by 1840 there were almost no slaves in the North.

[3] One interpretation is that Southern Whites were nervous about providing any education to Blacks because they would become more dissatisfied with their situations and might be in a better position to plan a revolt against Whites. The Gabriel [Prosser] revolt in Richmond, Virginia in 1800, the Denmark Vesey's plan for "the rising" up against White Charlestonians in 1822, and Nat Turner's bloody revolt in Southampton County, Virginia fueled this fear.

[4] A Kentucky law passed in 1904 segregated all education. It was upheld by the Supreme Court in 1907. Berea only resumed admitting African Americans in 1954.

The Civil War opened many opportunities for African-American education. During the war, schools were founded for Black students in some regions of Virginia, North Carolina, South Carolina, and Kansas after Union troops had taken control. The Thirteenth Amendment, passed in 1865, abolished slavery and overturned the Dred Scott ruling by the same court just 8 years earlier, which had denied citizenship to slaves and made slavery legal throughout the United States.[5]

In the first decade after the Civil War, between 1865 and 1875, 24 Black colleges opened. Schools founded during these years included many of the ones that remain leading colleges and universities today: Atlanta, Fisk, Hampton, Howard, Morehouse, Morgan State, Johnson C. Smith, and Tougaloo. By 1890, more than 200 Historically Black Colleges and Universities (HBCUs) had been founded, although only a few were advanced enough to offer a bachelor's degree.[6] Many of these schools were founded by either the African Methodist Episcopal Church or the American Missionary Association.

In 1865, Congress passed legislation creating a new federal agency known as the Bureau of Freedmen, Refugees, and Abandoned Lands (Freedman's Bureau), which lasted until 1872. It was responsible for the early Reconstruction efforts in aiding freed slaves. The Freedman's Bureau created many public and trade schools, as well as a few colleges. In 1867, Congress appropriated funds to open Howard Normal and Theological School for the Education of Teachers and Preachers (later Howard University) in Washington, DC under the direction of General O.O. Howard, who was the commissioner of the Freedmen's Bureau.

However, the Reconstruction Era was over by 1877. Reconstruction had mainly been supported by the Northern states and imposed on the Southern states. As memories of the Civil War faded, Northerners became less concerned about Reconstruction while Southern states began to return to Democratic control, the South was set back into a principally agricultural economy, and Whites regained control of Southern society through political means and intimidation. African Americans in the South quickly became second-class citizens with limited political and other rights. The Reconstruction Era, which at first looked promising for the advancement of African Americans, quickly and resoundingly faded.

Nevertheless, the last quarter of the nineteenth century witnessed a number of key events in African American educational history. The West Point military academy became integrated. Tuskegee Institute was founded with instruction given by the famous educators Booker T. Washington and George Washington Carver. Congress passed the second Morrill Act in 1890, which required states with segregated higher education systems (i.e. the Southern and the North–South border states) to create and provide ongoing support for land-grant colleges to serve the

---

[5] The Thirteenth Amendment was bolstered by passage of the Fourteenth Amendment in 1868, giving equal protection under the law, and the Fifteenth Amendment in 1870, giving African Americans the right to vote.

[6] According to Allen et al. (2007), as of 1900 only 58 of 99 HBCUs taught a college-level curriculum and 90 % of the enrollment on these campuses was pre-college students.

Black higher educational system as well as the White system.[7] The second Morrill Act stimulated the opening of 16 new public HBCUs by 1900 – most of them offering vocational training – at new institutions such as North Carolina A&T and Florida A&M; and more generally this legislation provided much needed funds to support Black higher education.[8] However, the second Morrill Act also reinforced the notion of separate but equal higher education. By the turn of the century, 55 % of the African-American population was literate (up from 5 % in 1865).

The situation was worst for African-American education in the South, where some 90 % of African Americans then lived. Southern Whites remained generally opposed to African-American education. At the time, there was a widespread belief among Whites that African Americans were mentally inferior and Southern Whites worried that broadening the educational system would be another way for Northerners and their beliefs to gain a foothold in the South. Between 1875 and 1930, Southern state legislatures pushed for segregated education and emphasized industrial rather than liberal arts education in the African-American schools.[9] The growth in urban, Southern, Black high schools fueled a demand for new teachers that was largely filled by graduates of the HBCUs.

In the Progressive Era of the early twentieth century, the National Association for the Advancement of Colored People (1909) and the National Urban League (1910) were founded.[10] By 1932, there were 117 African-American higher educational institutions in the United States. Of these, 36 were public. The overwhelming majority of the 81 private schools were church-affiliated. The 1930s were also a turning point in African-American enrollment. In 1900, African American college students had been overwhelmingly male. In 1930 there were still perhaps four Black men for

---

[7] The first Morrill Act, passed in 1862, had established the system of land-grant colleges, which are public universities whose mission included teaching agriculture and the mechanical arts.

[8] By 1900, three quarters of the HBCUs that exist today had been founded.

[9] The Southern states were not keen on educating poor Whites either. In 1915, the North Central states spent $28 per White pupil on public education, while South Carolina spent $14 per White child and $1.13 per African American child. Fewer African American children attended public school than White children because the African American schools were more widely scattered and little school transportation was available. The student-teacher ratio was approximately twice as high at African American schools as at White schools. The average length of the public school year in South Carolina was 173 days for White students and 114 days for African American students. (Bechtel 1989)

In higher education, the model of industrial education was developed at Hampton Institute in Virginia and replicated at Tuskegee Institute in Alabama by the great educator Booker T. Washington, a graduate of Hampton. The liberal arts model of education for African Americans was championed by another great educator, W.E.B. DuBois, who graduated from Harvard and taught at Atlanta University. Atlanta, Fisk, and Howard universities were the Historically Black Universities with the strongest liberal arts tradition.

[10] To provide a more complete account than is possible here, one might want to not only tell the stories of the National Urban League and the NAACP, but also the history of the National Association for Equal Opportunity in Higher Education (NAFEO) – formed in 1969 – and perhaps some other institutions such as the Office for Advance of Public Black Colleges (part of the National Association of State Universities and Land Grant Colleges) and the National Urban League.

every Black woman in college. However, by 1935, Black women were outnumbering Black men in college by a 4 to 3 ratio.

Funding to higher education generally and HBCUs specifically decreased significantly during the early 1940s, with federal funds being diverted to the war effort. In 1944 the United Negro College Fund (UNCF) was established by 39 HBCUs to improve the quality of these institutions and provide financial aid to their students. UNCF turned out to be a much more effective fundraiser than the individual colleges and universities had been.

Much of the advancement of African-American education since the Second World War has come through federal legislation and court decisions in the U.S. Supreme Court. Let us begin with the legislation and court decisions affecting K-12 education before turning to higher education. The most famous Supreme Court decision was *Brown v. Board of Education* (1954), which overturned separate but equal education on the grounds that the Black schools did not have the resources to provide equal education and that segregation in and of itself was psychologically damaging to African-American children.[11] In 1957 President Eisenhower sent federal troops to enforce integration of Central High School in Little Rock, Arkansas.

A number of actions took place as part of President Johnson's Great Society Program. Title IV of the 1964 Civil Rights Act addressed school segregation at both the K-12 and higher education levels, while Title VI required federal agencies to fund schools in a non-discriminatory manner. President Johnson's Executive Order 11246 explicitly required non-discriminatory hiring and employment practices for government contractors. The Head Start program, launched in 1965, provided assistance to disadvantaged preschoolers. The Elementary and Secondary School Act of the same year gave the federal government the right to intervene in state education, and it was applied to improve K-12 education for African-American students, especially in the South.[12] The Higher Education Act of 1965 provided funds to HBCUs for faculty and curricular improvements, student services, and administrative improvements.

In the late 1960s and the early 1970s, disputes over these federal K-12 initiatives were fought out in the courts. In 1968, the U.S. Supreme Court ruled in *Green v. County School Board of New Kent County* that policies that gave parents freedom of choice where to send their children to school were unconstitutional because they perpetuated dual systems of education; the court mandated more proactive steps to dismantle dual systems. In 1971 the court ruled in *Swann v. Charlotte-Mecklenburg Board of Education* that gerrymandering in order to maintain segregation was

---

[11] *Brown* overturned, at least for public education, the U.S. Supreme Court decision *Plessy v. Ferguson* (1896), which had upheld the Constitutionality of separate but equal public facilities (in this case, for seating on railway cars). For a detailed discussion of *Brown*, see Kluger (1975).

[12] The influential Coleman Report, *Equality of Educational Opportunity*, published in 1966 stimulated busing to fight segregation. However, the recommendation in the report to track students on the supposedly objective grounds of testing led to segregation within schools, with large numbers of African Americans directed into vocational tracks and large numbers of Whites directed into college preparatory tracks.

unconstitutional.[13] In a ruling unfavorable to minorities, *San Antonio Independent School District v. Rodriguez* (1973), the U.S. Supreme Court ruled that education is not a fundamental right under the Fourteenth Amendment and that it is permissible for wealthier communities to spend more money per student on public school education than do schools in poor neighborhoods.

White flight from inner-city neighborhoods to new suburbs increased segregation in the schools in the 1980s. The 1994 Improving America's Schools Act substantially increased the funding the federal government was paying out to help economically disadvantaged school children under Title I of the Elementary and Secondary School Act.

Legal engagement in higher education issues concerning African Americans began in 1935, when the NAACP won a suit, *Murray v. Pearson*, in the Maryland Court of Appeals requiring the University of Maryland to admit Black students to its law school. Of more significance, since it was a broader ruling that applied nationally, was the 1938 ruling by the U.S. Supreme Court in *Gaines v. Canada* that Missouri could not require that an African American seeking entry to law school go to an out-of-state school for his legal education. If a state was providing a certain kind of education to Whites, the court ruled, it must also provide that same kind of education to Blacks – either by allowing the Black students to attend the White school or by establishing a separate Black institution within the state.

In the late 1940s and through the 1950s, there were a number of court actions affecting higher education for African Americans. In 1949, in *Johnson v. Board of Trustees of the University of Kentucky*, the federal courts overturned the Day Law that segregated public higher education in Kentucky. Lyman Johnson, an African-American schoolteacher, was one of 30 graduate students admitted to the university that year.[14] In 1950, in *McLaurin v. Oklahoma State Regents*, the U.S. District Court allowed George McLaurin, a doctoral student in education, to attend the University of Oklahoma because there was no Black institution in the state. The president of the university admitted McLaurin but forced him to sit in an adjacent classroom and made him sit in segregated sections of the school library and cafeteria. The U.S. Supreme Court ruled that this segregation was unconstitutional.

That same year, an African-American man, Heman Marion Sweatt, was denied admission to the University of Texas at Austin law school on the grounds that the Texas state constitution did not allow integrated education. Sweatt sued and the State District Court continued the case for 6 months, long enough to allow the state to open a Black law school in Houston at Texas State University for Negroes (today the Thurgood Marshall School of Law at Texas Southern University). In *Sweatt v. Painter* the U.S. Supreme Court found that the separate facility was unequal because

---

[13] In *Swann* the court also allowed busing from majority to minority schools and use of racial quotas in each school as tools to implement a single integrated school system. The courts, however, pointed out clearly that racial quotas were not an end in themselves.

[14] The admission of Black students to the University of Kentucky led to cross burnings on campus. Lyman dropped out of school after 1 year but was always proud of his role in integrating the university. Admission of African-American undergraduates began 5 years later.

of the comparative sizes of the two faculties and of their law libraries, and on other factors that affected "substantive equality"; so the court allowed Sweatt admission to the law school in Austin.

In a case that is widely cited as one demonstrating the lengths that segregationists would go to avoid integration, Virgil Hawkins was denied admission to the University of Florida law school in 1949 because he was African American. Hawkins sued in 1950 and his case went through both the state and federal courts multiple times. In 1954, the U.S. Supreme Court ruled in the case *Hawkins v. Board of Control* that the University of Florida must admit African Americans even if the action might be the cause of "public mischief" as the defendants claimed would happen, but the state courts continued to drag their heels and the university imposed a new law school entrance exam intended to screen out "undesirable" candidates. It was not until 1956 that the courts compelled the University of Florida to admit African Americans. The first African-American law student was admitted in 1958, but it was not Hawkins, who had gone north for his law degree by then.

In 1952 Autherine Lucy, an African American woman, applied for admission to the graduate library school at the University of Alabama. In 1955 the NAACP helped to secure an order for the university not to deny her admission on the grounds of race, upheld in *Lucy v. Adams* (1955), although the university barred her entrance to all dormitories and dining halls. She began school in 1956, but on the third day of classes a mob prevented her from reaching class and the police had to be called to campus. That evening, the university suspended her on the grounds that they could not ensure her safety. Lucy and the NAACP could not prevent this suspension, and she was subsequently expelled from the university for alleged "false and outrageous" statements about the university. This expulsion was eventually overturned in 1988, and Lucy earned a master's degree in elementary education there in 1992.

The 1960s were a tumultuous time for African Americans and higher education. Tensions were already high throughout the South because of the 1960 sit-in at the Woolworth's department store lunch counter in Greensboro, NC and subsequent sit-ins across the South. In riots at the University of Georgia in 1961, two African-American students, Charlayne Hunter and Hamilton Holmes, were suspended; these suspensions were only lifted after a federal court order. The following year there were riots at the University of Mississippi, when James Meredith tried to attend classes as the first African American student at the university. The university had originally denied his admission on the grounds of race, but this decision was overruled by the U.S. Supreme Court. U.S. Attorney General Robert Kennedy sent in U.S. Marshalls, and President John Kennedy sent in U.S. Army troops to quell the riots and allow Meredith to attend classes when the Governor and Lieutenant Governor were unwilling to protect Meredith. Meredith eventually completed his undergraduate degree elsewhere. He was the focal point of another Civil Rights activity when he was shot (not fatally) during a freedom march in 1966. In a similar story, in 1963 Governor George Wallace tried to block entrance into the administration building of two African-American students at the University of Alabama, but

President Kennedy called in the National Guard to protect the students and assure entry.[15]

As a result of the Civil Rights Act of 1964, White colleges and universities were forced to admit African-American students. The HBCUs, which had historically offered admission to all students and had hired a number of Jewish professors fleeing Nazi Germany in the 1930s, had been the only higher educational institutions open to African Americans in the South. The courts did not pressure the HBCUs to increase their White enrollment. The period saw a rapid growth in the number of African Americans enrolled in college – from 83,000 in 1950 to 666,000 in 1975. In 1950 the vast majority of African American college students attended HBCUs. By 1970 there were approximately equal numbers of African American students on predominantly Black as on predominantly White campuses, and by 1975 almost three-quarters of the African American students were attending Predominantly White Institutions (Allen et al. 2007).

In 1969, on Parents Day at Cornell University, African American students took over the student center in protest of a cross burning in front of the Black women students' cooperative and other acts of racism on campus. The students brought guns to the student center to defend themselves after a fraternity tried to retake the building, and police from across the state were brought in and were being assembled for an assault. However, the students and university administration resolved the issues peacefully after 36 h of negotiation. In 1970, a student protest on the campus of Ohio State University demanding admission of more African American students resulted in the governor calling out the National Guard to restore order. The following year a student protest about low African American enrollments on the University of Florida campus was suppressed by the university administration; and when no changes resulted, about 100 African American students withdrew from the university.[16]

In 1972, in *Adams v. Richardson*, the NAACP sued the Department of Health, Education, and Welfare for not enforcing Title VI of the Civil Rights Act of 1964 by continuing to provide funds to ten states that did not have desegregation plans in place. The NAACP won this case in U.S. District Court, and the states were required to develop desegregation plans so that (primarily non-HBCU) public colleges had better plans for recruitment and retention of minority students and improved their racial balances among faculty and staff. The court also demanded that the states provide metrics for the improvement of facilities and programs at HBCUs so that they would be more attractive to White students. This led to increased state expenditures on HBCUs.

In 1973 and again the following year, Allan Bakke applied for admission to the medical school at the University of California at Davis. In both years, he was rejected even though he had a higher admission rating than a number of students

---

[15] For an interesting perspective because it was written so close to the time of these events, see Ballard (1973).

[16] For a more general account of African Americans and higher education in the turbulent times of the late 1960s and early 1970s, see Rogers (2012).

who had been admitted through a special admission process for disadvantaged students. Bakke sued and the case eventually made its way to the U.S. Supreme Court. A highly divided court ruled that race could be one of several factors considered in admission but that specific quotas for admission of minorities, as the University of California at Davis had used, were not acceptable. Bakke was admitted to the medical school. This case introduced the concept of reverse discrimination, which was to become prevalent later in the century.

With the support of the NAACP, several Black litigants filed suit in 1975, claiming that the state of Mississippi was still operating a segregated public university system with three Black and five White universities, in which the Black institutions had substantially poorer academic programs, staff, and facilities. The U.S. government took up the battle on behalf of the litigants, and in 1992 the Supreme Court ruled in *U.S. v. Fordice* that Mississippi had to do more to integrate its Black and White institutions. The court also ruled that the state must find "educational justification for the continued existence" of the dual systems of education. This ruling had the side effect of generating substantial public concern about whether the operation of HBCUs would continue to be legal.[17]

In 1997 the NAACP sued the federal Department of Health, Education, and Welfare (*Adams v. Califano*) because it had not enforced its 1969 and 1970 requirement, in keeping with Title VI of the Civil Rights Act of 1964, that North Carolina and five other states file plans for how they were going to desegregate their public systems of higher education. The NAACP asked that federal funds be withheld from these schools until they submitted plans for how they were going to enroll more African Americans at the predominantly White public colleges and universities and to provide more equitable funding for HBCUs compared to the predominantly White institutions. The court ruled in the NAACP's favor, and these Southern and Border states were required to develop plans to implement these changes.[18]

In 1988 and 1989 there were again protests over race on campuses. 200 minority students convened a 6-day sit-in of the Africa House on the University of Massachusetts campus. The administration agreed to demands for a more multicultural curriculum and harsh punishment for students convicted of racial violence. Student protestors at Howard University succeeded in getting the Chairman of the Republican National Committee, whose views were regarded by the students as antithetical to the interests of minorities, to resign from the Howard University board of trustees.

---

[17] During the 1980s, White enrollment at HBCUs increased by 41 % (on a small base) while Black enrollment increased only 9 %; and at least two HBCUs, Bluefield State University and West Virginia State University, became majority White. (Allen et al. 2007) White enrollment in an HBCU can be a difficult sale. In one study reported in Minor (2008), White high school students in Mississippi believed the HBCUs had poor academic quality, anticipated they would feel social discomfort and experience racial discrimination if they attended an HBCU, and expected their parents would disapprove.

[18] This case is often discussed in connection with *Adams v. Richardson* (1973), which ruled that higher education was to be part of the desegregation actions of the U.S. Department of Education.

In 1992 the U.S. Supreme Court decided in *U.S. v. Fordice* that when race-neutral policies and practices were insufficient to dismantle a dual system of public universities for Blacks and Whites in the state of Mississippi, additional affirmative actions were required. These included changing admissions practices and, in some cases, eliminating duplicate programs at predominately Black or White universities so that anyone wishing to study in a given academic field at a public university in Mississippi would be required to attend the one integrated program that remained in the state, whichever university might happen to offer it. One impact of this was to give the HBCUs a new opportunity to build strong academic programs. The U.S. Supreme Court also instructed the states to either find educational justification for the HBCUs or dismantle them. This ruling was followed by a number of impassioned educational justifications of the HBCUs by educators, scholars, and legislators; and generally there has been continuous strong public support for HBCUs up to this day.

In 1995 the U.S. Supreme Court ruled in *Adarand Constructors, Inc. v. Peña* that federal affirmative action programs (and this applied to higher education) must be reviewed by the courts under "strict scrutiny" (the most exacting level of review); and, following normal legal practice, there must be a compelling government interest involved and the remedy must be narrowly tailored to meet that interest. In this case, the U.S. Department of Transportation had awarded a highway construction contract in Colorado to Mountain Gravel and Construction Company. Mountain Gravel sublet a contract for guardrails to Gonzales Construction (a disadvantaged business under the rules of the Small Business Administration) rather than to the lower bidder, Adarand Construction. Adarand Construction sued the Department of Transportation and the U.S. Secretary of Transportation, Federico Peña.

In 1996, California voters passed the California Civil Rights Initiative, better known as Proposition 209.[19] According to this proposition, state institutions were prohibited from considering race, ethnicity, or sex in public education as well as in public employment or public contracting. This eliminated race-based affirmative action at California's public universities. As a result, African-American enrollment dropped significantly at the University of California. An example of a re-entry program in computer science at the University of California at Berkeley that was closed by this proposition is discussed in Chap. 10.

In 1996 Cheryl Hopwood and three additional applicants to the University of Texas at Austin law school sued when they were not admitted, arguing that they had better LSAT scores and grades than a number of African-American and Hispanic applicants who were admitted. The judge in District Court ruled in favor of the university, but the Appellate Court reversed the decision and the U.S. Supreme Court refused to hear the case because the university had changed its admissions policies from those that had been in effect when Hopwood had applied. This was the first case in which a university's admissions policy had been overturned in court since

---

[19] There were similar efforts to California's Proposition 209 in Florida, Michigan, and other states. The person principally behind these activities across the nation was Ward Connerly, first as the chairman of the California Civil Rights Initiative and later as the founder of the American Civil Rights Institute.

Bakke. In order to avoid the ban on affirmative action as it was posed in *Hopwood*, in 1997 the Texas legislature passed Texas House Bill 588, better known as the Top 10 % Rule, which assured admission at a Texas public institution of higher learning to any student who graduated in the top 10 % of his or her graduating class in a Texas public high school. This provided ethnic and racial diversity as well as geographic coverage of the entire state.

The impact of *Hopwood*, *Adarand*, and Proposition 209 was felt immediately in higher education. There was a drop of 20 % between 1996 and 1997 in African American applications to graduate school in the STEM disciplines, whereas the admission variation from year to year had not been more than three percent in any year in the prior decade. This occurred at a time when the undergraduate pool of African American students – the most likely applicants to graduate school – had been trending upwards. Texas and California, which were directly affected by these laws, essentially stopped considering race in either admission or financial aid. Administrators at universities in other states were uncertain what affirmative action efforts, if any, were now legal. A study group that visited some of the institutions that experienced drops in graduate enrollments of African Americans (and, to a lesser extent, in Hispanic enrollments) observed:

- greater emphasis on foreign graduate student recruitment;
- lukewarm attention to minority recruitment and retention;
- changes in policy to emphasize GRE scores in graduate admissions;
- lack of focused institutional effort (no special office concerned with tracking or retention);
- decentralization of minority recruitment efforts; and
- no express policies to encourage faculty commitment to minority graduate education. (Malcom et al. 1998)

The issue of whether race could be a factor in admission decisions was further considered by the Supreme Court in *Grutter v. Bollinger* (2003) in which student Barbara Grutter sued the University of Michigan law school when she was denied admission on the grounds of reverse discrimination based on race. The District Court ruled in favor of Grutter. The Appeals Court reversed the decision. The Supreme Court decided to hear the case and ruled 5–4 in favor of the university. The court argued that the university has a compelling interest in creating a diverse student body; and use of race as a factor is permitted under the law, so long as it used for narrowly defined purposes such as creating a critical mass of minority students in the academic program and is a temporary measure until color-blind admissions can produce a suitably diverse class. One impact of *Grutter* was to abrogate the *Hopwood* decision and allow race to be used as a factor in the same way it was allowed in *Grutter*.[20]

One of the most remarkable changes in African American education was the rapid rise of educational attainment in the second half of the twentieth century. In 1940 African American males aged 30–34 years old had on average 3.5 years less

---

[20] For another account of legal influences on HBCUs, see Jackson and Nunn (2003, Chapter 3).

formal education than White males of the same age. By 1985 this gap had shrunk to 0.5 years, and by 1995, there was close to the same level of educational attainment. Economic analysis (Bauman 1998) has shown that while vast improvements in the public schools attended by African Americans during the 1940s and 1950s, especially in the South, as well as the civil rights legislation of the 1960s, had some bearing on this change, these two factors do not wholly explain the rapid rise in African American educational attainment. Bauman, following some other scholars, speculates that the high value placed on education in the African American community may also be a factor.

3.2   African-Americans and STEM Education and Careers – A Brief History

This section surveys some of the social science literature concerning STEM education and careers for African Americans.[21] The first part discusses STEM education generally, while the last part focuses on computing as a particular example of a STEM disciplines.

---

[21] On the effect of Black English vernacular on the way that public school students learn quantitative relations, its impact on learning of math and science by African American students, and as a case study for the Sapir-Whorf hypothesis, see Orr (1987). On various aspects of African-American STEM students in baccalaureate programs, see Cook (2014), Gochenaur (2005), Freeman et al. (2008), Taylor et al. (2008), Palmer et al. (2010b), Green and Glasson (2009), Smith et al. (2008), National Research Council (2014) and Bonner et al. (2009). On African-American community college students in STEM, see Jackson (2013, 2014). On African-American graduate students in STEM, see Malcom et al. (1998) and Bush (2014). On comparisons and relations between HBCUs and PWIs with regards to the STEM disciplines, see Essien-Wood and Wood (2013), which contrasts the role of faculty members in helping students at HBCUs integrate their academic and social lives, whereas faculty at PWIs are sometimes negative influences on Black students through "microaggressions" such as questioning students' academic ability, suggesting they change to a less demanding major, or avoiding contact with students while on campus. Newman and Jackson (2013) reviews the advantages to students, HBCUs, and PWIs of programs in which African American students can spend their first 3 years at an HBCU and two additional years at a PWI, earning STEM degrees from both schools. On female African American STEM students, see Parker (2013), Galloway (2012), Wilkins (2014), Perna et al. (2009), Borum and Walker (2012) and Holmes (2013). On male African American STEM students, see Lundy-Wagner (2013).

In addition to the social science literature on African Americans and the STEM disciplines, there is some historical literature on this topic. See the various writings, for example, of Rayvon Fouche, Evelynn Hammonds, Kenneth Manning, and Willie Pearson. Williams (2003) compiles materials from the Blacks at MIT History Project. Harding (1993) includes a number of relevant articles on the racial economy of science.

On African American students in the computing disciplines, see DiSalvo et al. (2011), Quesenberry et al. (2013), Lopez et al. (2006), Lopez and Schulte (2002) and Williams et al. (2007) – in addition to the sources cited later in the book where broadening participation organizations in computing focused specifically on African Americans are discussed. Van Sertima (1983) has a list of African American inventors of information technologies, prepared by Kirstie Gentleman.

African Americans are underrepresented at all stages of STEM education and careers. They represent 12.3 % of the total U.S. population (2012), 13.4 % of the total undergraduate enrollment (2011), and 14.8 % of all 18–24 year olds in the United States (2012). However, they represent only 4.0 % of all engineering bachelor's degrees (2011), 3.6 % of the engineering workforce (2010), and 2.5 % of engineering faculty (2013). (NACME 2014) Percentages are similarly low in other science, mathematics, and computing disciplines. African-American students attending predominantly White institutions are equally likely to declare a STEM major as White students, but they are more likely to switch out of a STEM major or otherwise not complete their degree.

In 2012, 48 % of African Americans attending college were enrolled in community colleges. (American Association of Community Colleges 2014) These colleges provide the first 2 years of the bachelor's degree for many first-generation college students, including African Americans, but they also provide terminal degrees to prepare students for direct entry into lower-level positions in the workplace.[22] In one small qualitative study of a STEM workforce development program at Minneapolis Community and Technical College (Jackson 2014), the African-American students indicated that they did not have friends and family members who had attended college, so it was difficult for them to obtain good advice about their academic choices. Consequently, they found making choices in course offerings burdensome and preferred to be told exactly which courses to take. They also found it difficult to find role models at school. All of the students either had internships or on-the-job training, and they believed that this was more important than their community college degree in locating a STEM job.

Another small qualitative study (Jackson 2013) was conducted with female African-American STEM majors at HBCUs who had transferred from community college. One common theme was the importance of consistent information from the community college and the HBCU to ease both anxiety and the transition itself from one type of school to another. A second common theme was the importance to the students in being told about the career options their STEM education provided, including STEM careers that the students found to be "feminine", "nurturing", or "caring". A third common theme were the positive feelings about how safe an environment the students found the HBCU to be as they began to form a STEM identity and find ways to make it work together with their identity as an African-American female.

There is a significant social science literature on African-American STEM college majors. In one study of engineering majors at two HBCUs (Bonner et al. 2009), successful students pointed to support from their family, pride in being able to advance beyond the educational achievements of friends and family, African-American teachers as role models, participation in student-led support communities that have both academic and social functions, and religion/spirituality as among the

---

[22] For an extended comparison of community colleges and HBCUs, see Hughes (2012). On African American students and community colleges, see Blau (1999) and Freeman (1999).

most important factors of success.[23] Both students and faculty members at these schools believed that academic giftedness goes beyond good grades and test scores to include personal characteristics such as "resourcefulness, self-directedness, leadership potential, aptitude and passion for learning, a certain degree of maturity, and pride in education." (Bonner et al. 2009)

At predominantly White institutions, science, math, and computing departments sometimes have weed-out courses early in the curriculum. African-American students have been disproportionately weeded out of majors through these courses. This may be because of poorer high school preparation.[24] Other factors that are sometimes postulated as inhibiting African American persistence are "academic and cultural isolation, motivational and performance vulnerability in the face of negative stereotypes and low expectations for performance,[25] peers not supportive of academic success, and actual discrimination." (Maton et al. 2000, as quoted in Green and Glasson 2009)

One of the most successful examples of a program to prepare African American college students for STEM careers is the Meyerhoff Scholars Program, started in 1988 at the predominantly White University of Maryland, Baltimore County. In 2000, 95 % of the Meyerhoff scholars were completing STEM degrees, and 89 % continued on to graduate or professional school. The Meyerhoff program included using STEM researchers from campus or the local community as mentors; a formal program for counseling students on both academic and personal matters; summer research internships; and a special program that engaged the faculty in the recruitment of these students and in working with them on an ongoing basis while they are in college. (Cook 2014)

There have been few studies of African American men pursuing STEM education and careers. More numerous are studies of African American women pursuing STEM education and careers. Let us first consider gender without taking race into consideration. As of 2010, considering bachelor degrees awarded, agriculture (52.5 %) and the biological sciences (59.0 %) were predominantly female, while computer sciences (18.2 %), physical sciences (41.3 %), engineering (18.4 %), and

---

[23] Other studies have also pointed to the salutary effects of learning communities – in which students take responsibility outside of class to help each other master course material – for African-American students. For example, Freeman et al. (2008), in a study of four HBCUs (Howard, Jackson State, Talladega, and Xavier) showed that learning communities had a positive influence on student attitudes, educational experiences, and motivation. The learning community of these four HBCUs is also studied in Taylor et al. (2008). Smith et al (2008) provides another account of the learning communities at Howard University.

[24] Some HBCUs have specifically addressed lack of academic preparation of freshmen STEM students by offering an online course or a precollege summer institute to bolster math and critical thinking skills, developing web-based tutoring systems to help students while they are studying for their difficult math courses, or providing upper-class students to tutor and mentor freshmen. (Palmer et al. 2010b)

[25] In the mid-1990s, Claude Steele and Joshua Aronson at Stanford University conducted the first studies on how stereotype threat could affect academic performance of African Americans. Their work became well known beyond the academy through a series of articles in *Atlantic Monthly*. (Steele 1999) Stereotype threat is now a widely studied area of social psychology.

mathematics (43.1 %) had females as a minority. In the case of every one of these STEM disciplines, the percentage of African American women receiving the degree was higher than that for all women: agriculture (52.5 % women among all graduates vs. 59.0 % women among African American graduates), biological sciences (59.0 % vs. 70.0 %), computer sciences (18.2 % vs. 31.9 %), physical sciences (41.3 % vs. 57.5 %), engineering (18.4 % vs. 26.1 %), and mathematics (43.1 % vs. 50.6 %). (Parker 2013, using NCES data)

The only study of African-American males in STEM discussed here is Lundy-Wagner (2013). Her study compares graduation rates from HBCUs for males in STEM disciplines in the years 1981 and 2009. In agricultural, mathematical, and physical sciences, fewer African-American males graduated in 2009 than in 1981. There were small gains over time in the number of biological science and engineering undergraduate degrees awarded to African American males in HBCUs. The one field where there was tremendous growth in degrees for African American males was computer science – with 123 bachelors degrees in 1981 and 445 in 2009, representing a 262 % gain (while computer science degrees for African American women increased only 53 % during this same time period). This study found that African American males perceived less social support and more social barriers than women, both at home and at school, for pursuing STEM degrees.

African American women in STEM face a "double bind" of being women in a male culture and Black in a White Culture. (Malcom et al. 1976) A recent qualitative study (Galloway 2012) indicates that African American women in STEM found gender to be a more significant issue than race. A large quantitative study (Wilkins 2014), using the Cooperative Institutional Research Program Freshman and College Senior Surveys, found that father's education, the student's SAT scores, self-perception as a senior, and overall sense of community influenced degree completion in STEM disciplines for African American women. A qualitative study (Holmes 2013) of African American physics majors indicated that there were mismatches between the worldviews of these students and of the physics (and other STEM) departments; but that these students flourished in small class settings and in cases where the faculty took a personal interest in their academic success. Russell and Atwater (2005) point to the fact that African American women enter college with the disadvantage of fewer high school opportunities than others to learn about science and engineering through "STEM magnet programs, science fairs, co-curricular science organizations/programs, and rigorous experiences in math and science classrooms." (as quoted in the literature review by Parker 2013)

Parker (2013) notes in her review of the literature that various social scientists had found that African American women interested in STEM careers fare better at HBCUs than at predominantly White institutions because of a more nurturing academic environment with "less social isolation, alienation, personal dissatisfaction, and overt racism." Perna et al. (2009) reach similar conclusions in their study of African American women studying STEM disciplines at Spelman College:

> The following four themes emerged from the data analyses. First, participating students chose to attend Spelman College at least in part because of the institution's well-known success in promoting the success of Black women in STEM fields. Second, participating

students enter Spelman College with high STEM-related educational and occupational aspirations and maintain these aspirations while enrolled. Third, participating students and faculty acknowledge the academic, psychological, and financial barriers that limit the persistence of Black women in STEM fields. Fourth, most relevant to the guiding research question, the potential negative impact of these barriers on the attainment of Black women in STEM fields is mitigated by several institutional characteristics and practices, including: structural characteristics, the cooperative rather than competitive peer culture, the efforts of faculty to actively encourage and promote students' success, the availability and use of academic supports, and the availability of undergraduate research opportunities.

The HBCUs have been an important producer of African American undergraduate STEM majors who continue their education to earn a STEM doctorate. Many of these students pursue their graduate education in predominantly White institutions because of both capacity and quality.[26] Twice as many African American women earn doctorates as African American men; in fact, African American males represent only about 2% of the STEM doctorates earned in the United States. One recent study (Bush 2014) provides a qualitative analysis of African American males earning their bachelor degree at an HBCU and their doctorate at a predominantly White institution (PWI). Here is a summary of Bush's findings:

The findings from this study reveal that the participants generally felt prepared by their HBCUs for graduate study at a PWI. The participants also expected graduate school to be challenging both socially and academically. Though the participants felt inhibited by the pressures of being Black male doctoral students at PWIs and felt like imposters, they often persevered and stayed focus on their goal of completing their doctoral degrees. Overall, the participants described the climate at their PWIs as "chilly." They often mentioned being one of few Black graduate students within their environments and also felt isolated. Encounters with racism and perceived notions of inadequacy by their faculty and colleagues also contributed to the chilly climate. However, the participants took several actions to be successful as Black male doctoral students. One of the actions the participants took was practicing self-care by making their physical, cognitive, and spiritual needs a priority in their lives. They exercised, sought therapy, and solicited spiritual guidance from family and church members. Another action the participants took was being observant and understanding the culture of their PWIs. The participants also formed supportive relationships with faculty, staff, and colleagues on campus. Though using commonalities to bond and form relationships with their White counterparts, the participants were intentional in forming relationships with other Black persons as a means to feel supported and have colleagues that they could relate to culturally.

A study of 12 African-American females who had received doctorates in mathematics (Borum and Walker 2012) provides some comparison of their undergraduate experience at HBCUs versus PWIs. The seven women who attended HBCUs had a more positive undergraduate experience based on the encouragement from faculty and other students to pursue an advanced degree, a collaborative learning experience, and the general nurturing environment; while the students who attended PWIs mentioned the challenges of isolation and being singled out for their race or gender in the White, male environment of mathematics. While studying for their doctoral degrees, these women pointed to faculty mentoring and student support

---

[26] For information about doctoral programs at HBCUs, see Fountaine (2008), Duncan and Barber-Freeman (2008) and Fountaine (2012).

groups as positives that helped them complete their studies, while isolation, large weed-out classes at the beginning graduate course level, and faculty members who are indifferent about whether they complete their degree were the most trying elements.[27] Two of the 12 students changed graduate schools because of the hostile environment in their first program.

There is a small social science literature about African Americans and the study of computing disciplines, and we close this section with a brief discussion of these findings. We summarize four studies that are representative of issues of interest to social scientists. These studies do not give a complete picture of the situation for African American computing students; there are many other topics that deserve study. This sample does suggest, however, the inchoate state of research on African American computing students.

One study (DiSalvo et al. 2011) addresses the issue of getting high school age African Americans interested in computing. The authors note that many computer scientists have reported that they became interested in the subject because of their tinkering with computers before they come to college. The authors note that there is a widespread interest in video games among young African Americans, but that has not translated into an interest in computer science. The authors developed a job training program, called Glitch, in Atlanta in which a small number of young African American men were hired to work full time in the summers and part-time during the school year to debug pre-releases of games for Yahoo, Electronic Arts, and the Cartoon Network. Through this work, the young men learned about software engineering, for example learning the programming languages Alice and Jython (a cross between the Java and Python languages). (Epstein 2009) Early results indicated that peers regarded those who worked at Glitch to be technical resources. Whether this will translate into people from this program pursuing further education in the computing disciplines is not yet clear.

Glitch received a formal evaluation at the end of the project. While 20 % of the students expressed an interest in pursuing computing as a career before their Glitch experience, 65 % of the Glitch participants have pursued postsecondary computer studies of some kind. The students also reported significant increases in computing confidence, interest in information technology and computer programming, intent to pursue computing while in college, and plans to work as a computer scientist or programmer. (DiSalvo et al. 2013)

The second study concerns an intensive, 6-week service-learning program run by the information systems department at Carnegie Mellon University from 2004 through 2011. (Quesenberry et al. 2013) The students were recruited from 11 HBCUs. The students learned fundamental information about software development,

---

[27] Another study (Gray 2013) pointed to the important role of faculty members at HBCUs in promoting doctoral STEM education. Faculty had a positive influence on their students by serving as role models, maintaining high-level expectations of academic performance, providing undergraduate research experiences, providing positive feedback on student performance, being accessible, and encouraging students to become scientists. These factors, Gray argued, overcame the weaknesses in infrastructure and deficiencies in academic rigor in the HBCU STEM curricula.

web development, and database programming while practicing teamwork, project management, and professional communication in a project for a client from the Pittsburgh area. The researchers found that the students were exposed to real-world IT skills, gained a better understanding of work and graduate school, felt a stronger sense of belonging to the IT community, and built a stronger professional network.

The third study considered self-efficacy among undergraduate computer science majors at both HBCUs and PWIs. (Lopez et al. 2006; also see Lopez and Schulte 2002) The study surveyed 1208 computing majors drawn from 21 HBCUs and 21 PWIs, contrasting their work with a control group of 581 non-computing-discipline majors drawn from those same institutions. With regard to computer self-efficacy, not surprisingly the computing majors scored much higher than the control group. For Caucasian students, males scored significantly higher than females for computer self-efficacy. For African-American students, however, there was no significant difference in computer self-efficacy between males and females. The computing majors scored higher overall than the control group on math self-efficacy. There was no difference between males and females in math self-efficacy for either the Caucasian or African-American computing majors. Math self-efficacy was higher for students at the PWIs than at the HBCUs.

The fourth study concerns the value of collaborative pedagogy to African-American computing students. (Williams et al. 2007) The study was carried out through semi-structured interviews with a small number of third- and fourth-year students at North Carolina State University (a PWI), North Carolina A&T (an HBCU), and Meredith College (a women's college). The study found that the use of pair programming and agile software development created a collaborative environment that was attractive to both majority and minority students, and to both male and female students. The African-American students in particular found the collaborative environment better prepared them for real-world work environments.

3.3　Historically Black Colleges and Universities – A Brief History

This section takes a brief look at Historically Black Colleges and Universities from a social science perspective.[28] The HBCUs have traditionally been a large producer of African-American STEM baccalaureate degree recipients.

---

[28] There is a large social science literature on the topic of HBCUs. Provasnik et al. (2004), Esters and Strayhorn (2013), Bonner et al. (2009) and Hale (2006) provide general information. Brown and Davis (2001), U.S. Commission on Civil Rights (2010), Allen et al. (2007), Brown (2013) and Minor (2008) provide useful information about the history of HBCUs. Ellis and Stedman (1989), Matthews (2008), Brown and Burnette (2014), Palmer et al. (2011) and Lee (2010) discuss various issues related to the funding and politics of HBCUs. Centra et al. (1970), Ehrenberg and Roibstein (1993), Wenglinsky (1996), Kim and Conrad (2006), Irvine and Fenwick (2011) and Constantine (1995) provide information about the effectiveness – include economic effectiveness – of HBCUs. Some of this literature was written in response to the *Fordice* decision, after which HBCUs felt the

An Historically Black College or University (HBCU) is an institution founded prior to 1964 with primary concern for the education of African Americans. As of 2007, there were 89 four-year and 14 two-year HBCUs. In addition to the HBCUs, there are approximately 50 predominantly Black universities. These are institutions in which student enrollment is more than half African American. These schools were not founded with the express purpose of educating African Americans. (Brown and Davis 2001)

The HBCUs have been an important source in the production of doctorates, judges, and doctors. Although the HBCUs represent only about 3 % of all colleges and universities in the United States, they educate 14 % of African Americans pursuing a higher education and award 30 % of undergraduate engineering degrees, 44 % of undergraduate natural science degrees, and 17 % of African-American graduate STEM enrollments. 12 HBCUs have ABET-accredited engineering programs.[29] (Bonner et al. 2009)

Finances are one of the biggest challenges at the HBCUs. These schools have out-of-date technological infrastructures, little money to spend on academic program development, and low faculty and staff salaries. Because these schools typically have small endowments, small fundraising and public relations budgets, and limited externally funded research, the public HBCUs are highly dependent on state allocations for their operating budgets. When state budgets are tight, public HBCUs suffer more acutely than other higher education institutions. The *Fordice* decision in 1992 required a number of states to make extra funding available to public HBCUs to offset past discriminatory underfunding. In some cases, this funding was delayed and ended in additional court battles. For example, *Knight v. Alabama* (2006) led to an additional $7 million for Alabama A&M and $28 million to Alabama State University in capital funds.[30]

---

need to justify segregated Black institutions. Cole (2006), Shapiro (2008), Palmer et al. (2010a), Sydnor et al. (2010), Davis and Montgomery (2011), Haskell and Champion (2008) and Hughes (2012) explore various aspects of the academic programs at HBCUs. Van Camp et al. (2009), Lewis et al. (2008), Guiffrida (2005), Mawhinney (2011–2012) and Griffin (2013) discuss various social and cultural issues such as school selection, gender, and "othermothering". Fountaine (2008, 2012) and Duncan and Barber-Freeman (2008) explore issues of doctoral education and HBCUs. Sibulkin and Butler (2005), Kim (2011), McDonald (2011), Kynard and Eddy (2009), Renzulli et al. (2006), Kim (2002) and Strayhorn et al. (2010) provide comparisons between HBCUs and PWIs.

[29] The ABET-accredited engineering programs are located at Howard, North Carolina AT&T, Alabama A&M, Prairie View A&M, Southern University-Baton Rouge, Tennessee State, Morgan State, Florida A&M, Hampton, Tuskegee, Norfolk State, and Jackson State.

[30] Other examples include $503 million from the state of Mississippi to Mississippi Valley State University, Jackson State University, and Alcorn State University; $125 million from the state of Louisiana to Grambling University and Southern University; and $580 million from the state of North Carolina to Elizabeth City State University, Fayetteville State University, North Carolina A&T, North Carolina Central University, and Winston-Salem State University. See, for example, Matthews (2008) and Brown and Burnette (2014). Palmer et al. (2011) provides a case study of state funding to HBCUs in Maryland, while Lee (2010) provides a case study of state funding to HBCUs in Mississippi.

When the Supreme Court ruled in *Fordice* (1992) that states should demonstrate the effectiveness of HBCUs or integrate them into a single public higher education system, there was a renewed effort by social scientists to demonstrate the special effectiveness of HBCUs. Various claims had been made since the 1960s about the effectiveness of HBCUs in terms of educational outcome as compared to PWIs: higher graduation rates for Blacks who attended HBCUs than who attended PWIs, greater likelihood of HBCU students attending graduate school and entering professional careers than Black students at PWIs, stronger ability of HBCUs to develop Black leaders, a curriculum at HBCUs tuned to prepare students for engagement in their communities, and the effectiveness of HBCUs in preparing Black students for careers in social service organizations or engineering.

The problem was that the evidence from social science research studies conducted from the 1960s to the 1980s was inconclusive; the studies showed weak differences between HBCUs and PWIs, contradicted one another, or were methodologically unsound. Armed with more powerful statistical methods and larger data sets, social scientists since *Fordice* have again studied these issues. Most of these common beliefs continue to be at best weakly supported by quantitative social science research, although a few studies have found educational justifications for HBCUs. For example, Wenglinsky (1996) found that African American students trained at HBCUs are more likely than those trained at PWIs to pursue a postgraduate education and enter a profession. Kim and Conrad (2006) take a different approach and pose an economic efficiency argument as the educational justification for the HBCUS: that HBCUs have equally strong outcomes as PWIs in the education of African-American students while providing that education for significantly lower cost per student.[31] Kim and Conrad also found that African American students at HBCUs were 50 % more likely to participate in research with their faculty while undergraduates than were undergraduate African American students at PWIs. Other research suggests that this may be a factor in the strong role of HBCUs in preparing Black students for STEM careers.

In addition to comparisons of educational outcome for African American students at HBCUs and PWIs, there have also been comparisons of institutional characteristics, student profile, and student experience for African American students at these two types of institutions. In terms of institutional characteristics, the HBCUs have fewer resources, lower faculty salaries, weaker technological infrastructures, smaller enrollments, lower student-to-faculty ratios, and higher student-faculty interactions.[32] In terms of student profiles, the HBCU students are more likely to come from poorer families and be less well academically prepared (as measured by

---

[31] Kim (2002) cites the instruction cost per student at an HBCU to be $6506, while at a PWI it is $8645.

[32] In terms of special characteristics of academic programs at HBCUs, see Cole (2006) on culturally sensitive curricula; Shapiro (2008) on learning communities generally and Haskell and Champion (2008) on learning communities specifically for STEM education; Sydnor et al. (2010) on community-based participatory research; and Davis and Montgomery (2011) on honors education.

standardized test scores and high-school grades). The HBCU students are less likely than African-American students at PWIs to feel a sense of isolation, racial discrimination, or lack of social or academic integration.[33] As McDonald (2011) summarizes the existing literature: "Black students purchase psychological well-being, cultural affinity, nurturing academic relations and happiness at the cost of limited physical facilities, fewer resources, and more restricted academic programs."

A term that is used sometimes in describing the nurturing role of the faculty members at HBCUs is 'othermothering.' It is adopted from a common practice in American slave history in which other female slaves would help the birth mother raise her child, perhaps because the birth mother had died or was sold away, or because the birth mother did not have the knowledge to teach her child in a community in which there was no formal education and slave children were instead taught in the home. In the HBCU setting, othermothering involves practices that some faculty in PWIs might regard as crossing professional boundaries. These practices involved paying close attention to the psychosocial as well as the academic needs of their students, advocating for their students, and demanding hard work and high achievement from the students.[34]

3.4 Fellowship Programs for African-Americans – A Brief History of the United Negro College Fund and the Thurgood Marshall College Fund

Fellowship support has been critical to African-American students – many of whom come from poor families – if they wish to attend and graduate from college and graduate school. We will briefly discuss two important organizations, the United College Negro Fund and the Thurgood Marshall College Fund, which support African-American students across all academic disciplines. In Chap. 7 we focus in greater detail on the National GEM Consortium, which provides fellowship support

---

[33] Strayhorn et al. (2010) provides an analysis of the role of the Black Cultural Center as a supportive environment for African American students on primarily White campuses.

[34] Faculty members at HBCUs often express satisfaction in the othermothering support they provide to students and regard it as one of the great values of an HBCU education; but othermothering does exact a heavy toll on the faculty in terms of time and emotional commitments to their students, disappointment and frustration when the students do not reciprocate by putting as much effort into the relationship, when the students expect favoritism, and the guilt that the faculty feel sometimes when they do not have the time or energy to practice othermothering to the highest standards or if the students fall short despite this othermothering effort. On othermothering, see, for example, Guiffrida (2005), Mawhinney (2011–2012) and Griffin (2013). Similar practices also occur sometimes at primarily White small liberal arts colleges, although the term 'othermothering' is not used. For the most part, however, mentoring of African American students at PWIs is much weaker than at HBCUs.

to African American and other underrepresented minority students who are pursuing graduate study in applied science and engineering.[35]

The United Negro College Fund was founded in 1944 with the express purpose of helping the HBCUs. Today, it is the largest scholarship provider to students of color, currently providing approximately 10,000 scholarships annually to students studying at more than 900 colleges and universities. African American students who receive UNCF scholarships are almost twice as likely as African American college students overall to graduate within 6 years (70–38 %). UNCF scholarship holders are more likely to study STEM disciplines at the undergraduate level than other African American students, with 12 % studying biological and biomedical sciences, 8 % studying social sciences, 6 % studying psychology, and 4 % each studying computer science and engineering. In addition to scholarships, UNCF provides operating expenses to 39 HBCUs.

The Thurgood Marshall College Fund (TMCF) was founded in 1987 by social scientist Joyce Payne, who had been a senior staff member in the Carter Administration and held various senior-level profit and non-profit positions in Washington. The purpose is to support public HBCUs. Students at almost all of its 47 member colleges and universities receive scholarships, with the largest number of scholarships provided to students at North Carolina A&T, Howard, and Central State universities. The scholarship recipients have an impressively high 97 % college graduation rate, according to a 2013 analysis. STEM is the most common category of major chosen by TMCF students, followed by education.

TMCF supports a number of activities in the STEM areas. It runs summer institutes for education and STEM majors who want to become teachers. In a recent year, three of the five summer institutes were focused on STEM teaching. TMCF also organizes a workshop for college presidents, provosts, and faculty members on retention, graduation, and career readiness in the STEM disciplines. A series of grants from Microsoft to TMCF over a decade has allowed it to help its member schools build up their IT infrastructures. TMCF also runs a Student Ambassador program jointly with the Centers for Disease Control, which helps prepare students for careers in public health. This program includes summer internships for approximately 80 interns. TMCF has also operated blended classrooms and online charter schools for middle-school and high-school students in New Orleans and Winston Salem, NC that are focused on college preparation.

References

Allen, Walter R., Joseph O. Jewell, Kimberly A. Griffin, and De'Sha S. Wolf. 2007. Celebrating the legacy of 'The journal'. *The Journal of Negro Education* 76(3): 263–280.

---

[35] In the science and technology fields, AT&T, Ford, IBM, Sloan, and Xerox have been historically important in providing significant fellowship support to African American students.

American Association of Community Colleges. 2014. *2014 fact sheet.* http://www.aacc.nche.edu/AboutCC/Documents/Facts14_Data_R3.pdf. Accessed 17 Nov 2015.

Ballard, Allen B. 1973. *The education of Black folk: The Afro-American struggle for knowledge in White America.* New York: Harper & Row.

Bauman, Kurt J. 1998. Schools, markets, and family in the history of African-American education. *American Journal of Education* 106(4): 500–531.

Bechtel, H. Kenneth. 1989. Introduction. In *Blacks, science, and American education,* ed. Willie Pearson Jr. and H. Kenneth Bechtel, 1–20. New Brunswick: Rutgers University Press.

Blau, Judith R. 1999. Two-year college transfer rates of Black American students. *Community College Journal of Research and Practice* 23(5): 525–531.

Bonner, F. Arthur, Mary V. Alfred, Chance W. Lewis, Felicia M. Nave, and Sherri Frizell. 2009. Historically Black Colleges and Universities (HBCUs) and academically gifted Black students in science, technology, engineering, and mathematics (STEM): Discovering the alchemy for success. *Journal of Urban Education* 6(2): 122–135.

Borum, Viveka, and Erica Walker. 2012. What makes the difference? Black women's undergraduate and graduate experiences in mathematics. *The Journal of Negro Education* 81(A): 366–378.

Brooks, F. Erik, and Glenn L. Starks. 2011. *Historically Black colleges and universities: An encyclopedia.* Santa Barbara: Greenwood.

Brown II, M. Cristopher. 2013. The declining significance of historically Black colleges and universities: Relevance, reputation, and reality in Obamamerica. *Journal of Negro Education* 82(1): 3–19.

Brown, Walter A., and Daarel Burnette. 2014. Public HBCUs' financial resource distribution disparities in capital spending. *The Journal of Negro Education* 83(2): 173–182.

Brown, M. Christopher, and James Earl Davis. 2001. The historically Black college as social contract, social capital, and social equalizer. *Peabody Journal of Education* 76(1): 31–49.

Bush, Antonio. 2014. *These are my keys to success: The experiences of African American male HBCU graduates in STEM doctoral programs at PWIs.* Doctor of Philosophy dissertation, Educational Research and Policy Analysis, North Carolina State University.

Centra, John A., Robert L. Linn, and Mary Ellen Parry. 1970. Academic growth in predominantly Negro and predominantly White colleges. *American Educational Research Journal* 7(1): 83–98.

Cole, Wade M. 2006. Accrediting culture: An analysis of tribal and historically Black college curricula. *Sociology of Education* 79: 355–388.

Constantine, Jill M. 1995. The effect of attending historically Black colleges and universities on future wages of Black students. *Industrial and Labor Relations Review* 48(3): 531–546.

Cook, Laurie. 2014. *Mentor/mentee relationships: The experience of African American STEM majors.* Doctor of Education dissertation, Morgan State University.

Cowan, Tom, and Jack Maguire. 1995. History's milestones of African-American higher education. *The Journal of Blacks in Higher Education* 7: 86–90.

Davis, Ray J., and Soncerey L. Montgomery. 2011. Honors education at HBCUs: Core values, best practices, and select challenges. *Journal of the National Collegiate Honors Council* 12(1): 73–80.

DiSalvo, Betsy, Sarita Yardi, Mark Guzdial, Tom McKlin, Charles Meadows, Kenneth Perry, and Amy Bruckman. 2011. *African American males constructing computing identity.* CHI 2011, May 7–12. Vancouver, BC, Canada.

DiSalvo, Betsy, Mark Guzdial, Charles Meadows, Tom McKlin, Kenn Perry, and Amy Bruckman. 2013. *Workifying games: Successfully engaging African American gamers with computer science.* ACM SIGCSE'13, 317–322.

Duncan, Bernadine, and Pamela T. Barber-Freeman. 2008. A model for establishing learning communities at a HBCU in graduate classes. *The Journal of Negro Education* 77(3): 241–249.

Ehrenberg, Ronald O., and Donna S. Roibstein. 1993. *Do historically black institutions of higher education confer unique advantages on black students: An initial analysis.* Working Paper #4356, May, Cornell University, Institute for Labor Relations.

Elam, Julia C. 1989. *Blacks in higher education: Overcoming the odds*. National Association for Equal Opportunity in Higher Education. Lanham: University Press of America.

Ellis, William W., and James. B. Stedman. 1989. *Historically Black colleges and universities and African-American participation in higher education*. CRS Report for Congress, 89–588 RCO. Washington, DC: Congressional Research Service.

Epstein, Jennifer. 2009. Computing surge in Georgia. *Inside Higher Ed*, October 6.

Essien-Wood, Idara, and J. Luke Wood. 2013. Academic and social integration for students of color in STEM: Examining differences between HBCUs and non-HBCUs. In *Fostering success of ethnic and racial minorities in STEM*, ed. Robert T. Palmer, Dina C. Maramba, and Gasman Marybeth, 116–129. New York: Routledge.

Esters, Lorenzo L., and Terrell L. Strayhorn. 2013. Demystifying the contributions of public land-grant historically Black colleges and universities: Voices of HBCU presidents. *The Negro Educational Review* 64(1–4): 119–134.

Fitch, Nancy Elizabeth, and Gillian Johnson. 1988. *The development and history of education for the African American, 1704–1954: A selected bibliography of secondary sources*. Monticello: Vance Bibliographies.

Fleming, Jacqueline. 1984. *Blacks in college*. San Francisco: Jossey-Bass.

Fountaine, Tiffany P. 2008. *African American voices and doctoral education at HBCUs: Experiences, finances, and agency*. PhD dissertation, Morgan State University.

Fountaine, Tiffany Patrice. 2012. The impact of faculty-student interaction on Black doctoral students attending historically Black institutions. *The Journal of Negro Education* 81(2): 136–147.

Freeman, Kimberley Edelin. 1999. *African American Education: Just the facts*. Vol. 1, no. 2 Fall. Frederick D. Patterson Research Institute, Fairfax, VA.

Freeman, Kimberley E., Sharon T. Alston, and Duvon G. Winborne. 2008. Do learning communities enhance the quality of students' learning and motivation in STEM? *The Journal of Negro Education* 77(3): 227–240.

Galloway, Stephanie Nicole. 2012. *African American women making race work in Science, Technology, Engineering, and Math (STEM)*. Ph.D. Dissertation, School of Education, UNC Chapel Hill.

Gochenaur, Deborah L. 2005. *African Americans and STEM: An examination of one intervention program*. Doctor of Philosophy dissertation, American University.

Gray, Shannon. 2013. Supporting the dream: The role of faculty members at historically Black colleges and universities in promoting STEM PhD education. In *Fostering success of ethnic and racial minorities in STEM*, ed. Robert T. Palmer, Dina C. Maramba, and Gasman Marybeth, 86–101. New York: Routledge.

Green, Andre, and George Glasson. 2009. African Americans majoring in science at predominantly white universities (A review of the literature). *College Student Journal* 43(2): 366–374.

Grier-Reed, Tabatha, John Ehlert, and Shari Dade. 2011. Profiling the African American student network. *TLAR* 16(1): 21–30.

Griffin, Kimberly A. 2013. Voices of the 'Othermothers': Reconsidering Black professors' relationships with Black students as a form of social exchange. *The Journal of Negro Education* 82(2): 169–183.

Guiffrida, Douglas. 2005. Othermothering as a framework for understanding African American students' definitions of student-centered faculty. *The Journal of Higher Education* 76(6): 701–723.

Hale Jr., Frank W. 2004. *What makes racial diversity work in higher education*. Sterling: Stylus.

Hale Jr., Frank W. 2006. *How Black colleges empower Black students: Lessons for higher education*. Sterling: Stylus.

Harding, Sandra (ed.). 1993. *The 'racial' economy of science: Toward a democratic future*. Bloomington: Indiana University Press.

Harper, Shaun R., and Christopher B. Newman. 2010. In *Students of color in STEM*, New directions for institutional research no. 148, ed. Shaun R. Harper and Christopher B. Newman. Hoboken: Wiley Periodicals.

Haskell, Deborah H., and Timothy D. Champion. 2008. Instructional strategies and learning preferences at a historically Black university. *The Journal of Negro Education* 77(3): 271–279.

Holmes, Kimberly Monique. 2013. *The perceived undergraduate classroom experiences of African American women in science, technology, engineering, and mathematics (STEM)*. Doctoral dissertation, University of Maryland, College Park.

Hughes, Bob. 2012. Complementarity between the transfer goals of community colleges and historically Black colleges and universities: Learning from history. *Community College Journal of Research and Practice* 36(2): 81–92.

Irvine, Jacqueline Jordan, and Leslie T. Fenwick. 2011. Teachers and teaching for the new millennium: The role of HBCUs. *The Journal of Negro Education* 80(3): 197–208.

Jackson, Cynthia L. 2001. *African American education: A reference handbook*. Santa Barbara: ABC CLIO.

Jackson, Diraitra Lynette. 2013. A balancing act: Impacting and initiating the success of African American female community college transfer students in STEM into the HBCU environment. *The Journal of Negro Education* 82(3): 255–271.

Jackson, Cynthia L., and Eleanor F. Nunn. 2003. *Historically Black colleges and universities: A reference handbook*. Santa Barbara: ABC CLIO.

Jackson, Tina Marie. 2014. *Pathways to fast tracking African American community college students to STEM careers*. PhD Dissertation, University of Texas at Austin.

Kim, Mikyong Minsun. 2002. Historically Black vs. White institutions: Academic development among Black students. *The Review of Higher Education* 25(4): 385–407.

Kim, Mikyong Minsun. 2011. Early career earnings of African American students: The impact of attendance at historically Black versus White colleges and universities. *The Journal of Negro Education* 80(4): 505–520.

Kim, Mikyong Minsun, and Clifton F. Conrad. 2006. The impact of historically Black colleges and universities on the academic success of African-American students. *Research in Higher Education* 47(4): 399–427.

Kluger, Richard. 1975. *Simple justice*. New York: Vintage Books.

Kynard, Carmen, and Robert Eddy. 2009. Toward a new critical framework: Color-conscious political morality and pedagogy at historically black and historically white colleges and universities. *CCC* 61(1): W24–W44.

Lewis, Nicole, Henry T. Frierson, Terrell L. Strayhorn, Chongming Yang, and Raymond Tademy. 2008. Gender and college type differences in black students' perceptions and outcomes of a summer research program. *The Negro Educational Review* 59(3–4): 217–229.

Lopez, Antonio M., Jr., and Lisa J. Schulte. 2002. *African American women in the computing sciences: A group to be studied*. SIGCSE'02. February 27-March 3. Covington, KY.

Lopez, Antonio M., Jr., Marguerite S. Giguette, and Lisa J. Schulte. 2006. *Large dataset offers view of math and computer self-efficacy among computer science undergraduates* . ACM SE'06. March 10–12. Melbourne, FL.

Lundy-Wagner, Valerie C. 2013. Is it really a man's world? Black men in science, technology, engineering, and mathematics at historically Black colleges and universities. *The Journal of Negro Education* 82(2): 157–168.

Malcom, Shirley Mahaley, Paula Quick Hall, and Janet Welsh Brown. 1976. *The double bind: The price of being a minority woman in science*. Washington, DC: American Association for the Advancement of Science.

Malcom, Shirley M., Virginia V. Van Horne, Catherine D. Gaddy, and Yolanda S. George. 1998. *Losing ground: Science and engineering graduate education of Black and Hispanic Americans*. Washington, DC: American Association for the Advancement of Science.

Maton, K., F. Hrabowski, and C. Schmitt. 2000. African American college students excelling in the sciences: College and post college outcomes in the Meyerhoff Scholars Program. *Journal of Research in Science Teaching* 37: 629–654.

Matthews, Christine M. 2008. *Federal research and development funding at historically Black colleges and universities*. CRS Report for Congress, RL34435. Washington, DC: Congressional Research Service.

Mawhinney, Lynnette. 2011–2012. Othermothering: A personal narrative exploring relationships between Black female faculty and students. *The Negro Educational Review* 62 & 63(1–4): 213–232.

McDonald, Nicole L. 2011. *African American college students at predominantly White and Historically Black colleges and universities*. Ph.D. Dissertation in Leadership and Policy Studies, Vanderbilt University.

Michael Jr., John. 2010. United States v. Fordice: Mississippi higher education without public historically Black colleges and universities. *The Journal of Negro Education* 79(2): 166–181.

Minor, James T. 2008. A contemporary perspective on the role of public HBCUs: Perspicacity from Mississippi. *The Journal of Negro Education* 77(4): 323–335.

National Action Council for Minorities in Engineering (NACME). 2014. http://www.nacme.org Accessed 1 Oct 2014.

National Research Council. 2014. *Review of army research laboratory programs for historically Black colleges and universities and minority institutions*. Washington, DC: National Academies Press.

Newman, Christopher B., and M. Bryant Jackson. 2013. Collaborative partnerships in engineering between historically Black colleges and universities and predominantly White institutions. In *Fostering success of ethnic and racial minorities in STEM*, ed. Robert T. Palmer, Dina C. Maramba, and Gasman Marybeth, 181–191. New York: Routledge.

Orr, Eleanor Wilson. 1987. *Twice as less*. New York: Norton.

Palmer, Robert T., Ryan J. Davis, and Dina C. Maramba. 2010a. Role of an HBCU in supporting academic success for underprepared Black males. *The Negro Educational Review* 61(1–4): 85–106.

Palmer, Robert T., Ryan J. Davis, and Tiffany Thompson. 2010b. Theory meets practice: HBCU initiatives that promote academic success among African Americans in STEM. *Journal of College Student Development* 51(4): 440–443.

Palmer, Robert T., Ryan J. Davis, and Marybeth Gasman. 2011. A matter of diversity, equity, and necessity: The tension between Maryland's higher education system and its historically Black colleges and universities over the Office of Civil Rights Agreement. *The Journal of Negro Education* 80(2): 121–133.

Parker, Ashley Dawn. 2013. *Family matters: Familial support and science identity formation for African American female STEM majors*. Doctor of Philosophy Dissertation, Curriculum and Instruction, University of North Carolina at Charlotte.

Perna, Laura, Valerie Lundy-Wagner, Noah D. Drezner, Marybeth Gasman, Susan Yoon Enakshi Bose, and Shannon Gary. 2009. The contribution of HBCUs to the preparation of African American women for STEM careers: A case study. *Research in Higher Education* 50: 1–23.

Provasnik, Stephen, Linda L. Shafer, and Thomas D. Snyder. 2004. *Historically Black colleges and universities, 1976 to 2001*. NCES 2004–062. Washington, DC: U.S. Department of Education, Institute of Education Sciences.

Quesenberry, Jeria, Randy Weinberg, and Larry Heimann. 2013. *Information systems in the community: A summer immersion program for students from Historically Black Colleges and Universities (HBCUs)*. SIGMIS-CPR'13. May 30-June 1, 2013. Cincinnati, Ohio.

Renzulli, Linda A., Linda Grant, and Sheetija Kathuria. 2006. Race, gender, and the wage gap: Comparing faculty salaries in predominantly White and historically Black colleges and universities. *Gender & Society* 20(4): 491–510.

Richards, D.A.R., B.K. Bridges, and J.T. Awokoya. 2013. *Building better futures: The value of a UNCF investment*. Washington, DC: Frederick D. Patterson Research Institute, United Negro College Fund.

Roebuck, Julian B., and Komanduri S. Murty. 1993. *Historically Black colleges and universities: Their place in American higher education*. Westport: Praeger.

Rogers, Ibram H. 2012. *The Black Campus Movement: Black students and the racial reconstitution of higher education, 1965–1972*. New York: Palgrave Macmillan.

Roscoe, Wilma J. (ed). 1989. *Accreditation of historically and predominantly Black colleges and universities*. National Association for Equal Opportunity in Higher Education. Lanham: University Press of America.

Russell, Melody L., and Mary M. Atwater. 2005. Travelling the road to success: A discourse on persistence throughout the science pipeline with African American students at a predominantly white institution. *Journal of Research in Science* 42(6): 691–715.

Shapiro, Nancy S. 2008. Powerful pedagogy: Learning communities at historically Black colleges and universities. *The Journal of Negro Education* 77(3): 280–287.

Sibulkin, Amy E., and J.S. Butler. 2005. Differences in graduation rates between young Black and White college students: Effect of entry into parenthood and historically Black universities. *Research in Higher Education* 46(3): 327–348.

Smith, Tori Rhoulac, Jill McGowan, Andrea R. Allen, Wayne David Johnson II, Leon A. Dickson Jr., Muslimah Ali Najee-ullah, and Monique Peters. 2008. Evaluating the impact of a faculty learning community on STEM teaching and learning. *The Journal of Negro Education* 77(3): 203–226.

Steele, Claude M. 1999. Thin ice: 'Stereotype threat' and Black college students. *Atlantic Monthly*, August. http://www.theatlantic.com/past/issues/99aug/9908stereotype.htm. Accessed 1 Mar 2016.

Strayhorn, Terrell L., Melvin C. Terrell, Jane S. Redmond, and Chutney N. Walton. 2010. A home away from home: Black cultural centers as supportive environments for African American collegians at white institutions. In *The evolving challenges of Black college students: New insights for practice and research*, ed. Terrell L. Strayhorn and Melvin C. Terrell, 122–137. Sterling: Stylus.

Sydnor, Kim Dobson, Anita Smith Hawkins, and Lorece V. Edwards. 2010. Expanding research opportunities: Making the argument for the fit between HBCUs and community-based participatory research. *The Journal of Negro Education* 79(1): 79–86.

Taylor, Orlando L., Jill McGowan, and Sharon T. Alston. 2008. The effect of learning communities on achievement in STEM fields for African Americans across four campuses. *The Journal of Negro Education* 77(3): 190–202.

Thurgood Marshall College Fund. 2013. Annual Report.

U.S. Commission on Civil Rights. 2010. *The educational effectiveness of historically Black colleges and universities*. Washington, DC: US Commissions on Civil Rights.

Van Camp, Debbie, Jamie Barden, Lloyd Ren Sloan, and Renee P. Clarke. 2009. Choosing an HBCU: An opportunity to pursue racial self-development. *The Journal of Negro Education* 78(4): 457–468.

Van Sertima, Ivan (ed.). 1983. *Blacks in science: Ancient and modern*. New Brunswick: Transaction Books.

Wenglinsky, Harold H. 1996. The educational justification of historically Black colleges and universities: A policy response to the U.S. Supreme Court. *Educational Evaluation and Policy Analysis Spring* 18(1): 91–103.

Wilkins, Ashlee N. 2014. *Pursuit of STEM: Factors shaping degree completion for African American females in STEM*. Thesis, Master of Science in Counseling, California State University, Long Beach.

Williams, Clarence G. 2003. *Technology and the dream: Reflections on the black experience at MIT, 1941–1999*. Cambridge, MA: MIT Press.

Williams, Laurie, Lucas Layman, Kelli M. Slaten, Sarah B. Berenson, and Carolyn Seaman. 2007. *On the impact of a collaborative pedagogy on African American millennial students in software engineering*. 29th International Conference on Software Engineering (ICSE'07).

Willie, Charles V., Antione M. Garibaldi, and Wornie L. Reed (eds.). 1991. *The education of African-Americans*. New York: Auburn House.

Chapter 4
Opening STEM Careers to Hispanics

Abstract This chapter discusses Hispanics, the largest ethnic minority group underrepresented in the science and technology disciplines today. The first section describes the history of higher education for Hispanics in the United States. The middle section describes the history of science and technology education for Hispanics. The concluding section presents a history of Hispanic-Serving Institutions, one of the most important places for the higher education of Hispanic students.

This chapter discusses Hispanics, the largest ethnic minority group underrepresented in the STEM disciplines today.[1] The Hispanic population has been the fastest growing sector of the U.S. population for at least 30 years, growing 58 % in the decade of the 1990s and an additional 43 % from 2000 to 2010. As of 2010, there were 50 million Hispanics in the United States – 16 % of the population. (NCES 2011) In 2000, Hispanics surpassed African Americans as America's largest minority group.

Hispanics are a diverse group. The largest number of Hispanics in the United States have their origins in Mexico. However, a number have origins in Central and South America, the Spanish-speaking Caribbean islands, and Spain. Thus, rather than being a homogeneous group, Hispanics include people who speak various dialects of Spanish, have different racial and cultural heritages, have resided in the

---

[1] This book uses the term "Hispanic" when discussing this group of Americans. This is the term that is used by the US Office of Management and Budget (Directive Nol 15, May 12, 1977) in reference to "a person of Mexican, Puerto Rican, Cuban, Central or South American or other Spanish culture or origin, regardless of race." Other terms can be used either for the entire group or to make distinctions within the group, as the following quotation from Marta, an academic administrator from California, indicates:

> My identity depends on who I'm talking to. It depends on which setting I'm in. If I'm writing, I call myself a Chicana. If I'm in a group of people who are in the community, who are the people who really are involved in community affairs like arts, those kinds of things, writers, literary people, *Chicana* is what I use so it's more politicized in those circles. At home and talking to other people I would say *Mexican American*, and with people who speak Spanish I would say *Mexicana*. Within the university, *Hispanic*, so these terms are used all the time. (as quoted in Ibarra 2001)

© Springer International Publishing Switzerland 2016
W. Aspray, *Women and Underrepresented Minorities in Computing*,
History of Computing, DOI 10.1007/978-3-319-24811-0_4

United States for varying periods of time, are from different countries of national origin, and live in different regions of the country (e.g. those of Mexican origin live primarily west of the Mississippi River).

While there are only a few schools that were historically established explicitly to be Hispanic-Serving Institutions (unlike the Historically Black Colleges and Universities), with the population shifts a number of colleges have come primarily to serve Hispanics or are well along on the way to doing so (the so-called "Hispanic-emerging institutions"). The largest Hispanic populations are in the Southwestern states, the West Coast, Florida, and greater Metropolitan New York. This geographic distribution has a bearing on STEM careers because there is a strong Hispanic cultural value in caring for the extended family, which means that many Hispanic students are reluctant to move far from home to attend college or assume a professional job.

In 1973 and 1974, three organizations were formed to advance the opportunities for Hispanic scientists and engineers, the Society for Advancement of Chicanos/Hispanics and Native Americans in Science (SACNAS), Latinos in Science and Engineering (MAES), and the Society of Hispanic Professional Engineers (SHPE). SACNAS is primarily oriented towards graduate education and research careers, whereas MAES and SHPE are primarily focused on professional degrees and engineering careers.

These organizations came into existence at a time when there were few Hispanics studying and working in the STEM disciplines. These developments followed closely on the civil rights movement, which began soon after the Second World War ended (e.g. the 1947 Supreme Court ruling in *Mendez v. Westminster* against racial segregation in California public schools; and in *Hernandez v.* Texas (1954), which provided equal protection under the 14th Amendment to the Constitution), but which heated up in the 1960s and early 1970s. Thus the creation of these STEM activities can be seen to be of a piece with better-known activities such as Cesar Chavez's organization of farm workers in the 1960s, the poor people's march on Washington, DC in 1967, the formation of the Mexican American Legal Defense and Educational Fund in 1968, and the formation of *La Raza Unida* political party for Mexican-American rights in 1970.

While Hispanics are still among the most underrepresented minority groups across the STEM disciplines, progress is being made. In 1979, of the 15,000 doctorates awarded in science and engineering disciplines, only 151 were awarded to Hispanics. Thirty years later, in 2009, the number of Hispanics receiving doctorates in the science and engineering fields had increased significantly – to 1131 – out of the 21,000 such doctorates awarded. While this is significant progress and represents a little more than 5 % of the total, today the Hispanic population represents approximately 17 % of the U.S. population; so Hispanics remain severely underrepresented in STEM doctorates.

4.1   Hispanics and Higher Education – A Brief History

At every stage, from elementary education through graduate school, there are significant challenges for Hispanic education in America.[2] Many Hispanics live in low-income communities, where the schools are underfunded. These schools often have a limited number of college-preparatory courses. Even where there are such courses, Hispanics are more often than Whites directed to vocational-track courses. There is often limited information available from counselors and others about the value in and process of going to college. Language and culture can both be an issue in gaining a good public education. High-school dropout rates are high (higher than for African Americans, for example), and many Hispanic students – especially males – drop out of school to get a job. College applications and admissions are low.

More than half of all Hispanic college students choose to attend 2-year community colleges over 4-year colleges. The community colleges typically cost less, are close to home, often provide open admissions that is attractive to students with poor high school preparation, and have night and weekend class schedules that are more convenient to the students who work large numbers of hours at the same time they are attending school.

Historically, community colleges have had the dual mission of providing lower-level vocational training as well as providing a pathway from high school to a 4-year college degree. However, many community colleges prioritize their vocational mission over their college preparation mission. In these colleges, there is often no transfer culture and inadequate counseling about the transfer process.[3] More than half of Hispanic community college students, even if their intention is a college-preparation

---

[2] The social science research literature about Hispanics and higher education is large. By far, the strongest literature is about 4-year college students. On the general background and characteristics of students, see especially Vélez-Ibáñez et al. (2013). Also see Darder and Torres (2014), HACU (1995), NCES (2011), Vernez (1997) and Vernez and Mizell (2001). On the history of policy efforts related to Hispanic education, see MacDonald (2004). On early formal education or K-12 education generally, see Leal and Meier (2011) and Moller et al. (2013a, b). On high school, including transition to college, see Brown et al. (1980), National Council of La Raza (1981), Duran (1983), Lopez (2013) and Boden (2011). On community college, see especially Saenz (2002). Also see Gross et al. (2014). On 4-year undergraduate education, see especially Solorzano et al. (2005), Oseguera et al. (2009), Arana et al. (2011), Kim et al. (2014) and Crisp et al. (2014). Also see Padilla and Montiel (1998), HACU (1994), Gonzales et al. (2014), Cavazos et al. (2010), Gross (2011), Montalvo (2012), Cano and Castillo (2010), Castellanos and Gloria (2007) and Valverde and Garcia (1982). On faculty and graduate school, see Delgado-Romero et al. (2007) and Ibarra (2001). These are the sources used to write this section.

[3] Ornelas (2002) highlights seven elements of successful conditions required for transfer: "(a) personnel at the college must be committed and must prioritize the transfer function; (b) institutions should provide programs with high expectations and accept responsibility for student transfer; (c) there should be an emphasis and availability of a transfer curriculum and articulation with 4-year colleges; (d) student progress to transfer must be continually monitored; (e) institutions should provide learning community programs so students can experience the transfer process in cohorts; (f) institutions should establish bridge and partnership summer programs with universities; and finally, (g) institutions must build on the assets and strengths of students, their families, and communities." (Also see Oseguera et al. 2009)

path, fail to transfer to a 4-year college. The most common reason is that they drop out to take a job. In fact, Hispanic students are much more likely to attain a college degree if they enroll in a 4-year college directly out of high school instead of taking the route through community college.

One successful program to address this issue of low rates of associate degree completion and transfer to 4-year colleges by Hispanics is the Puente program, a partnership between a network of California high schools and community colleges with the University of California system. The Puente program includes enhanced counseling, community mentors, and courses in Hispanic culture.[4]

Hispanic students are more likely not only to be paying for their own college education but also be contributing to family finances at the same time.[5] Thus it is not surprising that, for Hispanic students at community colleges, receiving financial aid of any sort is positively associated with degree completion – although social science research shows that this relationship attenuates over time. This caveat about attenuation is important because many Hispanic students take 3 or 4 years to complete their community college degrees, presumably because of work and family obligations. First-year community college students of Hispanic origin have been less likely to apply for grants than students from any other minority group. (Gross et al. 2014)

In addition to issues of adequate high school preparation and financial aid, Hispanic college students face other challenges to academic performance and persistence: "cultural and social isolation, negative stereotypes, low expectations from teachers and peers, and non-supportive educational environments." (Oseguera et al. 2009) Mentoring, financial aid, the ability to go to school full time, as well as community learning opportunities, have all been shown to enhance retention.

When Hispanic students do attend college, they find few Hispanic role models. Less than 5 % of faculty and college administrators are Hispanic. Even in Hispanic-Serving Institutions, only slightly more than half of the college presidents are Hispanic.[6] Hispanic students who come from homogeneous high schools experience more stress and alienation in college than those from heterogeneous high schools.

---

[4] Oseguera et al. (2009) identify some other national programs that assist Hispanic college students at the community college or baccalaureate level: The Student Support Services program – part of the TRIO programs of the U.S. Department of Education – helps with basic study skills, counseling, mentoring, and assistance in completing admission and financial applications to 4-year colleges; ENLACE, a program of the Kellogg Foundation, helps communities to advance Hispanic students through public school and into college; and the Adelante U.S. Education Leadership Fund – a joint venture of Miller Brewing Company and HACU – provides counseling, academic support services, and other features on select college campuses.

[5] White families on average are wealthier than Hispanic families. According to the Urban Institute, in 2010 the average White family had family wealth (assets minus liabilities) of $632,000 compared to $110,000 for Hispanic families. (Miller and Horrigan 2014).

[6] Duran (1983) identifies the following factors as effective predictors of college achievement of Hispanic students: type of college (e.g. size, location, gender composition, general mission, etc.); general institutional commitment to the education of Hispanics; ethnic composition of the faculty and staff; peer support systems (counseling, tutoring, other services); institution-operated support systems (special orientation programs, advisory services, remedial and tutorial opportunities,

Before turning to STEM education for Hispanics and to Hispanic-Serving Institutions, we give a brief history of the politics of Hispanic education in the United States since the 1960s.[7] The 1960s are an appropriate time to begin this story because Hispanic educational protests and other political actions were part of the activism of the 1960s and 1970s that also included the Civil Rights, Black Power, Free Speech, Women's Rights, and American Indian self-determination movements.

There was cause for complaint. The public education for Hispanics was much worse than it was for Whites in the 1960s. On average, a Hispanic student had 3 to 4 fewer years of schooling than a White student. The schools teaching most of the Hispanic students were crowded and had weak infrastructures, and Hispanic students were commonly channeled into vocational tracks independent of their wishes and their academic capabilities. Instruction was delivered only in English.

A number of Hispanic activist organizations that were engaged in a wider set of issues participated in these education reform efforts. These organizations included the United Farm Worker's Association, the Mexican American Youth Organization, *La Raza Unida*, *Alianza de los Pueblos Libres*, and even the Young Lords street gang. The most visible of these protests were the walkouts that occurred in four schools in East Los Angeles in 1968, where a total of 15,000 students joined the protests. The walkouts spread to Texas and Colorado.

The fight over better public education for Hispanic students was also fought in the courts. One battle concerned separate but equal education. Although *Brown v. Board of Education* (1954) had overturned as unconstitutional the practice of separate but equal education for African Americans, separate but equal education persisted for Hispanics because many school districts treated Hispanics as Whites. With funding from the NAACP and the Ford Foundation, the Mexican American Legal Defense and Education Fund (MALDEF) fought on behalf of Hispanic students in the courts in the 1960s and 1970s. The landmark case was *Cisneros v. Corpus Christi Independent School District* (1970) in which the court ruled that Hispanics had to be treated as a separate minority group from Whites, and hence separate but equal education was illegal. After this ruling, MALDEF turned its attention primarily to bilingual education. However, the separate but equal issue was not fully resolved at this point, and there was still some wiggle room in the legal system for local school districts to persist in separate but equal education. Thus there continued to be battles over this issue in the state and local courts for some time after *Cisneros*.

A number of improvements in Hispanic education came in the mid-1960s through President Johnson's War on Poverty. The Elementary and Secondary Education Act (1965) included funding for migrant education, adult instruction in

---

extracurricular activities for Hispanic students); housing (academic attitudes of other residents); financial aid (amount, type, information about); and sponsorship of the student by some organization that takes a personal interest in the individual student's career.

[7] While a number of additional studies discuss the history of the federal designation of Hispanic-Serving Institutions, the account here is drawn primarily from MacDonald (2004), Galdeano et al. (2012) and Valdez (2013).

the English language, and early childhood education. The Bilingual Education Act (1967) provided substantial federal funds for Spanish-language instruction in the public schools. However, many local school districts resisted federal inducements to provide bilingual education. The federal government fought back through its Office of Civil Rights, requiring under Title VI regulations that any public school district receiving federal funds (hence all school districts) file compliance plans that explain how children whose native language is not English were not being unfairly treated. This issue was only finally resolved when the Supreme Court ruled in 1974 in *Lau v. Nichols* that the civil rights of children who did not speak English were being violated if the schools did not address their linguistic needs as part of public instruction.

At the same time that the protests were taking place in Los Angeles and Texas, Puerto Rican students and parents were protesting in New York City and Hartford, Connecticut. In 1968, PS 25 in New York City became the first bilingual public school in the Northeast. ASPIRA and the Puerto Rican Legal Defense Fund battled in the courts against the New York City Board of Education. A 1974 consent decree mandated bilingual education in the New York City public schools.

Also in the 1960s and 1970s, there were movements to address Hispanic higher education. The first Hispanic student organization, the Mexican American Youth Organization, was founded in 1967 at St. Mary's College in San Antonio, Texas; soon a chapter was opened at the University of Texas at Austin. In Los Angeles, not long thereafter, the United Mexican American Students organization opened chapters at both UCLA and Loyola University, and the Mexican American Student Organization was organized at East Los Angeles Community College. The first campus protest occurred in 1968, when five students walked off the stage and several hundred people protested outside the graduation ceremony at San Jose State University. An important conference held the following year at the University of California at Santa Barbara resulted in *El Plan de Santa Barbara*, which set out reform demands that were widely influential across the nation. *El Plan* called, among other things, for more open university access for Hispanic students and the hiring of additional Hispanic faculty and administrators.

At about the same time, Puerto Rican student activist organizations formed in New York City. There were major student protests in 1969 and 1970. These actions resulted in a change to an open admission policy in 1969 at City College, the academically strongest school in the CUNY system, thus leading to a flood of new African-American and Hispanic enrollments; the formation of a department of Puerto Rican studies at City College in 1971; and the creation of a new Hispanic institution of higher education, Boricua College, in 1973. Other developments included the opening of centers for Puerto Rican studies at various campuses of CUNY and SUNY, as well as at elite private universities such as Princeton and Yale, and at the University of Illinois-Chicago and Wayne State in Detroit.

There was little, if any federal funding targeted specifically at higher education for Hispanic students in the 1960s and 1970s. However, private foundations including Rockefeller, Carnegie, and Danforth made large investments.

The politics of Hispanic education in the 1980s and 1990s was closely associated with the large increase in Hispanic population in the United States during this time period. During the 1990s, for example, the Hispanic population grew by almost 60 %, largely through immigration. There was a serious backlash that led to majority protests to both affirmative action and welfare rights for immigrants. Texas and several other states tried to withhold funding to school districts that accepted undocumented students, but this practice was ruled unconstitutional by the Supreme Court in *Plyler v. Doe* (1982). Three times in the early 1980s, Congress introduced legislation to make English the official language of the United States, but each time the bill failed to become law. Frustrated at federal inaction, 23 states passed English-only laws in the 1980s or early 1990s; though eventually all of these laws were struck down by the courts. In the 1990s, California passed a series of restrictive propositions, including Proposition 187 (the Save Our State Initiative, which denied health care and public education to illegal aliens and which was also eventually struck down by the courts) and Proposition 227, which restricted bilingual education and promoted English-language use (and is still in effect).

The first major federal study of Hispanic education was published in 1980 under the title *The Condition of Education of Hispanic Americans.* (Brown et al. 1980) In response, the U.S. Department of Education increased support for national programs that were beneficial to Hispanic students, mostly at the K-12 level but also at the college level.[8]

Lobbying efforts in the 1970s by the League of United Latin American Citizens to increase Hispanic access to higher education were largely unsuccessful. The Hispanic Higher Education Coalition (HHEC), formed in 1978, was somewhat more successful and became a powerful political force throughout the 1980s.[9] HHEC's principal goal was to broaden Title III (the Strengthening Developing Institutions program) of the landmark Higher Education Act of 1965 so as to include support for Hispanic students. HHEC testified five times between 1979 and 1985 on reauthorization of the Higher Education Act; and in the 1985 reauthorization bill, legislation was introduced to include $10 million in Title III funds that higher educational institutions enrolling at least 20 % Hispanic students could apply for. This amendment was passed by the House but not by the Senate, so it did not become law.

As a result of the inability to get Congress to act on behalf of Hispanic higher education, a new organization (the Hispanic Association of College and Universities, known as HACU) was formed by academic and business leaders in 1986.[10] For the

---

[8] These included Title IV funding under the Higher Education Act of 1965, which includes student aid and the TRIO programs.

[9] The initial members of the Hispanic Higher Education Coalition were ASPIRA of America, *El Congreso Nacional de Asuntos Colegiales*, League of United Latin American Citizens, Mexican American Legal Defense and Education Fund, National Association for Equal Educational Opportunities, National Council of La Raza, Puerto Rican Legal Defense and Education Fund, the Secretariat for Hispanic Affairs, and the U.S. Catholic Conference.

[10] The impetus for forming HACU came from Dr. Antonio Rigual and Sister Anne Sueltenfuss of Our Lady of the Lake University in San Antonio, TX, who approached Xerox about funding to

1992 authorization of the Higher Education Act, HACU wrote language included in the bill that was passed by Congress providing funds to a new entity known as "Hispanic-Serving Institutions" (HSIs), which were defined as those having a Hispanic student population of at least 25 %. HSIs enrolling at least 50 low-income students were able to apply for Title III funding from a \$45 million fund specifically set aside for HSIs.[11] While these funds were authorized in 1992, the first funds were not appropriated until 1995. In 1998, the funding for HSIs was moved to a new program under Title V of the Higher Education Act (the Developing Hispanic Serving Institutions Program), and the definition of an HSI was modified to require that the schools eligible for funding were accredited, degree granting, and nonprofit.

Another important political event in Hispanic education occurred when President Clinton, through Executive Order 12900, created a special committee to study Hispanic education with the task of addressing what federal, state, and local governments as well as private organizations could do to improve the situation. The Executive Order also called on each federal agency to examine its programs in support of education and identify ways to increase Hispanic participation in them. In 1996 the committee published its report, *A Nation on the Fault Line*. Increased funding was included in the Higher Education Acts of 1997 and 1998. The Clinton initiative also cleared the way for additional funding for a USDA HSI program under the reauthorization of the Farm Bill, as well as HSI funding through temporary funding programs from Housing and Urban Development and from the Department of Defense. The College Cost Reduction Act of 2008 provided \$200 million for HSIs for articulation agreements and STEM education. The Student Aid and Financial Responsibility Act of 2010 extended the HSI programs from the College Cost Reduction Act for an additional 10 years. Nevertheless, federal funds provide only two-thirds as much funding per student to HSIs as they do to colleges and universities overall.

---

establish a Center for Hispanic Higher Education at their school. They began meeting with representatives of Xerox and other HSIs, including Texas A&I Kingsville and New Mexico Highlands University. Eighteen colleges and universities were original members of HACU.

[11] Various percentages, from 20 to 40 %, were proposed over the years leading up to the 1992 law. The term "Hispanic colleges" had been used to designate these schools as early as 1979, but HSI has been uniformly used since 1992. The United College Negro Fund and some other organizations supporting federal support to African-American higher education, were resistant to expanding the Title III law to include other than Historically Black Colleges and Universities. Little known is that HHEC also lobbied for a larger allocation of the Title III funds for community colleges – a proposed increase from 24 to 40 % of the total funds – because of the large number of Hispanic students attending community colleges.

4.2 Hispanics and STEM Education – A Brief History

According to the U.S. Department of Education's Early Childhood Longitudinal Study, Hispanic children enter kindergarten with a significant educational achievement gap compared to Whites or Asians.[12] Preschool programs have been found to help to reduce that initial gap for Hispanic students and result in higher math scores by the age of 15. However, lower percentages of Hispanic children attend pre-school than any other major demographic group. Gandara (2006) cites the reasons given in the social science literature for this phenomenon:

> Latino parents maintain a strong sense of familism that runs counter to the practices of many preschool programs. In addition to simply retaining young children at home with family members for a longer period of time, many parents in their study remarked that teaching traditional values (such as respect for elders and authority figures) and use of the primary language were important concerns that they did not see supported in mainstream preschool programs. Parents may worry that their children will not be understood if caregivers and teachers do not speak Spanish, and they may also want to extend the familial language as long as possible before children enter the predominantly English-only world of school.

Many Hispanic students – perhaps as many as half in California – enter public school learning English as a second language. Those students tend to receive lower grades and be channeled into low-level and remedial courses as opposed to college preparatory courses. A few successful programs have been targeted at Hispanic students to encourage them to stay in school and prepare them for college.[13] One

---

[12] There is a great deal of overlap in the discussions in the social science literature about Hispanics and education generally and in the literature on Hispanics and STEM education. This section focuses on topics and results that are particular to STEM. On early public education and its effects on STEM trajectories for Hispanic students, see Gandara (2006). On high school and its effects on STEM career plans for Hispanic students, see especially Riegle-Crumb and Grodsky (2010) and Crisp and Nora (2012). Also see Zimmerman et al. (2011). On community colleges and Hispanic STEM education, see Gilroy (2012) and Malcom (2010). On Hispanic students and undergraduate degrees, see especially Camacho and Lord (2011). But also see Carpi et al. (2013), Crisp et al. (2009), Malcom (2008), Vasquez (2007), Cole and Espinoza (2008), Camacho and Lord (2013) and Fifolt and Searby (2010). Supplemental instruction is one of the common interventions employed in community colleges and baccalaureate programs in the STEM fields; on the impact of supplemental instruction programs on Hispanic students, see Rabitoy (2011), Meling (2012) and Meling et al. (2013). On STEM graduate education and Hispanics, see Millett and Nettles (2006) and Baker (2000). There is a large body of literature on the broad topic of STEM education and Hispanics, often also covering the more general issues of Hispanic education described in the previous section. See especially Rochin and Mello (2007), Taningco et al. (2008) and Miller and Horrigan (2014). Also see U.W. White House (2014), Landivar (2013), Anon. (2012), Flores (2011), Torres et al. (2014) and Chapa and De La Rosa (2006). On Hispanics and the STEM workforce, see Taningco (2008) and San Miguel et al. (2014). There is little general literature on Hispanics and computing education. See McFarland (2004) and Yau (2013).

[13] Successful programs include Achievement for Latinos Through Academic Success, which is targeted at middle schoolers; High School Puente and Achievement Via Individual Determination, which is targeted at high schoolers; and Project GRAD, which follows students from first grade through high school. (Gandara 2006)

notable program of this sort having a STEM orientation is the College Board's Equality 2000 program, which enhances high school math instruction as preparation for college STEM majors and STEM careers. The program has been a moderate success in low socioeconomic status and high-minority population schools.

Concern over the low U.S. scores in the International Math and Science Survey (TIMSS), plus a demand for a well-trained workforce from the technology sector in the 1990s, resulted in a call for enhanced math education in the public schools.[14] This has led to a significant increase in Hispanic students taking advanced math courses in high school,[15] but nevertheless significant achievement gaps continue to exist between Hispanic students and White students in those classes; and those gaps are the greatest in the highest level math courses: pre-calculus and calculus. Social scientists have identified some possible reasons for this phenomena: socioeconomically advantaged parents have greater knowledge of the higher educational system and can inform their children better and with more certainty; these parents tend to have higher educational and occupational expectations of their children; and these parents may also feel more confident in engaging the school and lobbying for their children's access to advanced courses. There are also institutional and other reasons: schools in wealthier communities have greater resources to devote, especially to "frills" such as advanced courses when they are being required to meet the No Child Left Behind standards; teachers, especially math teachers, of Hispanic students are likely to have fewer years of experience and be teaching outside of their field of expertise; and especially in these advanced courses, Hispanic students may fail to perform as well as they might on account of stereotype threat.[16] (Riegle-Crumb and Grodsky 2010)

Hispanic culture places faith in teachers as experts, and thus it is particularly effective when teachers direct Hispanic students towards STEM careers. Working-class and lower-class Hispanic parents are particularly deferential to teachers. One study suggests that teachers are particularly influential in the case of schools where there is a collaborative teaching environment (Moller et al. 2013b; also see Moller et al. 2013a).

Families can have a negative impact on Hispanic students attending college. Some of the reasons include parents holding gender-based stereotypes, being reluctant to take on the burden of loans for educational purposes, prioritizing spend-

---

[14] Elementary and middle school preparation in math and science has been shown to affect not only a student's academic preparation for high school STEM courses, but also student interest in taking such courses and in pursuing a STEM major in college. For a review of this literature, see Crisp and Nora (2012).

[15] 3 % of Hispanic high school graduates from the Class of 1982 had completed a pre-calculus course, compared to 15 % in 2004.

[16] Another study (Crisp and Nora 2012) points to literature that Hispanic students have lower feelings of self-efficacy than Whites in math and science, and that this type of self-efficacy is a strong predictor of a choice of a college STEM major. This same paper cites literature on gender and its impact on Hispanic choice of a STEM major, arguing: "as early as junior high school, at which time Latina students may be more hesitant to ask questions during class discussions, less likely to report that they are looking forward to taking eighth grade mathematics classes, and are the least likely of any group to have STEM career aspirations".

ing on other expenses than education, and pressuring their children to take jobs as soon as they graduate from high school. (Taningco 2008)

There has been little study of Hispanic high school students and their interest in computer science.[17] One study (Zimmerman et al. 2011) compares to other nearby schools a charter school in East San Jose, CA with a curriculum designed to encourage Hispanic students to attend college and increase their proficiency in English. The computer science study found that the Hispanic students were especially influenced by job compensation (they generally come from low-income families); knowing someone in the computing field; and gaining programming, robotics, or electronics experience through exposure to computers. The study also found what many other studies say about girls: Hispanic girls were interested in how computing applies to other fields.

Gilroy (2012) points out the advantages of community college for Hispanic students interested in STEM careers. Class sizes are small, so there is more chance for the student to interact with professors, obtain hands-on laboratory experience, and gain access to tutoring centers – all benefits that may be in short supply at a large public university. These advantages are regarded as a way of helping Hispanic students to persist in the face of what many students regard as challenging courses. Community colleges often enroll a diverse student population, which may be more comfortable for Hispanic students. Many community colleges are responsive to the needs of local employers, so many of the STEM students may find the curriculum more relevant to the outside world.

Social science research has found that, in both community colleges and 4-year colleges, supplemental instruction has been an effective means for improving academic performance and retention, particularly among Hispanic STEM students. Supplemental instruction was pioneered at the University of Missouri at Kansas City in the early 1980s and has been adopted by more than 800 colleges. It is typically used in connection with gateway courses in math and science – for example, college algebra, or introductory college-level physics and inorganic chemistry. These are courses that the students find difficult and for which there is high incidence of low or failing grades. Under this initiative, students meet outside of class in small study groups with a supplemental instruction leader to review basic material from the course and hone their study skills. It is a cooperative, participatory environment in which the students set the agenda of material to be covered and are responsible for helping each other to learn. In a study of a Hispanic-Serving Institution in Texas, participation in supplemental instruction for these gateway math and science course – no matter whether the student's level of participation was slight or great – led to increased course completion and higher course grades for Hispanic undergraduates. (Meling 2012)

---

[17] There are similarly few studies of Hispanic students in college degree programs in computing disciplines. See, for example, McFarland (2004) on the management information science and computer science curriculum at Western New Mexico University. Also see the discussion of the computing programs at the University of Texas at El Paso and some of its partner institutions in the NSF Broadening Participation in Computing Alliance known as CAHSI, as discussed in Aspray (2016).

Both male and female Hispanic students persist in their college education in engineering at a similar rate to Whites overall (slightly over half are still enrolled by the time of their eighth semester). However, unlike fields such as business, a lower percentage of Hispanics are recruited into engineering majors and careers than Whites. Thus the problem for Hispanic students in the STEM disciplines is one of recruitment rather than retention. This occurs despite the fact that engineering is the most intended major among male Hispanic high school students planning on entering a STEM discipline (and the third most common, behind social and behavioral sciences, and biological and agricultural sciences, for Hispanic females). Since the 1990s, the numbers but not the percentages of Hispanics graduating with engineering degrees have increased.

One of the challenges for Hispanic undergraduate engineering majors is a feeling of isolation. The professional societies MAES and SHPE (both described in Chap. 7) have worked on building community among Hispanic engineers and engineering students. Because of the heavy demands of science and engineering degrees, it often takes students longer to complete these majors. As a result, an issue of particular concern is financial aid. Another research finding is that minority student interest in STEM is correlated to the percentage of minority students at the college who are STEM majors. This may be one reason that more than half of the Hispanic students earning STEM degrees attend Hispanic-Serving Institutions, and that the HSIs are the fastest growing producer of STEM degrees for Hispanics. Another reason may be that the HSIs are less likely to teach their introductory science and math courses in large, impersonal lecture classes – as is often the case in large majority institutions. HSIs are also more likely to offer more culturally sensitive programming. Another possible reason is that the HSIs have more role models for students, and the social science literature shows a correlation between the presence of role models and student self-efficacy in their STEM major.

One issue faced by Hispanic students attending a majority institution is finding a mentor who is from the same cultural background. Fifolt and Searby (2010) explain some of the issues:

> Faculty and other professionals need to be prepared for potential issues that can arise as a result of cross-cultural mentoring relationships. They may see that students possess cultural mistrust … based on personal history and experiences of racism or mistreatment by the majority race or a reluctance to establish a relationship with the cross-cultural mentor for fear of having the appearance of betraying one's own cultural group. Mentors may find students who feel inhibited based on hierarchical power structures, both real and perceived, and who are less willing to participate fully in the mentoring relationship based on concerns of cultural stereotyping by the mentor. Finally, the mentor may also possess cultural mistrust, negative cultural biases, and fears about being successful in relating to someone from another culture based on a negative experience or a lack of experience with a culture that is different from his or her own.

There are only a few studies of Hispanic graduate students in the STEM fields. Based on data that is now 15 years old, Hispanics have a lower completion rate than Whites in doctoral studies in engineering. This is despite the fact that Hispanic and White doctoral students in the STEM disciplines were equally likely to hold a teaching or research assistantship or to have a mentor. (Millett and Nettles 2006) Using data from 1983 to 1997, another study (Baker 2000) showed that in the STEM disciplines, White and Hispanic doctoral students had similar distributions in their area of study, with the biological sciences highest, the physical sciences next highest, and engineering lowest of all. The Hispanic students were much more likely than the White students to have mothers and fathers who had received only a high school education or less. Most of the Hispanic students earned their doctorates at public universities with Research I Carnegie classifications, and most of those students studied in regions of the country in which there were large Hispanic populations.

Many Hispanic doctoral students have received their undergraduate education in Texas, New Mexico, California, Florida, or Puerto Rico. Hispanic-Serving Institutions have been among the top producers of baccalaureates for Hispanic students who went on to study for their doctoral degree. For doctoral completion, a study by Baker (2000) identified the most commonly mentioned factors: "a caring faculty, a supportive environment facilitated by the use of strong intervention strategies, good advising and counseling services, mentoring, peer tutoring, and the availability of role models in the community."

The number of doctoral degrees awarded to Hispanics in all fields produced annually in the United States remained low from 1976 to 1994: fewer than 1000 in each of those years and typically representing no more than 2% of all doctorates awarded. However, doctoral degrees earned by Hispanic students began to pick up after 1994, and in 2004 there were more than 1600 Hispanic recipients of doctoral degrees. From 1976 through 1998, Hispanic males received more doctorates than Hispanic females, but from that year forward, more Hispanic women than men were earning doctorates, and the gap is widening each year.

4.3 Hispanic-Serving Institutions – A Brief History

Hispanic-Serving Institutions[18] differ from HBCUs or tribal colleges in that the HSIs – in all but a few cases[19] – did not have a historic mission to serve Hispanic students and many of them do not have an identity centered on serving these

---

[18] The general literature on HSIs consulted here includes U.S. House of Representatives (2003), Stearns et al. (2002), Murphy (2013) and Laden (2001, 2004). Literature focused on administrative and institutional issues – usually for 4-year HSIs but sometimes for all HSIs – includes Garcia (2012), Espinoza and Espinoza (2012), Lu et al. (2014), Godoy (2010), Mulnix et al. (2002), De Los Santos and Cuamea (2010, 2008), Torres and Zerquera (2012) and Santiago (2012). Literature taking a student-centered approach (attraction of an HSI, sense of belonging, persistence, etc.) include Musoba and Krichevskiy (2014), Butler (2010), Núñez and Bowers (2011), Cuellar (2014) and Crisp and Cruz (2010). Literature focused specifically on HSI community colleges includes Núñez et al. (2011), Perrakis and Hagedorn (2010) and Gastic and Nieto (2010). Rudolph et al. (2014) focuses on graduate school and Ginther et al. (2011) discusses biases in research funding.

[19] The only universities, other than those in Puerto Rico, with a historical mission to serve Hispanics are: National Hispanic University (California), St Augustine's College (Illinois), Boricua College (New York), Northern New Mexico University, and Hostos Community College (New York).

students. There is a wide variation across HSIs as to whether they provide specific services to address the needs of Hispanic students, and whether the school identifies itself as an HSI.

Although the number of HSIs nationally changes each year with fluctuations in enrollments, the numbers have been steadily climbing as the Hispanic population increases, and as the numbers of Hispanic students attending community and 4-year colleges also increases. The numbers have grown rapidly because of the Civil Rights movement, increasing availability of financial aid for college, waves of immigration from Latin America since 1980, and relocation around the country in search of employment. The number of HSIs has grown from 242 in 2003 to 370 in 2013; and there are an additional 277 "emerging" HSIs, in which Hispanics make up between 15 and 24% of the student population. 59% of Hispanic undergraduates attend HSIs. 52% of HSIs are 2-year colleges. 38% offer some kind of graduate degree. 18% offer a doctorate. The typical HSI has fewer than 2000 Hispanic students enrolled. 61% of the HSIs have open admissions policies (compared to 38% of all US higher education institutions). Although the largest concentrations of HSIs still are found in Florida, New York, and the states bordering Mexico, HSIs are increasingly appearing in many rural and urban areas widely scattered across the United States. (Dervarics 2014)

HSIs are dependent to a large degree on state and federal funding for their operations, and as a result they have limited funds to spend on institutional advancement activities such as fundraising, public relations, alumni affairs, marketing, enrollment management, and government relations. Using data from 1998, which nevertheless appears to still be at least somewhat representative of today's situation, total revues of HSIs are 42% less per full time equivalent student than an other universities; endowment funds at HSIs are 91% less than at other institutions; and HSIs spend 43% less on instruction, 51% less on academic support functions, and 27% less on student services per FTE student than other higher educational institutions. (Mulnix et al. 2002)

More than half of HSIs are community colleges. Common attributes of a Hispanic student in an HSI community college include being male, first generation to attend college, working, responsible for a family, attending school part-time, older, and never having attained a high school diploma. These characteristics present risks of not completing college. Hispanics make up about half of the students enrolled in the typical 2-year HSI, but these schools also enroll significant numbers of other minorities. In some cases, the students enroll not because they are aware that the school is Hispanic-Serving but instead because it is located near home or work, allows part-time enrollment, offers classes at times convenient to work schedules, or because the costs are low. Hispanic students in the 2-year HSIs are less likely to graduate with an associate's degree or a certificate than students overall.

Not only are Hispanic students often unaware that they are attending an HSI, some of the emerging HSIs themselves are also not paying close attention to the fact that their Hispanic enrollments are growing to a point that they are approaching or have actually attained the status of HSI. In a content analysis of the websites of 19 HSIs, 8 of these schools appeared to be unaware of their HSI status, 6 were aware but not building programs specifically for their Hispanic students, and 5 appeared committed to building programs that served the Hispanic community.[20] (Torres and Zerquera 2012)

There is a significant body of social science literature concerned with the administration and institutional characteristics of HSIs. For example, in a survey administered in 2007, HSI presidents and chancellors identified (in order) their five greatest challenges: lack of funding, poor academic preparedness of students, low student retention and completion rates, hiring a diverse faculty that was adequately prepared, and keeping college tuition affordable. (De Los Santos and Cuamea 2010) Institutional resources as well as selectivity of the institution have been correlated in multiple studies to graduation rates for Hispanic students. (See Garcia 2012)

The costs of operating an HSI are generally higher than those for operating a predominantly White school because of the perceived need to provide the many nontraditional Hispanic students with mentoring, tutoring services, career counseling, exposure to cultural events, supplemental instruction learning assistance centers, and information literacy instruction at higher rates than is deemed necessary for White students. The challenges are great for these HSIs because the majority of them are public institutions where there are calls from legislators for efficiency in the delivery of instruction (measured as decreases in state support per student), while at the same time typically having to manage rapid growth in student numbers. There are increasing calls from politicians to measure the effectiveness of these schools (often in terms of graduation rates), but it is hard to determine appropriate metrics for evaluating the education of these nontraditional students. The provision of adequate services to educate and graduate non-traditional students is often in conflict with efforts to build up a school's institutional prestige – it is often a competition between excellence and access.

There is also a substantial body of social science literature on the experiences of students at HSIs. One of the topics covered thoroughly in the literature is persistence. (See Musoba and Krichevskiy 2014; Musoba et al. 2013; Maestas et al. 2007; Cuellar 2014) What factors relate to the persistence through an undergraduate degree for students at an HSI? Elevated high-school grades (but not elevated SAT scores), elevated grades in the first math and first English courses in college (but not elevated first-semester GPA), academic integration, developing a higher sense of academic self-concept over the course of college,[21] having a sense of belonging,

---

[20] The content analysis of the institutional websites focused on eight characteristics in determining the institutional readiness to be an HSI:

- Institutional mission;
- Emphasis on local community;
- Approach to diversity issues;
- Institutional plans posted on web site;
- Marketing strategies for enrollment;
- Student support program, especially for students of color and Latinos;
- Stated approach to serving the local community; and
- Any additional mention of Latino/a in the web site. (exact quotation, Torres and Zerquera 2012).

[21] Factors that were found to be positively correlated to gaining a better academic self-confidence include engaging in tutoring, interacting outside of class with faculty, doing homework, and having conversations with peers about their studies. (Cuellar 2014)

higher family income, and the receipt of financial aid all correlate to persistence through the degree.

One issue that has been studied is why Hispanic students so often attend HSIs. One theory (Braddock's perpetuation hypothesis) argues that racial segregation is perpetuated across institutional settings. This theory would indicate that Hispanic students who attended principally Hispanic high schools might select HSIs. However, another study (Butler 2010) showed that proximity of the nearest 2- or 4-year college explained most of the selection phenomena of an HSI and that it was not about perpetuation of racial segregation. The average distance in first college choice for a Hispanic student coming from a predominantly Hispanic high school is 84 miles (compared to 196 miles for African-American students from predominantly Black high schools).

Many Hispanic students come from lower socioeconomic backgrounds. One study (Núñez and Bowers 2011) found that a lower socioeconomic background is correlated both with attending an HSI instead of a predominantly White institution and with selecting a 2-year HSI over a 4-year HSI. Hispanic students who come to college from a high school with a high student-to-faculty or high student-to-counselor ratio are more likely to select a 2-year HSI. Núñez and Bowers speculate this is because high schools with these characteristics are "less likely to offer access to a curriculum that is 'constrained' to college preparatory classes." Interestingly, 4-year HSIs are more likely to enroll student bodies that are less well prepared in mathematics; nevertheless, these 4-year HSIs award high numbers of STEM degrees. This suggests to Núñez and Bowers that "high-performing HSIs successfully cultivate 'talent development' among students with varied academic preparation."

References

Anonymous. 2012. Excelencia in education reveals top 25 colleges producing Latino grads in STEM fields in 2009-10. *The Hispanic Outlook in Higher Education* 23(1): 26.

Arana, Renelinda, Carrie Castañeda-Sound, Steve Blanchard, and Teresita E. Aguilar. 2011. Indicators of persistence for Hispanic undergraduate achievement: Toward an ecological model. *Journal of Hispanic Higher Education* 10(3): 237–251.

Aspray, William. 2016. *Participation in computing: The national science foundation's expansionary programs.* London: Springer.

Baker, Maricel Quintana. 2000. *The baccalaureate origins of Latino doctorates in science and engineering, 1983-97.* Ph.D. Dissertation, Education, American University.

Boden, Karen. 2011. Perceived academic preparedness of first-generation Latino college students. *Journal of Hispanic Higher Education* 10(2): 96–106.

Brown, George H., Nan L. Rosen, Susan T. Hill, and Michael A. Olivas. 1980. *The condition of education for Hispanic Americans.* National Center for Education Statistics. Washington, DC: Government Printing Office.

Butler, Donnell. 2010. Ethno-racial composition and college preference: Revisiting the perpetuation of segregation hypothesis. *The Annals of the American Academy of Political and Social Science* 627(1): 36–58.

Camacho, Michelle Madsen, and Susan M. Lord. 2011. Quebrando fronteras: Trends among Latino and Latina undergraduate engineers. *Journal of Hispanic Higher Education* 10(2): 134–146.

Camacho, Michelle Madsen, and Susan M. Lord. 2013. Latinos and the exclusionary space of engineering education. *Latino Studies* 11: 103–112.

Cano, Miguel Ángel, and Linda G. Castillo. 2010. The role of enculturation and acculturation on Latina college student distress. *Journal of Hispanic Higher Education* 9(3): 221–231.

Carpi, V., Darcy M. Ronan, Heather M. Falconer, Heather H. Boyd, and Nathan H. Lents. 2013. Development and implementation of targeted STEM retention strategies at a Hispanic-serving institution. *Journal of Hispanic Higher Education* 12(3): 280–299.

Castellanos, Jeanett, and Alberta M. Gloria. 2007. Research considerations and theoretical application for best practices in higher education: Latina/os achieving success. *Journal of Hispanic Higher Education* 6(4): 378–396.

Cavazos Jr., Javier, Michael B. Johnson, and Gregory Scott Sparrow. 2010. Overcoming personal and academic challenges: Perspectives from Latina/o college students. *Journal of Hispanic Higher Education* 9(4): 304–316.

Chapa, Jorge, and Belinda De La Rosa. 2006. The problematic pipeline demographic trends and Latino participation in graduate science, technology, engineering, and mathematics programs. *Journal of Hispanic Higher Education* 5(3): 203–221.

Cole, Darnell, and Araceli Espinoza. 2008. Examining the academic success of Latino students in science technology engineering and mathematics (STEM) majors. *Journal of College Student Development* 49(4): 285–300.

Crisp, Gloria, and Irene Cruz. 2010. Confirmatory factor analysis of a measure of "mentoring" among undergraduate students attending a Hispanic serving institution. *Journal of Hispanic Higher Education* 9(3): 232–244.

Crisp, Gloria, and Amaury Nora. 2012. *Overview of Hispanics in science, mathematics, engineering and technology (STEM): K-16 representation, preparation, and participation.* White paper. Original version, 2006. Hispanic Association of Colleges and Universities.

Crisp, Gloria, Amaury Nora, and Amanda Taggart. 2009. Student characteristics, pre-college, college, and environmental factors as predictors of majoring in and earning a STEM degree: An analysis of students attending a Hispanic serving institution. *American Educational Research Journal* 46(4): 924–942.

Crisp, Gloria, Amanda Taggart, and Amaury Nora. 2014. Undergraduate Latina/o students: A systematic review of research identifying factors contributing to academic success outcomes. *Review of Educational Research.* Advance online publication. doi:10.3102/0034654314551064.

Cuellar, Marcela. 2014. The impact of Hispanic-Serving Institutions (HSIs), emerging HSIs, and non-HSIs on Latina/o academic self-concept. *The Review of Higher Education Volume* 37(4): 499–530.

Darder, Antonia, and Rodolfo D. Torres (eds.). 2014. *Latinos and education: A critical reader*, 2nd ed. New York: Routledge.

Delgado-Romero, Edward A., Angela Nichols Manlove, Joshua D. Manlove, and Carlos A. Hernandez. 2007. Controversial issues in the recruitment and retention of Latino/a faculty. *Journal of Hispanic Higher Education* 6(1): 34–51.

Dervarics, Charles. 2014. Hispanic-serving institutions continue growth with more poised to join the ranks. *Diverse: Issues in Higher Education*, February 25. http://diverseeducation.com/article/60920/. Accessed 23 Feb 2015.

Duran, Richard P. 1983. *Hispanics' education and background: Predictors of college achievement.* New York: College Entrance Exam Board.

Espinoza, Penelope P., and Crystal C. Espinoza. 2012. Supporting the 7th-year undergraduate: Responsive leadership at a Hispanic-serving institution. *Journal of Cases in Educational Leadership* 15(1): 32–50.

Fifolt, Matt, and Linda Searby. 2010. Mentoring in cooperative education and internships: Preparing protégés for STEM professions. *Journal of STEM Education* 11(1-2): 17–26.

Flores, Glenda M. 2011. Latino/as in the hard sciences: Increasing Latina/o participation in science, technology, engineering and math (STEM) related fields. *Latino Studies* 9: 327–335.

Galdeano, Emily Calderón, Antonio R. Flores, and John Moder. 2012. The Hispanic association of colleges and universities and Hispanic-serving institutions: Partners in the advancement of Hispanic higher education. *Journal of Latinos and Education* 11(3): 157–162.

Gandara, Patricia. 2006. Strengthening the academic pipeline leading to careers in math, science, and technology for Latino students. *Journal of Hispanic Higher Education* 5(3): 222–237.

Garcia, Gina A. 2012. Does percentage of Latinas/os affect graduation rates at 4-year Hispanic Serving Institutions (HSIs), emerging HSIs, and non-HSIs? *Journal of Hispanic Higher Education* 12(3): 256–268.

Gastic, Billie, and David Gonzalez Nieto. 2010. Latinos' economic recovery: Postsecondary participation and Hispanic-serving institutions. *Community College Journal of Research and Practice* 34(10): 833–838.

Gilroy, M. (2012). Community colleges growing importance in STEM education benefits Hispanics. *The Hispanic Outlook in Higher Education*, April 23.

Ginther, Donna K., Walter T. Schafer, Joshua Schnell, Beth Masimore, Faye Liu, Laurel L. Haak, and Raynard Kington. 2011. Race, ethnicity, and NIH research awards. *Science* 333(6045): 1015–1019.

Godoy, Cuauhtemoc. 2010. *The contribution of HSIs to the preparation of Hispanics for STEM careers: A multiple case study.* University of Pennsylvania, Doctor of Education dissertation.

Gonzales, Sandra M., Ethriam Cash Brammer, and Schlomo Sawilowsky. 2014. Belonging in the academy: Building a 'Casa away from casa' for Latino/a undergraduate students. *Journal of Hispanic Higher Education.* Advance online publication. doi:10.1177/1538192714556892.

Gross, Jacob P.K. 2011. Promoting or perturbing success: The effects of aid on timing to Latino students' first departure from college. *Journal of Hispanic Higher Education* 10(4): 317–330.

Gross, Jacob P.K., Desiree Zerquera, Brittany Inge, and Matthew Berry. 2014. Latino associate degree completion: Effects of financial aid over time. *Journal of Hispanic Higher Education* 13(3): 177–190.

Hispanic Association of Colleges and Universities and The Institute for Higher Educational Policy (HACU). 1994. *Student financial aid: Impact on Hispanics and Hispanic-serving institutions.* San Antonio: Hispanic Association of Colleges and Universities.

Hispanic Association of Colleges and Universities and The Institute for Higher Educational Policy (HACU). 1995. *Enhancing quality in higher education: Affirmative action and the distribution of resources in US Department of Education Programs.* San Antonio: Hispanic Association of Colleges and Universities.

Ibarra, Robert A. 2001. *Beyond affirmative action: Reframing the context of higher education.* Madison: University of Wisconsin Press.

Kim, Young K., Liz A. Rennick, and Marla A. Franco. 2014. Latino college students at highly selective institutions: A comparison of their college experiences and outcomes to other racial/ethnic groups. *Journal of Hispanic Higher Education* 13(4): 245–268.

Laden, Berta Vigil. 2001. Hispanic-serving institutions: Myths and realities. *Peabody Journal of Education* 76(1): 73–92.

Laden, Berta Vigil. 2004. Hispanic-serving institutions: What are they? Where are they? *Community College Journal of Research and Practice* 28: 181–198.

Landivar, Liana Christin. 2013. *Disparities in STEM employment by sex, race, and Hispanic origin.* American Community Survey Reports. Washington, DC: U.S. Department of Commerce, Economics and Statistics Administration, U.S. Census Bureau.

Leal, David L., and Kenneth J. Meier (eds.). 2011. *The politics of Latino education.* New York: Teachers College Press.

Lopez, J. Derek. 2013. Differences among Latino students in precollege multicultural exposure and the transition into an elite institution. *Journal of Hispanic Higher Education* 12(3): 269–279.

Lu, Ming-Tsan Pierre, Hsuying C. Ward, Terry Overton, and Yousun Shin. 2014. The synergetic approach to effective teachers' research education: An innovative initiative for building educational research capacity in a Hispanic-serving institution. *Journal of Hispanic Higher Education* 13(4): 269–284.

MacDonald, Victoria Maria. 2004. *Latino education in the United States: A narrated history from 1513–2000.* New York: Palgrave Macmillan.

Maestas, Ricardo, Gloria S. Vaquera, and Linda Muñoz Zehr. 2007. Factors impacting sense of belonging at a Hispanic-serving institution. *Journal of Hispanic Higher Education* 6(3): 237–256.

Malcom, Lindsey Ellen. 2008. *Accumulating (Dis)Advantage? Institutional and financial aid pathways of Latino STEM baccalaureates*. PhD dissertation, Education, University of Southern California.

Malcom, Lindsey E. 2010. Charting the pathways to STEM for Latina/o students: The role of community colleges. In *Students of color in STEM. New directions for institutional research, number 148*, ed. Shaun R. Harper and Christopher B. Newman, 29–40. Hoboken: Wiley Periodicals.

McFarland, Ronald D. 2004. Effective design considerations to address Hispanic learners in the MIS/CS curriculum. *Journal of Computing Sciences in Colleges* 20(1): 314–322.

Meling, Vanessa Bogran. 2012. *The role of supplemental instruction in academic success and retention at a Hispanic-serving institution*. Doctor of Education dissertation, Texas A&M University-Kingsville and Texas A&M University-Corpus Christi.

Meling, Vanessa B., Marie-Anne Mundy, Lori Kupczynski, and Mary E. Green. 2013. Supplemental instruction and academic success and retention in science courses at a Hispanic-serving institution. *World Journal of Education* 3(3): 11–23.

Miguel, San, M. Anitza, and Mikyong Minsun Kim. 2014. Successful Latina scientists and engineers: Their lived mentoring experiences and career. *Journal of Career Development*. Advance online publication. doi:10.1177/0894845314542248.

Miller, Joseph S., and John B. Horrigan. 2014. *STEM urgency: Science, technology, engineering, and mathematics education in an increasingly unequal and competitive world*. Washington, DC: Joint Center for Political and Economic Studies.

Millett, Catherine M., and Michael T. Nettles. 2006. Expanding and cultivating the Hispanic STEM doctoral workforce: Research on doctoral student experiences. *Journal of Hispanic Higher Education* 5(3): 258–287.

Moller, Stephanie , Roslyn Arlin Mickelson, Elizabeth Stearns, Neena Banerjee, and Martha Cecelia Bottia. 2013a. Collective pedagogical teacher culture and mathematics achievement: Differences by race, ethnicity, and socioeconomic status. *Sociology of Education* 86(2): 1745-1794

Moller, Stephanie, Neena Banerjee, Martha Cecilia Bottia, Elizabeth Stearns, Roslyn Mickelson, Melissa Dancy, Eric Wright, and Lauren Valentino. 2013b. *Moving Latino/a students into STEM fields: The role of teachers and professional communities in secondary schools*. Roots of STEM. Working Paper No. 104, UNC Charlotte.

Montalvo, Edris J. 2012. The recruitment and retention of Hispanic undergraduate students in public universities in the United States, 2000–2006. *Journal of Hispanic Higher Education* 12(3): 237–255.

Mulnix, Michael William, Randall G. Bowden, and Esther Elena Lopez. 2002. A brief examination of institutional advancement activities at Hispanic serving institutions. *Journal of Hispanic Higher Education* 1(2): 174–190.

Murphy, Joel. 2013. Institutional effectiveness: How well are Hispanic serving institutions meeting the challenge? *Journal of Hispanic Higher Education* 12(4): 321–333.

Musoba, Glenda Droogsma, and Dmitriy Krichevskiy. 2014. Early coursework and college experience predictors of persistence at a Hispanic-serving institution. *Journal of Hispanic Higher Education* 13(1): 48–62.

Musoba, Glenda Droogsma, Charlene Collazo, and Sharon Placide. 2013. The first year: Just surviving or thriving at an HIS. *Journal of Hispanic Higher Education* 12(4): 356–368.

National Center for Education Statistics (NCES). 2011. *Achievement gaps: How Hispanic and White students in public schools perform in mathematics and reading on the National Assessment of Educational Progress*. Statistical analysis report. Washington, DC: Institute of Education Sciences.

National Council of La Raza. 1981. *The Alpha Project: An NCLR program to increase the number of Hispanics in science, mathematics, and engineering careers*. Washington, DC: NCLR.

Núñez, Anne-Marie, and Alex J. Bowers. 2011. Exploring what leads high school students to enroll in Hispanic-serving institutions: A multilevel analysis. *American Educational Research Journal* 48(6): 1286–1313.

Núñez, Anne-Marie P., Johnelle Sparks, and Eliza A. Hernández. 2011. Latino access to community colleges and Hispanic-serving institutions: A national study. *Journal of Hispanic Higher Education* 10(1): 18–40.

Ornelas, A. 2002. *An examination of the resources and barriers in the transfer function and process: A case study analysis of an urban community college.* Unpublished Ph.D. Dissertation, University of California, Los Angeles

Oseguera, Leticia, Angela M. Locks, and Irene I. Vega. 2009. Increasing Latina/o students' baccalaureate attainment: A focus on retention. *Journal of Hispanic Higher Education* 8(1): 23–53.

Padilla, Raymond V., and Miguel Montiel. 1998. *Debatable diversity: Critical dialogues on change in American Universities.* Lanham: Rowman & Littlefield.

Perrakis, Athena, and Linda Serra Hagedorn. 2010. Latino/a student success in community colleges and Hispanic-serving institution status. *Community College Journal of Research and Practice* 34(10): 797–813.

Rabitoy, Eric. 2011. *Supplemental instruction in STEM-related disciplines on a community college campus: A multivariate path-analytic approach.* Doctor of Education dissertation, California State University, Fullerton.

Riegle-Crumb, Catherine, and Eric Grodsky. 2010. Racial-ethnic differences at the intersection of math course-taking and achievement. *Sociology of Education* 83(3): 248–270.

Rochin, Refugio I., and Stephen F. Mello. 2007. Latinos in science: Trends and opportunities. *Journal of Hispanic Higher Education* 6(4): 305–355.

Rudolph, Bonnie A., Carlos P. Castillo, Vanessa G. Garcia, Alina Martinez, and Fernando Navarro. 2014. Hispanic graduate students' mentoring themes: Gender roles in a bicultural context. *Journal of Hispanic Higher Education.* Advance online publication. doi:10.1177/1538192714551368.

Saenz, Victor B. (2002, September). *Hispanic students and community colleges: A critical point for intervention.* ERIC Digest. Los Angeles: ERIC Clearinghouse for Community Colleges.

Santiago, Deborah A. 2012. Public policy and Hispanic-serving institutions: From invention to accountability. *Journal of Latinos and Education* 11(3): 163–167.

Santos, De Los, G. Alfredo Jr., and Irene I. Vega. 2008. Hispanic presidents and chancellors of institutions of higher education in the United States in 2001 and 2006. *Journal of Hispanic Higher Education* 7(2): 156–182.

Santos, De Los, G. Alfredo Jr., and Karina Michelle Cuamea. 2010. Challenges facing Hispanic-serving institutions in the first decade of the 21st century. *Journal of Latinos and Education* 9(2): 90–107.

Solorzano, Daniel G., Octavio Villalpando, and Leticia Oseguera. 2005. Educational inequities and Latina/o undergraduate students in the United States: A critical race analysis of their educational progress. *Journal of Hispanic Higher Education* 4(3): 272–294.

Stearns, Christina, Satoshi Watanabe, and Thomas D. Snyder. 2002. *Hispanic serving institutions statistical trends from 1990 to 1999.* NCES 2002–051. National Center for Educational Statistics. Washington, DC: U.S. Department of Education, Office of Educational Research and Improvement.

Taningco, Maria Teresa V. 2008. *Latinos in STEM professions: Understanding challenges and opportunities for next steps.* Los Angeles: Tomas Rivera Policy Institute.

Taningco, Maria Teresa V., Ann Bessie Mathew, and P. Harry. 2008. *STEM professions: Opportunities and challenges for Latinos in science, technology, engineering, and mathematics: A review of literature.* Los Angeles: Tomas Rivera Policy Institute.

Torres, Vasti, and Desiree Zerquera. 2012. Hispanic-serving institutions: Patterns, predictions, and implications for informing policy discussions. *Journal of Hispanic Higher Education* 11(3): 259–278.

Torres, Luis, Christopher Peña, Karina Camacho, and Priscilla Silva. 2014. *STEM education: A bridge for Latinos to opportunity and success*. League of United Latin American Citizens. Washington, DC: Time Warner Cable Research on Digital Communications.

U.S. House of Representatives. 2003. *Expanding opportunities in higher education: Honoring the contributions of America's Hispanic serving institutions*. October 6. Subcommittee on Select Education, Committee on Education and the Workforce, Edinburg, Texas. Washington, DC: U.S. Government Printing Office.

U.S. White House. 2014. *Hispanics and STEM education. White House Initiative on Educational Excellence for Hispanics*. Washington, DC: U.S. Department of Education.

Valdez, Patrick Lee. 2013. *Hispanic-serving institution legislation: An analysis of policy formation between 1979 and 1992*. Doctoral dissertation, The University of Texas at Austin.

Valverde, Leonard A., and Shernaz B. Garcia. 1982. *Hispanics in higher education: Leadership and vision for the next twenty five years*. Proceedings of the invitational symposium, April 29 to May 1, 1982. Austin: Office for Advanced Research in Hispanic Education, University of Texas at Austin.

Vasquez, Philip L. 2007. *Achieving Success in Engineering: A phenomenological exploration of Latina/o student persistence in engineering fields of study*. Thesis, Master of Science, Education, Iowa State University.

Vélez-Ibáñez, Carlos, Elsie Szecsy, and Courtney Peña. 2013. The impact of the Ford fellowship program in the creation of Latina/o academic generations. *Journal of Hispanic Higher Education* 12(2): 109–137.

Vernez, Georges. 1997. *Education's Hispanic challenge*. The Jerome. Working paper No. 228. Annandale-on-Hudson: Levy Economics Institute of Bard College.

Vernez, Georges, and Lee Mizell. 2001. *Goal: To double the rate of Hispanics earning a bachelor's degree*. Center for Research on Immigration Policy. Santa Monica: RAND.

Yau, Margaret. 2013. Engaging Hispanic/Latino(a) youth in computer science: An outreach project experiencing report, consortium for computing sciences in colleges. *Journal of Computing Science in Colleges* 28(4): 113–122.

Zimmerman, Thomas G., David Johnson, Cynthia Wambsgans, and Antonio Fuentes. 2011. Why Latino high school students select computer science as a major: Analysis of a success story. *ACM Transactions on Computing Education* 11(2): 1–17.

Chapter 5
Opening STEM Careers to American Indians

Abstract This chapter provides an account of science and technology education for American Indians. It begins with a section on the history of higher education for American Indians and is followed by a section on American Indian higher education today. Both of these sections include material about the tribal colleges and universities. The chapter closes with sections on science and technology education for American Indians, and on computing at the tribal colleges.

This chapter provides an account of American Indians, the third of the three underrepresented minorities in STEM covered in this book. The number of American Indians is smaller than the number of Hispanics or African Americans, and the challenges they face are greater than those of the other two minority groups. This chapter begins with a section on the history of higher education for American Indians and is followed by a section on American Indian higher education today. Both of these sections include material about the tribal colleges and universities. We close with sections that discuss STEM education for American Indians and computing at the tribal colleges.

The tribal colleges play a central role in this story. They are typically much younger than the Historically Black Colleges and Universities or the Hispanic-Serving Institutions.[1] The first tribal colleges were formed in the late 1960s as an outgrowth of the civil rights and American Indian self-determination movements. Today there are 37 tribal colleges, spread across 14 states.[2] They are members of the American Indian Higher Education Consortium (AIHEC), described in Chap. 7. An important goal of the tribal colleges has been to provide higher education without the need to assimilate into White culture. While enrollments at the tribal colleges more than doubled between 1990 and 2005, still less than one-fifth of all American Indians enrolled in college attend tribal colleges. Most of the students enrolled in

---

[1] For a general history of American Indian education, see AIHEC (1999), Deloria (1991), Freeman and Fox (2005), Goulding (1995), Huff (1997), and Nee-Brenham and Stein (2003).

[2] The tribal colleges often play more than a higher educational role on the reservations. For example, the libraries and computer labs are often open to all members of the reservation, and the tribal colleges serve as a locus for community activities such as business incubators and events to promote good health to the entire population.

© Springer International Publishing Switzerland 2016
W. Aspray, *Women and Underrepresented Minorities in Computing*,
History of Computing, DOI 10.1007/978-3-319-24811-0_5

tribal colleges are from the first generation in their family to attend college. The average age of these students is 31, significantly older than at most U.S. colleges. Many of these students work and attend school part-time, and a significant number are single parents. About two-thirds of these students are female. Some of the tribal colleges offer only certificates and 2-year degrees, but an increasing number of them offer bachelor's degrees and a few offer masters and doctoral degrees. (NAE 2006; Varma 2009a, b)

The tribal colleges face a number of challenges. The reservations on which they are typically located have high levels of unemployment and low average income, and the tribal colleges are underfunded. Congress first passed the Tribally Controlled College or University Assistance Act in 1981, appropriating $2831 per Indian student for the operation of the tribal colleges.[3] But the actual federal appropriation must be negotiated each year, making it difficult for tribal college administrators to engage in long-term planning. Some programs on the tribal college campuses are also funded out of other federal agencies. Despite significant improvements over the past quarter-century, preparation for college among American Indians is well below the national average in terms of K-12 reading and mathematics proficiency; and American Indian students take fewer advanced science and mathematics courses in high school than Whites, Asians, African Americans, or Hispanics. When American Indian students reach college, they have the lowest recruitment and retention rates of all minorities. However, the tribal colleges are somewhat more successful than other colleges at retaining and graduating American Indian students, especially when the students enter college directly out of high school. It has also been difficult to recruit faculty to teach at the tribal colleges. There are limited numbers of American Indians themselves prepared to teach at the college level, and it is difficult to attract others to teach there. Salaries are low. Teaching and counseling loads are heavy (often five courses per semester). The schools are mostly located in remote places, making it difficult for the faculty to gain additional training to advance their academic credentials beyond the bachelor or masters degree that these faculty members typically hold. (NAE 2006; Varma 2009a, b; Boyer 2014)

---

[3] In 1994, Congress passed a law making the tribal colleges and universities land-grant colleges, similar to the Morrill Acts of 1859 and 1890, which created the land-grant colleges in many states. Under the 1994 act, some of the tribal colleges (e.g. Salish Kootenai and Turtle Mountain) received funding for their initiatives in e-learning through the Tribal Colleges Equity Education Grants Program administered by the US Department of Agriculture. In 1996, President Clinton signed an executive order for federal agencies to do more to support the tribal colleges and universities.

5.1   Higher Education for American Indians – A Brief History

For context, we will briefly discuss the history of higher education for Americans Indians since the first settlements of Europeans in the United States.[4] The story of European-style higher education for American Indians begins soon after the first permanent European settlement is made in Virginia. In 1611 a settlement was opened in Henrico, on the James River near the present-day city of Richmond. A Royal land grant was given in 1618 to form the first European-style higher-education institution in America, and by 1622 a school for American Indians had opened on the site. However, much of the Henrico settlement was destroyed in a battle between the settlers and the American Indians that same year, and the following year the British government closed the college and re-appropriated the land and remaining assets.[5]

Between the 1620s and the American Revolutionary War in the 1770s, a few major colleges opened on the eastern seaboard, and several of them included the education of indigenous people as part of their charters. Harvard College was founded in 1636 and opened an "Indian college" on its campus in 1654. Of the 20 American Indians enrolled in the Latin grammar school that was the feeder program into Harvard, only two graduated with a bachelor's degree.[6] In 1693 the College of William and Mary opened, with one of its founding goals being to Christianize American Indians. In 1769 Dartmouth College was founded, and funds were raised in both Britain and New England for the education of American Indian students. By the time of the Revolutionary War, however, only 47 American Indians had matriculated in these colleges and only four had graduated. There was a high price to be paid by those few American Indian students who attended these colleges, for

---

[4] There is a sizable body of literature on the history of American Indians and higher education. For general purposes, this section pulled material from Monette (1995), Pavel et al. (1998), Machamer (2000), Martin (2005), and McClellan, Fox, and Lowe (2005). On the history of tribal colleges and the politics of establishing them, see Olivas (1981), Stein (1988), Pease-Windy Boy (1994), Young (1998), and Pease-Pretty on Top (2003). Some examples from the large literature on the history of K-12 education for American Indians, see Adams (1971), Szasz (1977), and Belgarde (2004). Also see Fischbacher (1967), Adams (1971), Boyer (1989), DeJong (1993), Carney (1999), Hale (2002), Reyhner and Eder (2004), and Klug (2012).

Native American Studies programs were established at Cornell, Dartmouth, Michigan, Stanford, and other universities. These programs sometimes had significant value in enabling American Indian students to persist in principally White institutions. On the history of Native American studies programs, see Cook-Lynn (1997), Nelson (1997), Champagne and Strauss (2002), Krupat (2002), and Kidwell (2005).

[5] The College of William and Mary argues that it is, in fact, the college that was supposed to be built at Henrico.

[6] The low graduation rate was apparently caused in part by death to European diseases, loneliness, and cultural issues. To add further indignity, one of the two graduates died in a shipwreck on the return to campus for the graduation ceremony.

afterwards they had trouble being welcomed into either European or American Indian society.

The 1790 Trade and Intercourse Act brought all business transactions with American Indians under federal rather than state or private control. Between 1794 and 1871, the federal government signed treaties with various American Indian nations for support of American Indian precollege education, often in exchange for land. Perhaps the most significant development in American Indian education prior to the Civil War was the establishment in 1825, with federal and missionary funds, of the Choctaw Academy in Kentucky. It provided the most advanced education offered to American Indians at the time, including both liberal arts and vocational training, with a heavy dose of Christianity. While the majority of students were Choctaw, students attended from eight Indian nations. In an 1830 treaty, the federal government agreed to pay for school buildings and provide scholarships for up to 20 Choctaw youth to attend college. However, forced removal of Choctaw Indians began the following year as part of the plan to relocate all indigenous people to places west of the Mississippi. In 1842, when enough Choctaws had been relocated to Oklahoma, the Choctaw nation severed its ties with the Choctaw Academy and established six boarding schools in Oklahoma.

In 1860 the federal government established the first boarding school for American Indians, on the Yakima Reservation in Washington. In 1879 the federal government opened the Carlisle Indian School in Pennsylvania, the first federal boarding school for American Indians not located on a reservation. The school taught mostly industrial and agricultural arts. By the 1880s, there were more than 80 American Indian boarding schools. As a means of assimilation and control, federal law passed in 1893 allowed the government to withhold food supplies from American Indian families who would not send their children to boarding schools or day schools on the reservation; and many families, facing hunger, reluctantly complied.

In the late nineteenth century, three educational institutions were formed that eventually brought higher education to American Indians. In 1878 the Sitka Industrial and Training School opened in Alaska for Alaskan Natives. It received early funding through the efforts of a Presbyterian Minister, Sheldon Jackson, and was renamed in his honor in 1910. In 1880 Almon Bacone, a Baptist missionary teacher, opened an American Indian School in Oklahoma, and in 1885 it moved to Muskogee, Oklahoma, where the Creek tribe had donated land for the school. In 1910 it was renamed Bacone College, and it is still one of the few private colleges focused on American Indian education. In 1887 the North Carolina state government funded the Croaton Normal School for students from the Cherokee Nation. It began granting 4-year degrees in 1939 and is now part of the state university system and known as the University of North Carolina at Pembroke.[7] Hampton Institute in Virginia, a Historically Black college founded by the American Missionary

---

[7] The school went through a number of name changes: Indian Normal School of Robeson County (1911), Cherokee Indian Normal School of Robeson County (1911), Pembroke State College for Indians (1941), Pembroke State College (1949, which began admitting White students in 1953), and Pembroke State University (1969); and then it was made a part of the University of North

Association in 1868 to teach freed slaves, admitted American Indians from 1878 to 1923. Throughout the nineteenth century, however, all of these schools were primarily for secondary and vocational education rather than higher education.

The Rockefeller Foundation funded a study entitled *The Problem of Indian Administration*, better known as the Meriam Report after its principal author. The report, prepared by an organization that later became the Brookings Institution, was delivered to Congress in 1928. It was the first Congressional study of American Indian affairs since the Schoolcraft study in the 1850s.[8] Among many other indictments of the Interior Department for its management of American Indian affairs, the report criticized the boarding schools and missionary schools catering to American Indians for their poor education, harsh living conditions, and coerced enrollments. The Meriam Report was influential during the Hoover and Roosevelt Administrations. The federal government passed the Indian Reorganization Act in 1934, which provided tribes with many rights of self-government.[9] The bill included provision for loans for American Indians to attend college. Two years later, Congress passed the Johnson-O'Malley Act, which authorized federal contracts to states to provide higher education to American Indians.

The Second World War had a significant impact on education for American Indians. 25,000 American Indian GIs returned to their reservations after the war, either from the war fronts or from urban areas where they had relocated during the war to work in war-related factories. This experience gave the veterans a wider view of the world, expectations that the federal government would support them (rather than control them as the Bureau of Indian Affairs typically tried to do), and a regard for education as a principal way to achieve a better life for themselves and their families. Some American Indians took advantage of the GI Bill of Rights (Veterans Readjustment Act, 1948) to attend college or vocational school. The Bureau of Indian Affairs started a scholarship program that same year. By the late 1950s, more than 20 tribes offered college scholarships to their members. Even so, there were only about 2000 American Indians in college in the late 1950s and only about 3500 in college in the late 1960s.[10]

Carolina university system in 1972. It may be the only Historically Indian college that is now minority American Indian.

[8] It was not, however, the last federal report on American Indian education. Another report with similarly critical things to say as the Merriam Report was the Kennedy Report of 1969, *Indian Education: A National Tragedy – A National Challenge* (U.S. Senate Committee on Labor and Public Welfare 1969). For a discussion of the Merriam and Kennedy reports, see DeJong (1993).

[9] College scholarships for American-Indian students became available in 1948 through a Bureau of Indian Affairs scholarship program. The Bureau of Indian Affairs had taken over federal responsibility for higher education of American Indians in 1921 with the passage of the Snyder Act.

[10] One might have thought that the civil rights movement of the 1960s would be highly supportive of American Indian needs, including educational needs. However, civil rights legislation was driven principally by the demands of the African American community; and the legislative goal of integration of Whites and Blacks ran counter to the desire of the American Indians to create segregated educational institutions. For a comparison of American Indian and African American education, see Carney (1999, Chapter 5).

In the 1950s and early 1960s there were a few new venues offering higher education to American Indians. Arizona State University opened the Arizona State Indian Education Center in 1954, providing both undergraduate and graduate instruction. The Santa Fe Boarding School renamed itself the Institute of American Indian Arts in 1962 and began offering college-level instruction in the arts. But other than the handful of schools mentioned so far, most American Indians had to attend majority institutions if they wanted a higher education.

The tribal college movement began with the Navajo nation, which owned the largest amount of land and had the largest population of any tribe. There had been talk of self-control over both pre-college and college education on the Navajo reservation since the early 1950s. The first major action was taken in 1963, when Navajo leaders applied for grant funds to take over the Bureau of Indian Affairs school in Lukachuki, Arizona. This attempt failed, but it led to the creation by the Navajos of the Rough Rock Demonstration School in Rough Rock, Arizona for precollege education of their children. The same individuals who were involved in founding the Rough Rock School led the effort in 1968 to create Navajo Community College (today Dine College) in Tsaile, Arizona. The federal government provided support for both capital and operating expenses through the Navajo Community College Act of 1971.

The creation of Navajo Community College showed the way for the creation of a number of other colleges under tribal control. In 1970 the Haskell Institute, which had been founded in 1884 as an off-reservation boarding school in Lawrence, Kansas, changed its name to Haskell Indian Junior College (now Haskell Indian Nations University) and began offering postsecondary education. Other early colleges were Turtle Mountain Community College (North Dakota), Fort Berthold Community College (North Dakota), Standing Rock College (now Sitting Bull College, North Dakota), Oglala Lakota College (South Dakota), and Sinte Gleska College (now University, South Dakota). The numbers steadily grew, and today there are 37 tribal colleges and universities. There are more than 10 times as many tribes as colleges, however, so there is opportunity for growth in the number of tribal colleges in the coming years.

Many of these schools received federal support under Title III of the Higher Education Act of 1965. One funding requirement was that these institutions be accredited. None of the early tribal colleges were accredited, so the tribal colleges formed working partnerships with accredited majority institutions, which typically were the grant holders.[11] However, since the passage of the Tribally Controlled

---

[11] Efforts to achieve regional accreditation at the tribal colleges began in earnest in the 1970s, with efforts to accredit Navajo Community College, Oglala Sioux Community College, Sinte Gleska College, Turtle Mountain Community College, D-Q University, and Standing Rock Community College. In 1976, Navajo Community College was the first tribal college to receive accreditation to offer associate degrees. In 1983 Oglala Lakota College and Sinte Gleska College were the first two tribal colleges accredited to offer bachelor degrees. In 1989 Sinte Gleska was the first tribal college accredited to offer masters degrees. Today, most of the tribal colleges are accredited as institutions to offer 2-year degrees and many are also accredited for 4-year degrees. These early relationships between the tribal colleges and mainstream colleges and universities have developed over time to

College Assistance Act in 1978, most of the tribal colleges have received their funding directly from the federal government.[12]

In the 1990s there was ongoing federal concern about both precollege and college education for American Indians. In 1991 the U.S. Department of Education released a report entitled *Indian Nations at Risk: An Educational Strategy for Action* (U.S. Department of Education 1991), which identified many problems with precollege education on the reservations and presented various solutions. A White House Conference on Indian Education resulted in 114 recommendations. At the college level, in 1994 Congress granted land grant status (created under the Morrill Act of 1862) to the tribal colleges, which provided them with substantial ongoing funding. In 1996 President Clinton signed an Executive Order calling upon all the federal agencies to provide better access for American Indians and the tribal colleges to the various educational programs operated by federal agencies.

5.2 Higher Education for American Indians Today

American Indians[13] face many challenges in their higher education.[14] They are approximately twice as likely to live below the poverty line as other Americans. Many American Indians, especially the one third who live on reservations, grow up

---

include many other types of programs such as articulated agreements for transfer from the tribal college to complete a baccalaureate, distance learning programs, and access for tribal college students to certain kinds of science and technology facilities and faculty that are typically stronger at the mainstream college or university.

[12] There was a slew of federal legislation concerning American Indian education in the 1970s. In 1972 Congress passed the Indian Education Act, which recognized Indian culture and provided grants to support its incorporation into the curriculum; this Act also provided support for Indian teacher training and Indian adult education. In 1975 Congress passed the Indian Self-Determination and Education Assistance Act, which (in order to enhance self-governance over educational matters) authorized the federal government to contract with tribes and Indian organizations for the tribal operation of programs. The most important of these laws was the Tribally Controlled Community College Act (1978), which provided legal recognition to and funding for the operation and improvement of the tribal colleges.

Three schools received funding directly from the Bureau of Indian Affairs through the Snyder Act of 1923: Haskell Indian Junior College, the American Indian Art Institute, and Southwest Indian Polytechnic Institute.

Generally speaking, the state governments did not provide support for the tribal colleges.

[13] There are sizable social science and policy literatures on American Indians and higher education. Fuchs and Havinghurst (1972), and Larimore and McClellan (2005) provide general overview material. Pavel et al. (1998); Institute for Higher Education Policy (2000, 2007); DeVoe and Darling-Churchill (2008), and AIHEC (2009) provide general statistical information on the topic. More specialized studies include ones on persistence [Jackson et al. (2003); Mosholder and Goslin (2013–2014)], cultural issues [Kirkness and Barnhardt (1991), Frey and Pewewardy (2004), Lundberg (2007), and Tippeconnic and Tippeconic Fox (2012)], and policy [NAS (2011)].

[14] This paragraph, like those of so many other researchers, dwells on the challenges for American Indian students. It should not be forgotten that there are many excellent American Indian students.

in isolated areas where educational and employment opportunities are limited. While educational attainment of American Indians has increased markedly over the past 40 years, Indians trail behind the White population and most other minority groups in educational attainment. More than a fifth of American Indians do not graduate from high school and less than half pursue a college education. As of 2002, only 13 % had attained a college or higher educational degree. Of the small number of American Indians who pursue advanced degrees, few study in STEM disciplines. As of 2007, approximately half of the master's degrees earned by American Indians were in the fields of education or business, and approximately half of the doctoral degrees were in education, history, or the social sciences.[15]

American Indian students have a different demographic profile from American college students overall. They are more likely to be older (more than a third are over age 30), they are more likely to be financially independent, and approximately two-thirds of them are women.[16] About half of them attend 2-year colleges, and they are most likely to attend a public community college or public 4-year college, followed by a tribal college. Less than 10 % attend private colleges. Child services, public transportation, and scheduling flexibility are particularly important to the large number of American Indian students who work and have a family while they attend college. More than 70 % of American Indian students receive financial aid, most commonly federal grants; this group is particularly vulnerable to reductions or delays in financial aid or increases in tuition.

Where do American Indians attend college? In addition to the 37 tribal colleges and universities, there are three religiously affiliated schools (Bacone College, Nazarene Indian Bible College, and American Indian College) and three colleges under federal control (Haskell Indian Nations University, Southwestern Indian Polytechnic Institute, and the Institute for American Indian Art). However, the majority of Indian students attend one of 85 colleges and universities with only small Indian student populations. Only a handful of these colleges have more than 500 Indian students enrolled.[17]

The National Native American Honor Society, created in 1981 by Frank Dukepoo, challenges American Indian students to achieve academic excellence and gives recognition to those – from fourth grade through graduate school – who do. (Dukepoo 2001)

[15] For more detailed statistical information, see DeVoe and Darling-Churchill (2008).

[16] For information on male American Indian college students, see Stuart (2012).

[17] While there is a vast literature on students who attend tribal colleges, the social science literature on American Indian students attending predominantly White institutions is thin. One example (Makomenaw 2012) discusses the experiences of American Indian students who transfer from tribal colleges to complete their undergraduate degree at predominantly White institutions. These students, Makomenaw found, sought out meaningful interactions with other American Indian students, faculty, and staff but were indifferent to their interactions with other populations on campus. Academic advising and financial aid were the university services most often mentioned as important to these transfer students, but their experiences with these services were less than uniformly positive. The students experienced ignorance or stereotypes both in the classroom and elsewhere on campus, and this contributed to a feeling of alienation. Also see the personal stories of American Indian students at predominantly White institutions as told in Huffman (2008).

Jackson et al. (2003) have identified factors and characteristics important to the persistence of American Indian students in higher education. While their study is only a small qualitative study (n = 15) of students who grew up on reservations, their findings are consistent with the wider social science literature. The authors identify nine factors that have an impact on American Indian academic persistence in college. Support from a parent, grandparent, or other relative is "almost an imperative to be academically successful." Participation in structured social activities, such as multicultural offices or clubs for American Indians, is regarded as beneficial, even though several of the students hesitated before joining them. Students are motivated by warm interactions with staff and faculty members because these interactions provide a support network and make the students feel more a part of the community. Exposure to college-level programs, e.g. Upward Bound, or college-educated professionals prior to entering college makes the students feel more comfortable about the college and career path they are planning. Every student these researchers interviewed had experienced some form of racism, and this was universally acknowledged to be a demotivating force. Every one of the students in this study also tracked a nonlinear path through their education, in every case involving at least three colleges and often accompanied by gaps of a year or two when they were out of school. Several of the students reported tension between the student's aim to be a good student and the aim to be a good member of his family or community – even when the family and community were supportive of his education.

An older study focused on a cultural understanding of persistence (Kirkness and Barnhardt 1991) points out that blame is generally placed on American Indian students for "high attrition, poor retention, weak persistence, etc." However, these issues, the authors argue, can be viewed from an alternative perspective:

> From the perspective of the Indian student, however, the problem is often cast in more human terms, with an emphasis on the need for a higher educational system that respects them for who they are, that is relevant to their view of the world, that offers reciprocity in their relationships with others, and that helps them exercise responsibility over their own lives.

Cultural explanations abound for the reasons that American Indian student persistence in tribal colleges is higher than in majority colleges, such as a more culturally relevant curriculum, less isolation, more personalized attention, and no pressure to assimilate into a mainstream culture.[18]

---

[18] Here are some other examples of cultural issues related to the higher education of American Indians: in a national study (n = 643) of American Indian students (Lundberg 2007), the students reported higher levels of learning when the school had a strong commitment to diversity; through a qualitative study of the Commanche tribe, Tippeconnic and Tippeconnic Fox (2012) analyzed how tribal values shape teaching, learning, research, and educational governance; Pewewardy and Frey (2004) have shown the problems (what they call "ethnic fraud") that arise when colleges allow students to self-declare as American Indians according to their own definitions; and Boyer (2008) discusses the cultural issues that arise when an intertribal college attempts to support various individual tribal cultures. Guardia and Evans (2008) lists core values of American Indians: sharing, cooperation, noninterference with others, present-time orientation, being versus doing, extended family orientation, respect, harmony and balance between humans and their environ-

Students who attend tribal colleges and universities generally have a strong positive sense of their institution.[19] They praise the faculty, staff, and curriculum.[20] They
value the small class size, the affordability, and the availability of such services as
daycare and transportation. They are less satisfied with the quality of the laboratory
and library facilities on campus.[21] They believe that college will lift them from a life
as a minimum wage earner or welfare recipient.[22] Compared to American Indian
students attending a principally White institution, the students at tribal colleges and
universities score higher on tribally linked variables. There is some evidence that
attending a tribal college (compared to another type of college) slightly increases
associate's degree attainment levels but moderately decreases bachelor's degree
attainment, especially for males.

A number of the students who attend tribal colleges are not prepared for college
as measured by traditional metrics such as standardized test scores and high school
grade point averages. Several social science studies have examined programs
intended to help prepare students with weak high school academic records and
poorly defined academic and vocational goals to succeed academically at tribal college and transfer successfully into a 4-year degree program at a majority institution.
One model program that has been adopted at six tribal colleges is called Breaking
Through, designed by two advocacy groups: Jobs for the Future and the National

---

ment, spiritual causes for illness and problems, decision-making by consensus, and importance of
the tribe.

[19] For general overview material on tribal colleges and universities, see Oppelt (1990), Boyer
(1997a, b), Machamer (2000), AIHEC, the Insitute for Education Policy, and Sally Mae (2000);
Benham and Stein (2003), Campbell (2003), Institute for Higher Education Policy (2006), Reyhner
(2006), and Abelman (2011). For general statistical information, see AIHEC (2008). For an
extended case study of one (government-controlled) school for American Indians (Southwestern
Indian Polytechnic Institute), see Khachadoorian (2010); for a case study of a church-related
Indian school (Bacone College), see Neuman (2013). Various studies address specific topics
related to tribal colleges and universities: accreditation (Putnam 2000); student persistence
(Kicking Woman 2011) and attainment (Reese 2011; Wright and Weasel Head 1990), American
Indian male students (Stuart 2012), transition from secondary to postsecondary education (Brown
2003; Gonzalez 2012), Intertribal colleges (Boyer 2008), tribal colleges and universities and their
relations to private foundations (Boyer 2000), local economic development (American Indian
Higher Education Consortium and The Institute for Higher Education Policy 2000), and policy
(Olivas 1981).

[20] For example, in one study (Institute for Higher Education Policy 2006) 86 % of students were
satisfied with the course in their major field of study; 83 % were satisfied with their contact with
faculty and administrators; 82 % were satisfied with the overall quality of instruction; and 78 %
were satisfied with the curricula on tribal culture. Another study of Montana's seven tribal colleges
offered consistent findings. (Wright and Weasel Head 1990). Although focused more on persistence, a different, independent study of Montana's seven tribal colleges (Kicking Woman 2011)
found similar results concerning student satisfaction.

[21] The severe infrastructural problems faced by tribal colleges and universities, and possible policy
remedies, are discussed in The Institute for Higher Education Policy (2000).

[22] This belief was largely true. Many tribal college students found jobs related to their studies on
the reservation, enabling them to have employment and give back to their tribe. (AIHEC et al.
Survey of Tribal College Graduates 2000)

Council for Workforce Education. This program provides comprehensive academic support services, remedial work in math and English that is embedded in required courses, and accelerated learning so that students are back in the workforce more quickly, given the pressure to support their families. While these programs are primarily targeted at vocational training in fields such as carpentry or electrical construction, the same characteristics have proved useful in giving students a strong foundation for transferring to a 4-year program at another institution.[23]

5.3   American Indians and STEM Education – A Brief History

One of the reasons there are few American Indians engaged in STEM education and careers[24] is because of the poverty at home and the low quality of many of the schools in which they are trained at the primary and secondary levels.[25] Math and reading scores at the fourth and eighth grades, percentages of families with 25 or more books at home, and percentage of homes with computers are all below those of Whites, Asians, Hispanics, and African Americans. By percentage, only half as many American Indian students achieve the ACT math and science readiness benchmarks compared to the national average. In 2010, only 51 % earned a regular high school diploma – the lowest of any demographic group. At the college level, the 6-year postsecondary graduation rate for American Indians is 39 % – the lowest of any group. While college attendance and college graduation numbers for American Indians have increased slowly over the past quarter century, graduation numbers in engineering remain stubbornly low: approximately 300 bachelor's degrees, 100 masters degrees, and 10 doctoral degrees per year for more than a decade. (NACME 2013)

Carroll et al. (2010) conducted a set of interviews with American Indian students in South Dakota, a state with a significant American Indian population. Three salient

---

[23] On Breaking Through, see Gonzalez (2012). On the success rate of transferring from a tribal college to a 4-year degree at the University of North Dakota, see Brown (2003).

[24] American Indians comprise 0.7 % of the population but only 0.4 % of engineering bachelor's degrees and only 0.3 % of the engineering workforce.

[25] There is a modest literature on American Indians and STEM education. For general information, see NACME (2013). Some of the more narrowly focused topics include student characteristics of American Indians studying STEM disciplines (Schmidtke 2010), recruitment of American Indians into STEM fields (McNeil et al. 2011; Popovics et al. 1974), K-12 STEM education (Richardson and McLeod 2011; Carroll et al. 2010; Kafai et al. 2014), educational attainment of American Indian students in STEM disciplines (James 2000), the tensions between science and Indian culture (James 2006; Garroutte 1999; Murry et al. 2013), culturally agreeable science curricula for American Indians (Riggs et al. 2007; Kostelnick et al. 2009), stepping out and persistence of American Indian STEM students (McAfee 2000), STEM doctoral education for American Indian students (Oguntoyinbo 2014), engineering programs at tribal colleges and universities (NAE 2005), and attracting American Indian students to computing (Varma 2009a, b).

points emerged from their study. First, many students select subject areas to study that will enable them to help their reservation. Thus many of the students choose to study education or nursing in college. Often, the high school students could not see how the study of science or mathematics would prepare them for a career that either they could pursue on the reservation or that would help their people. Second, there was rapid turnover in teachers and administrators in the reservation schools. The lack of administrative continuity led, for example, to difficulties with plans for adequate technological infrastructure for the schools and local communities.[26] Many of the teachers in the schools were supplied by Teach for America; and while these (mostly) young teachers were enthusiastic, they were generally not experienced teachers.[27] Third, a number of programs had been put in place to try to interest American Indian and other high school students in math and science, give them hands-on experience with math and science, or provide support services such as counseling and tutoring.[28]

The South Dakota study found that, at the college level, most of the American Indian students who majored in science or engineering fields:

- took a non-traditional pathway through college that involved learning technical skills, serving in the military, or working and starting a family before they entered college;
- were often reserved in character, which was sometimes mistakenly interpreted by faculty as disinterest;
- exhibited strong personal drive to succeed;
- faced a dual cognitive load of learning a new culture at the same time they were learning STEM material;

---

[26] Richardson and McLeod (2011) call for high school administrators at American Indian schools to step up to the role of technology leaders if they want their students to succeed in the modern world. James (2000) also discusses the poor technological infrastructure in American Indian schools.

[27] James (2000) points to the particular difficulties the Indian schools have in staffing math and science faculty positions with qualified instructors and how many of these schools do not teach any advanced courses in these areas. He also points to the lack of role models, not only among family and friends, but also among teachers; most American Indian students never took a class in which the teacher was an Indian. He also points out that stereotypes are common "that native children are less capable than Euro-children, especially in science and mathematics," and "that Indian children are more inclined toward arts and crafts than intellectual pursuits. Such condescendingly positive stereotypes can lead to lack of intellectual challenge and stimulation that helps mitigate against intellectual interest" and also to the Indian student's lack of self-confidence.

[28] Programs that attempted to interest American Indian (and other) high school students in math and science included the Todd County math contest, Girls Day at South Dakota School of Mines, the Knowledge Bowl held at Sinte Gleska College, the Build a Computer program at Central High School, and the Math & Science Initiative run by the University of South Dakota. American Indian high schools students were given hands-on math and science experience through the Build a Computer program at Central High School, as well as through activities coordinated by the AISES chapters and by the Math & Science Initiative Program at the University of South Dakota. Support services were provided through federal TRIO and state GEAR UP programs, the Academic Café at Central High School, and transition counselors at Todd County high school.

- performed best academically in small-group, hands-on learning environments and worst in large lecture classes; and
- commonly experienced racism.

When enrolled in a large, majority university, the authors found, the students did better when they participated in smaller communities on campus where they felt comfortable, and that the college AISES chapter often played that role for the science and engineering students. (See Chap. 7 about AISES.)

There is a small body of literature about recruitment, retention, and attainment of American Indian students in undergraduate STEM studies.[29] McNeil et al. (2011) looked at a program at the South Dakota School of Mines and Technology called Tiospaye (the Lakota word for extended family), which was designed to enhance recruitment, retention, and graduation of American Indian students in STEM disciplines. The program includes professional, cultural, and social as well as academic activities. The students are encouraged to enroll together in the same large math and science lecture courses as well as in the same recitation sections, so as to build a cohort.[30]

A graduate student is employed as both a mentor and tutor for the students. The mentor meets weekly with each student during the freshman year until both mentor and protégé agree that meetings can occur less often. The mentor provides academic advising to each of the Tiospaye scholars and also provides tutoring sessions in some of these early core math and science courses (e.g. calculus and introductory chemistry) that many of the scholars are taking. The students are also encouraged to make regular use of the Tech Learning Center, which is free and open to all students.

The program director meets monthly with the Tiospaye students and holds regular office hours to deal with other problems that arise, such as financial aid issues or problems with an individual faculty member. The program director also communicates with the Tiospaye scholars regularly about the availability of career development seminars, scholarships, and other opportunities that might interest them. Twice a month, the students meet with faculty for an informal meal; and professional development seminars are often attached to these meals – on topics such as communication skills, resume writing, career planning, and graduate school. The Tiospaye students are required to hone their professional skills by writing a resume

---

[29] Popovics et al. (1974) is an interesting artifact of an earlier era. It is remarkable for how early it appeared in trying to recruit American Indian students into engineering education and careers. The article was written by engineers, for an education journal of engineers (*IEEE Transactions on Education*). It is not well informed by social science research, and in some ways it is not culturally sensitive. It takes the position, often expressed by engineers, that engineering is inherently interesting and that American Indians (and others) would necessarily feel the same way if only the barriers for its study were removed.

[30] A cohort approach is also being used in the doctoral program in the School of Social Transformation at Arizona State University, in which two cohorts of Pueblo Indians were established to study together through online courses and in-person classes in their community. Topics included issues concerning cultural preservation and economic development in their community. (Oguntoyinbo 2014)

and attending the annual career fair, even before they are at a stage when they are entering the job market.

The fact that American Indian students do not follow the traditional pathway through college, i.e. attending college immediately after high school and graduating 4 years later, is often seen by critics as a shortcoming of either the American Indian students themselves or of the higher educational system. However, McAfee (2000), in a qualitative study of 76 American Indian STEM majors and 33 faculty members and administrators from nine colleges and universities, addressed this phenomenon not as a failure but simply as a pattern known as "stepping out." Students might take absences from college and change schools – times of stepping out – but they would eventually return to school with determination and perhaps a better understanding (called "progressive discovery" in McAfee's paper) of who they are and what they can get out of college. One of the students interviewed stated:

> It seems like every time I went back to school I was clearer about what I wanted to do. It was hard though, because every time I went back my life was more complicated with a bigger family and more debt and less money. But I always felt like I knew myself better. Finally I got a good hold of what I wanted to do so I just went for it.

Moves to enroll, spend time away from school, or change colleges were motivated by various factors including cultural identity, academic preparation, financial resources, family needs, motivation, and interaction with the "institutional interface." These students showed remarkable persistence, McAfee noted, in their long-term pursuit for an education.

One of the common approaches used by STEM educators to reach out to American Indian students is to teach them about topics that have cultural relevance to them.[31] For example, Purdue and several other universities organized the Sharing the Land program. It provides a way for precollege students to learn about Earth science through a Young Native Scholars summer bridge program and an Explorers Club outdoor education program for primary school students. (Riggs et al. 2007) Another example is a geographic information systems curriculum that has been established at Haskell Indian Nations University in partnership with the Center for Remote Sensing of Ice Sheets (an NSF-funded center at the University of Kansas), the U.S. Geological Survey, the Kansas Biological Survey, the geospatial technology company Western Air Maps, and the Kansas City Area Transportation Authority. (Kostelnick et al. 2009)

However, some scholars – notably including the sociologist Eva Garroutte and the psychologist Keith James – believe that there are fundamental tensions between

---

[31] In 2005 the National Academy of Engineering held a two-day workshop on the campus of Salish Kootenai College in Montana. The purpose was to provide advice to 11 tribal colleges and universities that offer engineering programs. The report of this workshop (NAE 2005) reflects the importance of making the engineering curriculum culturally relevant to students. The examples given in the report, however, were weak at integrating cultural relevance into the engineering, science, and math courses being taught in these engineering programs: "(1) using Native symbols in school logos and campus designs; (2) offering, and in some cases requiring, courses on tribal culture and language…; (3) involving tribal elders in classroom teaching; and (4) providing facilities that can be used for adult education, boys and girls clubs, health clinics,….

American Indian culture and mainstream science as it is generally taught in Europe and the United States; and that this tension undermines, or at least makes more difficult, the efforts to attract American Indians to STEM disciplines through programs that select content for cultural reasons. For example, James (2006) points out that some American Indians are distrustful of science because of the ways it has undermined the culture and livelihood of Indian peoples:

> US and Canadian history reveal instances when science has been intentionally used against Native peoples. In other cases, science has put itself in the service of achieving ends valued by the mainstream culture even when those ends have injured Indian people directly or violated their cultural values: the actions of anthropologists and archaeologists in removing and exploiting the cultural and spiritual materials and remains of Indian ancestors contributed to negative views of mainstream science in particular, as did psychologists and social workers who participated in efforts to break down Indian cultures or who assisted with programs that promoted adoption of Indian children by non-Indian parents. Engineers, chemists, and other scientists supported relatively frequent expropriations of Indian lands for resource extraction, dam building, and other purposes and failed to defend Indians against pollution of their lands and other health-damaging actions, all of which added more support for the impression that science did Indians more harm than good.

James goes on in the same article to argue that there is an "incompatibility" between American Indian cultural values and scientific values.[32] American Indian cultures share this set of values:

---

[32] In a different paper, James (2000) discusses ways in which science is not value free. He writes: "Scientists and engineers are socialized and trained to value objectivity, but there are at least two problems with how this value is generally put into practice. First, the norm of being objective in gathering and evaluating information related to developing scientific understanding of a specific issue or problem is often unnecessarily and destructively extended to mean that the resulting knowledge should be applied without regard to consideration of anything other than its scientific or technical accuracy. Second, whereas few would argue with attempting to consider information objectively as a worthwhile ideal, scientists and engineers often invoke this ideal as a talisman to confer a veil of sanctity on their work despite abundant evidence that the human mind, even when possessed by a scientist, is inherently subjective in all its operations....

"Subjectivity creeps into science in many subtle ways. For instance, the problems that scientists and engineers address, far from being the universal want of some amorphous general society, more typically reflect the issues that particular groups possessed of significant economic and political power desire to have addressed. Not surprisingly, the benefits of addressing those problems typically go more to those powerful groups, and the costs typically fall more on less powerful social groups. Scientists and engineers, far from being objective in this process, are often part of the very elites that benefit and, conversely, outside of the groups that pay the costs. A substantial body of research indicates that judgments of the relative merit of a particular course of action do tend to be significantly, typically, unconsciously distorted by the social group memberships and the relative social statuses of the judges, of those who will benefit, and of those who will bear the burden of costs..."

Another value in science that Keith identifies is *technological fix* mentality: "Technology is seen by many as capable of solving anything, and technical virtuosity is admired in and of itself. These values are so strong among many scientists and engineers that problems are often immediately defined in technological terms and technical solutions are sought regardless of the true nature of the issue."

The third and last of the science values that Keith describes here is specialization, compartmentalization, and reductionism, which he argues can be useful but is harmful in their extremes: "But

(1) an equal respect and valuation of nonhuman and human beings; (2) a belief that inevitable bonds exist between the well-being of humans and the well-being of nonhumans; (3) an emphasis on the importance of place and the uniqueness of each locality; (4) a perception that the spiritual and the material are in harmony with each other; (5) a belief that there are multiple ways of knowing, including the scientific and the spiritual, that are equally valuable and equally required for complete understanding; and (6) an orientation toward extended time frames for analyzing phenomena and weighing potential outcomes of actions.

On the other hand, science (or at least scientific education) holds a set of values that are incompatible with those of American Indians:

…scientists are oriented toward mastery of nature, priority to the technically advanced, progress (a better future), independence, and personal prestige and achievement. Science and science education are also inclined toward reductionistic approaches that treat topics and applied issues in isolation from each other.

Garroutte (1999) claims that the epistemology of mainstream science teaching differs from indigenous epistemologies. She argues that mainstream educators trying to present culturally relevant education to American Indians are not really adopting an indigenous mindset but instead are stripping away certain fundamental indigenous assumptions and placing other aspects of American Indian understanding into a nonscientific thinking category (given lower status), where it cannot conflict with scientific thinking. This, the author argues, is harmful to preserving indigenous ways of thinking.

Similarly to Keith, Garroutte contrasts the epistemology of science teaching with the American Indian understanding of the natural world. Describing the science teacher first, she states:

The conventionally trained science teacher is carefully schooled to impart to her students a model of inquiry – what I have referred to as "classroom science" – in which knowledge is generated only through intersubjectively verifiable, replicable, sensory observations. These observations are ideally capable of expression in laws that are causal, universal, and impersonal and that allow for the prediction and control of the natural world. This natural world is separate from and unaffected by the language used to describe it, and its exploration is separate from the domain of ethics. A central value governing descriptions of this world is the reduction of explanatory complexity. The results of inquiry yield knowledge, while the ability to apply scientific methods frequently equates with rationality, thought, and intelligence.

Later in the paper, Garroutte describes, by contrast, the very different American Indian ways of knowing:

In American Indian models of inquiry into the natural world, knowledge tends to be received from a variety of observations. Information is not necessarily excluded from consideration if it is gained from sources other than the five senses, from an unrepeatable experience, or

---

the skewed values of modern science (and of some societies) and the structures and systems of sciences as professions tend to inhibit integration and coordination. This has two mutually reinforcing negative outcomes. Practical problems result because analyses and judgments tend to have very narrow foci regardless of the breadth of the issue(s) at hand; and those whose values tend more toward integration and synthesis tend to be driven away from scientific and technical fields."

from events which are not, by scientific definitions, intersubjectively verifiable. Indeed, ideas of subjectivity and objectivity may be, in themselves, quite different from those assumed by non-Indian thinkers. Indian models of inquiry often find other patterns in the natural world than the law-governed and causal ones sought in typical science classrooms: a broader, more complex, more personalized order, which is rooted in responsible interrelationship and co-creative activity. Laws do not grind blindly away, and the prediction, control, and manipulation of the natural world are less pronounced expectations. Language may be seen as a powerful, active force in the ongoing process of creation, and seeking knowledge becomes a sacred activity through which inquirers begin to penetrate the fabulous complexity of the world. Native models of inquiry understand the methods they prescribe as means of generating dependable, accurate knowledge about the natural world, but do not require the conclusion that there are no other sources of knowledge.

Both Keith and Garroutte believe that these tensions can be overcome by science education that is truly sensitive to American Indian culture and epistemology. However, both remain skeptical about the simplistic efforts to marry science education and American Indian culture that have generally been attempted.[33]

5.4 Computing and the Tribal Colleges

The story of computing in the tribal colleges has several strands: the development of a computing infrastructure (computers and networking), the use of this infrastructure to build a virtual (digital) library used in the tribal colleges and distant education programs operated out of the tribal colleges, and the teaching of computer science and information courses and entire curricula. Although there were some computer courses and even complete programs offered in the tribal colleges as early as the end of the 1980s, computing-related activities in tribal colleges only took off in the second half of the 1990s – with many different activities taking place at about the same time. There may have been multiple reasons for this timetable: dawning recognition of the educational importance of the Internet, increased interest in funding broadening participation activities at the National Science Foundation, new sources of funding from private foundations (especially the Kellogg Foundation), the growing recognition of the importance of the computer as an economic driver and a work opportunity, and the newfound stability of the tribal colleges after they received land-grant status in 1994 and through President Clinton's Executive Order in 1996 requiring federal agencies to make their programs more readily available to the tribal colleges.

This timing is quite a bit later than computing programs at most mainstream colleges inasmuch as the first campus computing laboratories were created in the 1940s and half of the computer science departments operating today were founded in the

---

[33] Murry et al. (2013) describe a method of Vision Mapping, which they argue can be used to enable scientists and American Indian communities to work together on issues of sustainability.

James (2001) has also edited a volume from a conference that addressed the issue of reconciling science and American Indian culture and values.

10 years between 1965 and 1975. In fact, many tribal colleges did not begin to develop a computing curriculum until after 2000, and many of these programs are taught only at the 2-year level. For example, one computer science department chair remembers (as quoted in Varma 2009a):

> I began by offering a course called Introduction to Computers, which basically used some version of the Apple computer to teach word processing. From there, I worked my way into Microsoft Platform, DOS, and Windows. I created a 1-year certificate program in data processing and a 2-year associate degree program in the late 1980s.

As Varma (2009a) evaluates the state of computing programs in the tribal colleges:

> Most TCUs offer associate degrees and/or certificates in IT-related fields, such as business data processing, business computer operator, CS, computer office skills, computer support technology, e-commerce, information systems, information technologies, integrated office technology, internetworking specialist, graphic arts technology, microcomputer applications in business, microcomputer operations, and microcomputer management. Some have moved from certification and associate's degrees to a bachelor's program in CS.

Let us look back at these intertwined histories of computing-related activities in the tribal colleges. One of the earliest efforts was the Summer Institute in Computer Science. The goal was to encourage American Indian students to transfer from 2-year tribal colleges to 4-year research universities. The program began in 1991, when Lubomir Bic and others from the computer science department at the University of California at Irvine visited the Navajo Community College in Tsaile, Arizona.

The SICS program as it developed was a non-residential program primarily for students in southern California, advertised through local tribal colleges and American Indian community centers. For eight summers, beginning in 1992, American Indian students engaged in an 8-week summer program. It included intensive, 4-week university-level courses in computer science and telecommunications on the Irvine campus. Tuition for the summer courses was paid by industry. The students with the top performance in these summer courses had their school-year tuition paid for the following academic year at their home colleges. The 4 weeks of coursework were followed by a 4-week internship in industry, to learn about the technological workforce and build mentoring relationships. The summer internships in 1998, for example, were held at Boeing North America, Rockwell, Silicon Systems, Toshiba America, TRW, Unisys, and Xerox. Evaluation of the SICS program showed that more than 90 % of the students continued their college education, more than 80 % transferred to a 4-year college to pursue a bachelor's degree, and more than 50 % pursued some type of degree in a computing discipline. The program also included cultural and social events to help build a sense of community and provide opportunities for the students to meet informally with the faculty. (Bic n.d.)

As of 1995, there had been limited progress in developing computing activities on the tribal college campuses themselves. Tom Davis, who has held multiple higher-level administrative positions in various tribal colleges and who was actively involved in bringing computing to several of the tribal colleges, assessed the situation in the second half of the 1990s: a few schools had computing labs, including College of the Menominee Nation in Wisconsin, Salish Kootenai in Montana, and Turtle Mountain in North Dakota; "[b]ut overall, the programs were pretty weak, not much connectivity, not much of anything."[34] (Davis 2014)

One program that Davis noted as a particular early success was an e-learning program at Bay Mills Community College in Michigan's Upper Peninsula, created by Davis working together with Helen Scheirbeck, the head of the Indian Head Start program in Washington, DC. Scheirbeck was concerned that Congress was going to pass legislation requiring Head Start teachers to have earned at least an associate degree and that this requirement would be difficult to achieve for the teachers in some of the remotely located tribal college Head Start programs. (Indeed, Congress did pass legislation requiring that 50% of Head Start employees have an associate degree by 2003.) Davis knew Larry Smarr and some of the work being done out of the supercomputer center at the University of Illinois, and he was also familiar with some projects that were being funded at the time by the Sloan Foundation. Based on these models, he built an asynchronous, online associate degree program in early childhood education using an electronic bulletin board, PowerPoint, Java Chat, email, and telephone – with many of the students connecting to the system through America On Line. The program, which became available to students beginning in 1998, was highly successful. In its first 3 years it attracted students in Alaska, Florida, Michigan, Oregon, Texas, Washington state, and Wisconsin. This "virtual college" as it was called, was replicated at other universities. (Davis 2014; Tribal College Journal staff 2001)

Another project, known by some as the AIHEC Virtual Library Project, had its origins at about the same time in the late 1990s. Dan Atkins, a computer scientist who was the dean of engineering at the University of Michigan, was selected in 1992 by the university president James Duderstadt to remake the university's library school so that it had continuing relevance in an increasingly digital world. This effort, which led to the creation of the School of Information in 1996, was funded primarily through a multi-million dollar grant from the Kellogg Foundation, which is headquartered in the state of Michigan. The Kellogg Foundation, and especially one of its officers Gail McClure, built a strong rapport with Atkins.

In 1995 the Kellogg Foundation made a $30 million commitment to support higher education at the tribal colleges, known as the Native American Higher Education Initiative.[35] The Kellogg Foundation grant supported the American Indian College Fund, AIHEC, and AIHEC's Student Congress. But the bulk of the funds

---

[34] Davis served as president of Lac Courte Oreilles Ojibwa Tribal College and Little Priest Tribal College, as acting President and Chief Academic Officer at Fond du Lac Tribal and Community College, helped found College of the Menominee Nation, and was Provost at Navajo Technical University.

[35] The Kellogg Foundation's $30 million commitment to American Indian higher education influenced the Lilly Endowment's decision to invest $30 million in tribal campus buildings and also positively influenced contributions from the Packard, Fannie Mae, and several smaller foundations. (Boyer 2000).

were intended for some competitive grants targeted not only at individual tribal colleges but also intended to foster collaborations among several tribal colleges or partnerships between tribal colleges and majority institutions. In the end, these funds were used for a wide range of purposes including creation of new courses and programs, purchase of administrative software for student records, and development of a community wellness center.

Unfortunately, there was a clash between the foundation and tribal representatives at their first meeting. Kellogg had had success with the African American community by funding centers of excellence at the most elite Black colleges and universities (e.g. Clark Atlanta, Howard, and Spelman) on mutually agreed upon targeted topics such as math and science education or jazz studies; and the Kellogg staff planned on adopting this same approach with the tribal colleges. However, the tribal college representatives resisted this plan:

> "What's the splash?" challenged [Fort Peck Community College President Jim] Shanley. He meant, how much money are you proposing, and how long will you stay with the colleges? And when Sinte Gleska University President Lionel Bordeaux spoke, he turned the foundation's assumptions about responsible grant-making upside down. "It's all of us or none of us," he said. "That's the way it has always been with us." "Don't pick us off," [foundation officer] Johnson recalls being told. "Don't you decide who the best of us are. If you have $10, spread it among all of us." (Boyer 2000)

In the end, Kellogg backed off, making grants to all but the smallest and newest of the tribal colleges and allowing each college considerable input and latitude in the ways the funds were to be used – although the foundation required that the funds be used on projects within a few broadly specified categories such as curriculum development, development offices, distance learning, or social services.

There were additional problems. Some of the schools were too weak to make effective use of the funds. Assessment was also a point of disagreement:

> The foundation "is really attached to outcomes" Johnson explains. "We plan almost everything. We do external scans, and then we do internal scans, and then we will review all the literature, and then we develop concept papers. We develop goals, strategies, and action plans." Little is purposefully left to chance. The [tribal] colleges take a slightly different approach. "They want outcomes, too," Johnson says, "but they are open to outcomes; they are open to what will happen. They say, 'This is what we hope will happen.' Whereas Kellogg is more machine-like. It's more about predictability and control in its drive for excellence." (Boyer 2000)

One of the values that Atkins instilled in the Michigan information school was to ground and inform research by engaging in real-world problems and trying to do social good. He and his colleagues in the Alliance for Community Technology – a part of the information school – were at the cutting edge in digital library research. McClure wondered whether a digital library for the tribal colleges might be of value to their teaching. However, McClure had a "long history of challenges of actually getting these tribal colleges to work together," so she made the award for this project to Atkins and the University of Michigan to build a digital library for the tribal colleges.

The tribal leaders were unhappy that the grant was awarded to Atkins and the University of Michigan instead of directly to the tribal colleges themselves. Early in the project, when Atkins traveled to scheduled meetings with the tribal leaders, they would sometimes not show up. Atkins was discouraged about whether this project could succeed, but things began to turn around when he began to work through Karen Buller, the director of the National Indian Telecommunications Institute. Buller vetted Atkins with tribal leaders and told them that Atkins was someone they could work with. She helped Atkins to convene a meeting in Ann Arbor, and this meeting seemed to be a turning point in the success of the project. The initial day of the meeting was acrimonious, but at dinner that first evening in a fancy restaurant in Ann Arbor, Atkins saw the first evidence of the project coming together. As he related the story:

> I had been told by Karen and by Gail that when Native Americans start joking with you or even making fun a little bit of you or making jokes about you or on you or with you, that that was a sign that they were accepting you… We were ordering dinner, it was a fairly nice place, and all of the sudden one of the Native Americans said, "Aha, Dan, I see now why you brought us to this restaurant." And I said, "Oh yeah?" He said, "Yes, they have boiled redskins on the menu." Referring to potatoes of course. Everybody laughed and [this] actually lightened things up. That meeting apparently, instead of being the end of the project, was one where we turned the corner on being able to work together. (Atkins 2015)

To build a culturally appropriate digital library presented some challenges. For example, there were questions about how well existing classification systems would fit with the cultures of the various tribes – which themselves were not always in agreement. Atkins's team used a "strong user-centered design approach", working closely with the different tribal colleges in creating the design attributes of the system so that the digital library was culturally appropriate for them. The project was challenging in Atkins's estimation because of difficulties with "logistical support, the ability to work together, and basic technical competence." (Atkins 2015) There were weaknesses in network infrastructure that made distribution of materials online suboptimal; and the grant did not include funds to build better Internet connectivity. Individual variation across the tribal colleges was also an issue. Not only might the materials to be used in the tribal colleges be different from materials used for research and teaching in a majority institution such as the University of Michigan, but there were also variations in needs across the different tribal colleges, such as the language in which the materials were presented and the most effective examples to use in illustrating particular scientific and engineering principles.

The Kellogg Foundation was generally happy with the results of the project. There was cooperation among the tribal colleges, the project provided a widespread familiarity of the value of digital libraries to the teaching and research missions of the tribal colleges, and the project built some human capacity on the tribal college campuses. The success stimulated the Kellogg Foundation to hold an international workshop in Hawaii on the value of digital libraries as cultural repositories, with representation from native peoples of Australia, New Zealand, Sweden, Canada, and the United States.

The virtual library supplemented the meager library resources of most of the tribal colleges. Over time, it was supported by grants not only from the Kellogg Foundation, but also from IBM and the National Science Foundation. (U.S. White House 2000; Billy 2006) As touted by the White House:

> The virtual library homepage will link into a major national database that will catalog electronic books, magazines, journals and Internet documents from around the world. The virtual library will have a reference section, a database section, and local exhibitions for the 32 tribal colleges across the country. There will also be a technical support system which will answer student, faculty and community members' questions interactively. (U.S. White House 2000)

In another effort, in 1999, AIHEC formed a High Technology Committee to consider what to do about the worsening digital divide for American Indians. Jack Barden, one of the founders of Standing Rock Community College (now Sitting Bull College) in North Dakota, and Tom Davis were the co-chairs. The committee created various plans in its first 2 years:

> It has come up with a series of projects and activities designed to increase the number of tribal IT workers and experts, improve higher education IT programs, extend the ability of tribal colleges and universities (TCUs) to serve increasingly larger Native American student populations in tribal communities, build strong partnerships with major higher education institutions and research labs, and generate resources to cover a national effort. Parts of the initiative also recognize the importance of prekindergarten schools as pipelines for TCU students and of introducing technology into the American Indian Program Branch of Head Start. (Davis and Trebian 2001)

Barden and Davis, together with Carrie Billy, who at that time was leading the White House Initiative on Tribal Colleges and Universities working in the Clinton Administration and who is now the President of AIHEC, developed a plan for a Circle of Prosperity Conference held in Silicon Valley in 2000. The plan, which received endorsement and support from the office of Senator Jeff Bingaman (D-NM) and ARPA Deputy Director Lee Buchanan, was designed to bring government and industry leaders together with leaders from the American Indian community to address solutions to the digital divide for American Indians. The conference adopted a prosperity game format, a domestic version of a strategic war game produced by Sandia National Laboratory. It involved:

> interactive simulations that encourage creative problem-solving and decision-making. This "game" format explored the challenges and opportunities of the new information age economy as it relates to Indian country.
>
> During the conference, the attendees spent two days playing by a predefined set of rules that specify teams of players, allowed interactions, forced group reporting periods, and a method of assessing outcomes. Different interest groups were layered into teams and given objectives that ensured intense discussion and debate. Over the two days, teams adopted strategies and then interacted with competing teams' "moves" based on those strategies, leading to outcomes that may or may not have been consistent with individual attendees' goals but that led toward a plan with a high probability for real-world success. (Davis and Trebian 2001)

Davis was pleased with the high-level attendance at the conference and the fact that it gave "Carrie Billy and [me] a lot of fodder for going after the Clinton

Administration and getting them ... to support increasing the amount of money going into STEM programs of the tribal colleges and universities, [funded through] the National Science Foundation." (Davis 2014) Davis credits this as an important influence in motivating the National Science Foundation to create the Tribal Colleges and Universities Program (TCUP, see below).[36]

At about the same time, two existing NSF programs, AN-MSI and EOT PACI, came together to help bring better connectivity to the tribal colleges. In 1997 the National Science Foundation created a program known as EOT PACI in computational education, outreach, and training that brought together all of the education, outreach, and training programs that were being carried out by the National Computation Science Alliance (those institutions working with the supercomputer center at the University of Illinois at Champaign) and the National Partnership for Advanced Computational Infrastructure (those institutions working with the super-computer center at the University of California at San Diego).[37] In 1999, NSF initiated a major program with Black and Hispanic-Serving institutions as well as tribal colleges, through its Advanced Networking Project With Minority-Serving Institutions project (AN-MSI). This project involved a $6 million award to Educause, a non-profit association whose mission is to advance education through educational technology. It was part of NSF's strategy to take advantage of its national supercomputer centers to advance educational opportunities across the nation.

AN-MSI was a joint venture between Educause and EOT PACI, with Dave Staudt from Educause as the project leader. The goal was to improve "Internet connectivity, campus networks and their technical support, and advanced use of the networks." (Davis and Trebian 2001) By early 2000, 100 minority-serving colleges and universities had signed up to participate in AN-MSI. AIHEC's High Technology Committee took the lead role in working on the tribal college component of AN-MSI. Davis had some reservations about whether the program would achieve the stated aims of Educause and NSF officials – to bring the African American, Hispanic, and American Indian institutions closer together.[38] He pointed out differences: the American Indian schools were more remote than most Black and Hispanic-serving colleges and universities; the American Indian community was slower to reach consensus but more resolute in following that consensus once it was attained; the tribal colleges and universities were generally much smaller than the Black and Hispanic-serving institutions, which meant that they could not offer as many innovative programs; and the American Indian community had stronger

---

[36] For a summary of Clinton Administration efforts to address digital divide issues in the American Indian community, see U.S. White House (2000).

[37] On the impact of EOT-PACI see Alexander and Foertsch (2003).

[38] The summative evaluation of the project by Foertsch (2004) was more positive about the collaboration gained between the three minority communities through this project. On the other hand, as Foerstsch notes, given the substantial networking infrastructure needs at these institutions and the large number of participating institutions (100), and the small amount of funding available when $6 million was split so many ways from the beginning, there were concerns about how far any individual institution would be able to advance on funding from this project.

cultural issues to face than the Black or Hispanic communities.[39] But Davis was bullish about the impact this program could have on the tribal colleges and universities themselves.

In the end, Davis was right on both counts. While there were some useful connections made through the program, such as quarterly meetings of the technical support staffs of these various minority-serving institutions, the initiative did not lead to deeper connections between the institutions serving different minority communities. The tribal colleges faced the greatest challenges of these three types of minority-serving institutions: "distrust of specific new technologies, geographic remoteness, weak economic bases in tribal communities, lack of private investment on tribal lands, poor targeting of specific government policies for improving technology infrastructures in Native American communities, and lack of protection of Native American intellectual property rights over the Internet." (Davis and Trebian 2001)

Nevertheless, Davis's assessment of the long-term outcome was very positive:

...what that money really did was it allowed us to bring all of these technicians from Hispanic and Black and Tribal colleges together in quarterly meetings, ... And we developed a number of different programs all over the country with that [funding], trying to build technology infrastructure. Now, in Tribal college country, of course, it really worked. I mean, today I don't think you'll find a Tribal college in the country (even some of the really, really small ones) that [does not] have a pretty good technology curriculum as well as pretty decent networking, pretty decent Internet connectivity, and quite a bit of expertise in

---

[39] Regarding these cultural issues, Tom Davis (2014) stated: "...some of the differences have to do with culture within Indian country. I mean, just to try to give you an example, ... Carol Davis at Turtle Mountain, who was the Chief Academic Officer up there for a long time, wrote her doctoral thesis upon what the medicine people and the Anishinaabe community in the northern tier of states in the United States where the Anishinaabe are, what they thought of information technology because there's a lot of controversy about that. Really, should you give it sort of a violation of what Indian people are about and have historically been about in Indian culture, in Indian spirituality? And Carol's thesis was pretty interesting. And what it basically showed is that, yeah, there's a lot of concern about it. You don't want certain information to get out about Indian culture on the net, for instance. You want to make sure that women stories are told properly, winter stories are told properly, that sort of thing. But, overall, the medicine people, I think, said, "Okay, this is a new way that our people can make money and make a living for their families, so go ahead, go do it". And I think, so you have that element in all of the communities."

Davis (2014) also reported a story about cultural issues concerning Navajo Tech: "At Navajo Tech, one of the most interesting things was that there's a lot of trouble bringing technology labs to communities and then one of the librarians had this idea of getting the Navajo rug-weavers, some of the elders in New Mexico together, where some of the finest of the rug work is happening around Crownpoint, and she introduced them, I think, ... to [the software program] Paint to help them design patterns for their rug. And it was sort of difficult to get a group of them to come to that meeting, of these elderly women, primarily, from all over, the elderly women. But they came, we sent out a bus to pick them up or the group that brought them in sent us a bus out to pick those ladies up. They all came in and they got introduced to it and then the truck driver got into trouble because he couldn't get them to leave. They got so deeply involved in it. And were talking about the cultural aspects and how cool this was and all that kind of stuff. And so, in the end, it's a wonderful story. So you have some barriers within Indian country that I don't think you have with any of the Hispanics or the Black communities, at least, that I know of."

information technology.[40] And a lot of them have developed a pretty sophisticated e-learning program of one kind or another…. So anyway, we managed to pull off AN-MSI, which was sort of an important marker along the road as we tried to develop, strengthen, especially the tribal colleges and universities' performance in the whole area of information technology. (Davis 2014)

Another important NSF program, the Tribal Colleges and Universities Program (TCUP), was initiated in 2000 with funds specifically earmarked by the Clinton Administration.[41] The program continues today, and to date it has awarded approximately $150 million in grants. The TCUP programs aim to build capacity in the STEM disciplines. TCUP funds have been used for "upgrading technology, purchasing lab equipment, strengthening core courses, building partnerships with mainstream universities, and promoting K-12 outreach." (Boyer 2014) A theme running through these grants is the support for information technology to leverage STEM education. (Varma 2009a, b) Another common theme is building a research experience for undergraduates, often on projects of value to the reservation such as reservation water supplies, as a means to enhance student engagement. Although TCUP has had a significant impact, major facilities shortcomings remain in the tribal college classrooms and laboratories (Boyer 2014; Nee-Benham and Stein 2003; Varma 2009a, b; CEOSE 2004)

In 2009, the Obama Administration created the Broadband Technology Opportunities Program (BTOP) as part of the American Recovery and Reinvestment Act, operating through the U.S. Commerce Department's National Telecommunications and Information Administration. The objective is to accelerate broadband diffusion to underserved areas so as to create jobs and do other public good. Two BTOP projects are having an impact on American Indians. BTOP funds are allowing the Computer Center at the College of the Menominee Nation to provide Internet access – and training programs – to the entire local community, not only to the college. Under the ZeroDivide Tribal Digital Village Broadband Adoption

---

[40] However, this does not mean there was Internet connectivity in student homes. There is no economic incentive for the telecommunications companies to provide service in remote areas such as on Indian reservations. The federal government has funded connectivity infrastructure to some of these remote areas, such as through the Internet to the Hogan Project of the mid-2000s, which sought to increase connectivity to the Navajo Nation, a project of the San Diego Supercomputer Center with funding from NSF's TerraGrid program. The first stage was to build a fast Internet connection from Albuquerque to Navajo Technical College, and from then "Through an extended mesh of wireless broadband towers that will be built by students, faculty and community members, NTC will offer broadband connectivity to 31 community centers, and later to schools, clinics, hospitals, police departments, fire houses and homes." (Mueller 2007, also see Davis 2014).

[41] These were not the first NSF grants to the tribal colleges. Several grants were given as early as the 1970s, for example to Turtle Mountain Community College to build and equip laboratories for basic science courses. In the mid-1990s, NSF had provided funds through its Rural Systemic Initiatives Program to strengthen K-12 education in a number of poor communities, including various Indian reservations. Other major funders of STEM activities at tribal colleges and universities have been the Ford, Iannan, Kellogg, Gates, and Bush Foundations as well as the federal departments of Education, Agriculture, Interior, Housing and Urban Development, and National Aeronautics and Space Administration. (Boyer 2014; AIHEC 2014)

Program, 19 tribal communities in southern California are receiving Internet access and digital literacy training.[42]

The early success of Bay Mills Community College in distance education set the tone for future efforts by the tribal colleges. By 2013, 29 tribal colleges and universities were providing distance education. For example, as of 2011, Northwest Indian College in Washington State was teaching 75 % of its students through distance learning. United Tribes Technical College in North Dakota, which was the first tribal college to offer complete degree programs online, now provides eight associate of applied science degrees through distance education. Salish Kootenai College in Montana offers 125 courses through distance education. (Hampton 2013)

References

Abelman, Robert. 2011. The institutional vision of tribal community colleges. *Community College Journal of Research and Practice* 35(7): 513–538.

Adams, Evelyn C. 1971. *American Indian education: Government schools and economic progress.* New York: Arno Press/The New York Times.

Alexander, Baine, and Julie Foertsch. 2003. *The impact of the EOT-PACI program on partners, projects, and participants: A summative evaluation.* Madison: LEAD Center, University of Wisconsin-Madison. http://www.wceruw.org/publications/LEADcenter/SumEOT-PACI.pdf. Accessed 3 Nov 2014.

American Indian Higher Education Consortium (AIHEC). 1999. *Tribal colleges: An introduction.* Washington, DC: Institute for Higher Education Policy.

American Indian Higher Education Consortium (AIHEC). 2008. *Sustaining tribal colleges and universities and the tribal college movement: Highlights and profiles.* Alexandria: AIHEC.

American Indian Higher Education Consortium (AIHEC). 2009. *AIMS fact book 2007.* American Indian Measures for Success. Alexandria: AIHEC.

American Indian Higher Education Consortium (AIHEC). 2014. *Organizational website.* http://www.aihec.org. Accessed 5 Nov 2014.

American Indian Higher Education Consortium (AIHEC) and The Institute for Higher Education Policy. 2000. *Tribal college contributions to local economic develop*ment. A collaborative effort between the American Indian Higher Education Consortium and the American Indian College Fund. Alexandria: AIHEC.

American Indian Higher Education Consortium (AIHEC), The Institute for Higher Education Policy, and Rhe Sallie Mae Education Institute. 2000. *Creating role models for change: A survey of tribal college graduates.* Alexandria: AIHEC.

Atkins, Dan. 2015. Oral history interview by William Aspray. Charles Babbage Oral History Collection, January 14.

Belgarde, Mary Jiron. 2004. Native American charter schools: Culture, language, and self-determination. In *The emancipatory promise of charter schools: Toward a progressive politics of school choice*, ed. Eric Rofes and Lisa M. Stulberg, 107–124. Albany: State University of New York Press.

---

[42] For a revealing, earlier example of the positive impact that telecommunications infrastructure can create on an Indian reservation, see the story of Oglala Lakota College and Pine Tree Reservation as told in James (2000). For a profile of the IT infrastructure at various individual tribal colleges and universities, see AIHEC (2009).

Benham, Maenette K.P., and Wayne J. Stein. 2003. *The renaissance of American Indian higher education: Capturing the dream.* Mahwah: Lawrence Erlbaum Associates.

Bic, Lubomic. n.d. SICS. http://www.ics.uci.edu/~aisi/history.html. Accessed 22 June 2015.

Billy, Carrie. 2006. *Statement on tribal colleges & the cyberinfrastructure, by AIHEC to NSF advisory committee on cyberinfrastructure.* Originally presented on January 22, 2002, updated with recommendations. Ballston, VA.

Boyer, Ernest L. 1989. *Tribal colleges: Shaping the future of native America,* The Carnegie foundation for the advancement of teaching. Princeton: Princeton University Press.

Boyer, Paul. 1997a. *Native American colleges: Progress and prospects,* The Carnegie foundation for the advancement of teaching. San Francisco: Jossey-Bass.

Boyer, Paul. 1997b. First survey of tribal college students reveals attitudes. *Tribal College: Journal of American Indian Higher Education* 9(2): 36–41.

Boyer, Paul. 2000. Kellogg initiative: Rewriting the way foundations do business in Indian country. *Tribal College: Journal of American Indian Higher Education* 12(1): 14–18.

Boyer, Paul. 2008. End of what trail? Intertribal colleges support thriving cultures. *Tribal College: Journal of American Indian Higher Education* 19(3): 23–25.

Boyer, Paul. 2014. *Tribal colleges and universities program: 2014 leaders forum,* National Science Foundation. http://online.swc.tc/peec/wp-content/uploads/2014/08/TCUP-Leaders-Forum-Report.pdf. Accessed 4 Nov 2014.

Brown, Donna. 2003. Tribal colleges: Playing a key role in the transition from secondary to post-secondary education for American Indian students. *Journal of American Indian Education* 42(1): 36–45.

Campbell, Margarett H. 2003. *Leadership styles of successful tribal college presidents.* Dissertation, Doctor of Education, University of Montana.

Carney, Cary Michael. 1999. *Native American higher education in the United States.* New Brunswick: Transaction Publishers.

Carroll, Becky, Heather Mitchell, Pamela Tambe, and Mark St. John. 2010. *Supporting Native American students along STEM education pathways: Findings from an exploratory study of south Dakota's educational landscape.* Inverness: Inverness Research.

Champagne, Duane, and Jay Strauss (eds.). 2002. *Native American studies in higher education.* Walnut Creek: Altamira Press.

Committee on Equal Opportunities in Science and Engineering (CEOSE). 2004. *Broadening participation in America's science and engineering workforce.* The 1994–2003 Decennial & 2004 Biennial Reports to Congress. http://www.nsf.gov/od/iia/activities/ceose/reports/ceose2004report.pdf. Accessed 26 Nov 2014.

Cook-Lynn, Elizabeth (1997, Spring). Who stole native American studies?. *Wičazo Ša Review* 12(1): 9–28.

Davis, Thomas. 2014. Oral history interview by William Aspray. Charles Babbage Institute Oral History Collection, October 13.

Davis, Thomas, and Mark Trebian. 2001. Shaping the destiny of Native American people by ending the digital divide. *Educause Review,* January–February: 38–46.

DeJong, David H. 1993. *Promises of the past: A history of Indian education in the United States.* Golden: North American Press.

Deloria, V. 1991. *Indian education in America.* Boulder: American Indian Science and Engineering Society.

DeVoe, Jill Fleury, and Kristen E. Darling-Churchill. 2008. *Status and trends in the education of American Indians and Alaska Natives: 2008.* NCES 2008-084. Washington, DC: U.S. Department of Education.

Dukepoo, Frank. 2001. The Native American honor society. In *Science and Native American communities: Legacies of pain, visions of promise,* ed. Keith James, 36–44. Lincoln: University of Nebraska Press.

Fischbacher, Theodore. 1967. *A study of the role of the Federal Government in the education of the American Indian.* Ph.D. Dissertation, Arizona State University.

Foertsch, Julie. 2004. *Summative evaluation report for the AN-MSI project.* Madison: LEAD Center, University of Wisconsin-Madison. http://www.wceruw.org/publications/LEADcenter/ANMSI_2003.pdf. Accessed 3 Nov 2014.

Freeman, C., and M. Fox. 2005. *Status and trends in the education of American Indians and Alaska natives.* Washington, DC: U.S. Department of Education.

Frey, Bruce, and Cornel Pewewardy. 2004. American Indian students' perceptions of racial climate, multicultural support services, and ethnic fraud at a predominantly white university. *Journal of American Indian Education* 43(1): 32–60.

Fuchs, Estelle, and Robert J. Havinghurst. 1972. *To live on this earth: American Indian education.* Albuquerque: University of New Mexico Press.

Garroutte, Eva Marie. 1999. American Indian science education: The second step. *American Indian Culture and Research Journal* 23(4): 91–114.

Gonzalez, Jennifer. 2012. Tribal colleges offer basic education to students 'not prepared for college'. *Chronicle of Higher Education,* April 13. http://chronicle.com/section/About-the-Chronicle/83. Accessed 26 Mar 2015.

Goulding, A.R. 1995. *Battling assimilation: American Indians in higher education.* Minneapolis: University of Minnesota Press.

Guardia, Juan R., and Nancy Evans. 2008. Student development in tribal colleges and universities. *Journal of Student Affairs Research and Practice* 45(2): 409–436.

Hale, Lorraine. 2002. *Native American education: A reference handbook.* Santa Barbara: ABC CLIO.

Hampton, Ayasia. 2013. Tribal colleges and universities: Rebuilding culture and education through distance education. *Distance Learning* 10(4): 45–51.

Huff, D.J. 1997. *To live heroically: Institutional racism and American Indian education.* Albany: State University of New York Press.

Huffman, Terry. 2008. *American Indian higher educational experiences: Cultural visions and personal journeys.* New York: Peter Lang.

Institute for Higher Education Policy. 2000. *Options for a federal role in infrastructure development at tribal colleges & universities.* Washington, DC: The Institute for Higher Education Policy.

Institute for Higher Education Policy. 2006. *Championing success: A report on the progress of tribal college and university alumni,* On behalf of the American Indian College Fund. Washington, DC: The Institute for Higher Education Policy.

Institute for Higher Education Policy. 2007. *The path of many journeys: The benefits of higher education for native people and communities.* In collaboration with the American Indian Higher Education Consortium and the American Indian College Fund. Washington, DC: Institute for Higher Education Policy.

Jackson, Aaron P., Steven A. Smith, and Curtis L. Hill. 2003. Academic persistence among Native American college students. *Journal of College Student Development* 44(4): 548–565.

James, Keith. 2000. Social psychology: American Indians, science, and technology. *Social Science Computer Review* 18(2): 196–213.

James, Keith (ed.). 2001. *Science and Native American communities: Legacies of pain, visions of promise.* Lincoln: University of Nebraska Press.

James, Keith. 2006. Identity, cultural values, and American Indians' perceptions of science and technology. *American Indian Culture and Research Journal* 30(3): 45–58.

Kafai, Yasmin B., Kristin Searle, Cristóbal Martinez, and Bryan Brayboy. 2014. Ethnocomputing with electronic textiles: Culturally responsive open design to broaden participation in computing in American Indian youth and communities. *SIGCSE '14, Proceedings of the 45th ACM technical symposium on computer science education,* 241–246.

Khachadoorian, Angelle A. 2010. *Inside the Eagle's head: An American Indian college.* Tuscaloosa: University of Alabama Press.

Kicking Woman, Cheri Lynn. 2011. *The tribal college movement: Ensuring that Native American students successfully complete an associates degree and persist to earn a four-year degree.* Missoula: Doctor of Education Dissertation, University of Montana.

Kidwell, Clara Sue. 2005. *Native American studies*. Lincoln: University of Nebraska Press.

Kirkness, Verna J., and Ray Barnhardt. 1991. First Nations and higher education: The four rs – Respect, relevance, reciprocity, and responsibility. *Journal of American Indian Education* 30(3): 1–10.

Klug, Beverly J. 2012. *Standing together: American Indian education as culturally responsive pedagogy*. Lanham: Rowman & Littlefield.

Kostelnick, John C., Rex J. Rowley, David McDermott, and Carol Bowen. 2009. Developing a GIS program at a tribal college. *Journal of Geography* 108: 68–77.

Krupat, Arnold. 2002. *Red matters: Native American studies*. Philadelphia: University of Pennsylvania Press.

Larimore, James A., and George S. McClellan. 2005. Native American student retention in U.S. Postsecondary education. *New Directions for Student Services* 109: 17–32.

Lundberg, Carol A. 2007. Student involvement and institutional commitment to diversity as predictors of Native American student learning. *Journal of College Student Development* 48(4): 405–416.

Machamer, Ann Marie. 2000. *Along the red road: Tribally controlled colleges and student sevelopment*. Doctoral Dissertation in Education, UCLA.

Makomenaw, Matthew Van Alstine. 2012. Welcome to a new world: Experiences of American Indian tribal college and university transfer students at predominantly white institutions. *International Journal of Qualitative Studies in Education* 25(7): 855–866.

Martin, Robert G. 2005. Serving American Indian students in tribal colleges: Lessons from mainstream colleges. In *Serving native American students*, ed. Mary Jo Tippeconnic Fox, Shelly C. Lowe, and George S. McClellan, 79–86. San Francisco: Jossey-Bass.

McAfee, Mary E. 2000. From the voices of American Indians in higher education and the phenomenon of stepping out. *Research News on Graduate Education* 2(2). http://ehrweb.aaas.org/mge/Archives/5/Macafee.html. Accessed 4 April 2016.

McClellan, George S., Mary Jo Tippeconnic Fox, and Shelly C. Lowe. 2005. Where we have been: A history of Native American higher education. In *Serving Native American students*, ed. Mary Jo Tippeconnic Fox, Shelly C. Lowe, and George S. McClellan, 7–16. San Francisco: Jossey-Bass.

McNeil, Jacqueline, Carter Kerk, and Stuart Kellogg. 2011. Tiospaye in engineering and science: Inculcating a sustained culture for recruiting, retaining, and graduating American Indian students. *41st ASEE/IEEE Frontiers in Education Conference*, S3H1–S3H6.

Monette, Gerald Carty. 1995. *Follow-up study of the graduates of an American Indian tribally controlled community college*. Doctoral Dissertation in Education, University of North Dakota.

Mosholder, Richard, and Christopher Goslin. 2013–2014. Native American college student persistence. *Journal of College Student Retention* 15(3): 305–327.

Mueller, Peter. 2007 (January 25). San Diego supercomputer experts help Navajos build 'An Internet to the hogan'. *UCSD News Center*, January 25. http://ucsdnews.ucsd.edu/archive/newsrel/science/hogan07.asp. Accessed 4 Nov 2014.

Murry, Adam Thomas, Keith James, and Damon Drown. 2013. From pictures to numbers: Vision mapping and sustainability collaboration between Native American community members and mainstream scientists. *American Indian Culture and Research Journal* 37(4): 1–24.

National Action Council for Minorities in Engineering (NACME). 2013. NACME. http://www.nacme.org. Accessed 1 Oct 2014.

National Academy of Engineering (NAE). 2005. *Engineering studies at tribal colleges and universities*. Letter report from the steering committee for engineering studies at the tribal colleges. Washington, DC: The National Academies Press.

National Academy of Engineering (NAE). 2006. *Engineering studies at tribal colleges and universities*. Washington, DC: National Academies Press.

National Academy of Science (NAS). 2011. *Expanding underrepresented minority participation*. Committee on underrepresented groups and the expansion of the science and engineering workforce pipeline. Washington, DC: National Academies Press.

Nee-Benham, M.K.P., and W.J. Stein (eds.). 2003. *The renaissance of American Indian higher education*. Mahwah: Lawrence Erlbaum.

Nelson, Robert M. 1997. A guide to native American studies programs in the United States and Canada. *Studies in American Indian Literatures*, Series 2, 9(3): 49–105.

Neuman, Lisa K. 2013. *Indian play: Indigenous identities at Bacone college*. Lincoln: University of Nebraska Press.

Oguntoyinbo, Lekan. 2014. Experts: More focus needed on guiding Native Americans to doctoral programs. *Diverse Education*, July 30. http://diverseeducation.com/article/66060/. Accessed 4 Apr 2016.

Olivas, Michael A. 1981. The tribally controlled Community College Assistance Act of 1978: The failure of Federal Indian higher education policy. *American Indian Law Review* 9(2): 219–251.

Oppelt, Norman T. 1990. *The tribally controlled Indian college: The beginnings of self determination in American Indian education*. Tsaile: Navajo Community College Press.

Pavel, D. Michael, Rebecca Rak Skinner, Elizabeth Farris, Margaret Cahalan, John Tippeconnic, and Wayne Stein. 1998. *American Indians and Alaska natives in postsecondary education*. NCES 98-291. Office of Educational Research and Improvement. Washington, DC: U.S. Department of Education.

Pease-Pretty on Top, Janine. 2003. Events leading to the passage of the tribally controlled Community College Assistance Act of 1978. *Journal of American Indian Education* 42(1): 6–21.

Pease-Windy Boy, Janine. 1994. *The Tribally Controlled Community Colleges Act of 1978: An expansion of federal Indian trust responsibility*. Doctor of Education Dissertation, Montana State University.

Pewewardy, C., and B. Frey. 2004. American Indian students' perceptions of racial climate, multicultural support services, and ethnic fraud at a predominately White university. *Journal of American Indian Education* 43(1): 32–60.

Popovics, Sandor, Lea M. Popovics, and John C. Johnson. 1974. Attracting and motivating the American Indian student into engineering and technology. *IEEE Transactions on Education* E-17(1): 12–15.

Putnam, T. Elizabeth Mennell. 2000. *Tribal college and university accreditation: A comparative study*. Doctoral Dissertation, University of Texas at Austin.

Reese, Mitchell Jordan. 2011. *The impact of the tribal college on Native American educational attainment*. Master of Science in Applied Economics. Montana State University.

Reyhner, Jon Allan. 2006. *Education and language restoration*. Philadelphia: Chelsea House Publications.

Reyhner, Jon, and Jeanne Eder. 2004. *American Indian education: A history*. Norman: University of Oklahoma Press.

Richardson, Jayson W., and Scott McLeod. 2011. Technology leadership in Native American schools. *Journal of Research in Rural Education* 26(7): 1–14.

Riggs, Eric M., Eleanora Robbins, and Rebekka Darner. 2007. Sharing the land: Attracting Native American students to the geosciences. *Journal of Geoscience Education* 55(6): 478–485.

Schmidtke, Carsten. 2010. Math and science instructors' perceptions of their American Indian students at a sub-baccalaureate technical college: A Delphi study. *Journal of Career and Technical Education* 25(2): 8–23.

Stein, Wayne John. 1988. *A history of the tribally controlled community colleges: 1968–1978*. Doctor of Education Dissertation, Washington State University.

Stuart, Reginald. 2012. College bound: Efforts to recruit American Indian males are working. *Diverse,* November 22.

Szasz, Margaret Connell. 1977. *Education and the American Indian: The road to self-determination since 1928*. Albuquerque: University of New Mexico Press.

Tippeconnic III, John W., and Mary Jo Tippeconnic Fox. 2012. American Indian tribal values: A critical consideration in the education of American Indians/Alaska natives today. *International Journal of Qualitative Studies in Education* 25(7): 841–853.

Tribal College Journal. 2001. Bay Mills trains head start workers online. *Tribal College: Journal of American Indian Higher Education, Tribal College News,* February 15.

U.S. Senate Committee on Labor and Public Welfare. 1969. *Indian education: A national tragedy – A national challenge. 1969 Report of the Committee on Labor and Public Welfare, United States Senate. Made by Its Special Subcommittee on Indian Education.* Washington, DC: U.S. Government Printing Office.

U.S. White House. 2000. *The President's new markets trip: From digital divide to digital opportunity.* Office of the Press Secretary, April 17–18. http://clinton3.nara.gov/WH/New/New_Markets-0004/20000417-3.html. Accessed 19 Nov 2014.

Varma, Roli. 2009a. Bridging the digital divide: Computing in tribal colleges and universities. *Journal of Women and Minorities in Science and Engineering* 15: 39–52.

Varma, Roli. 2009b. Attracting Native Americans to computing. *Communications of the ACM* 52(8): 137–140.

Wright, Bobby, and Patrick Weasel Head. 1990. Tribally controlled community colleges: A student outcomes assessment of associate degree recipients. *Community College Review* 18(3): 28–33.

Young, Alvin. 1998. Tribal colleges: New land-grant institutions grow and thrive in America. *Engineering and Technology for a Sustainable World* 5(5): 6.

Part II
Case Studies

Chapter 6
Organizations That Help Women to Build STEM Careers

Abstract This chapter begins the second half of the book, which examines various case studies. The chapter is devoted to the descriptions and histories of four organizations that have worked to support increased numbers and better experiences for women in science and engineering careers. The chapter considers these organizations in the chronological order of their origins: the Society of Women Engineers (SWE 1950), the Association for Women in Science (AWIS 1971), the Women in Engineering ProActive Network (WEPAN 1990), and MentorNet (1997).

This chapter is devoted to the descriptions and histories of four organizations that have worked to support increased numbers and better experiences for women in science and engineering careers. We consider these organizations in chronological order of their origins. The Society of Women Engineers (SWE) was created in 1950, the Association for Women in Science (AWIS) in 1971, the Women in Engineering ProActive Network (WEPAN) in 1990, and MentorNet in 1997. All four of these organizations continue to be active today. Although this is a book focused on underrepresentation in computing, not on underrepresentation in the broader field of STEM, we include profiles of these four organizations for two reasons: before there were organizations devoted to underrepresentation in computing (most of which were formed in the late 1980s or the 1990s), the STEM organizations described below were the only ones that provided support to women in computing; and for the past 25 years, even though there have been specialty broadening-computing organizations, the organizations discussed here have continued to support women in computing. This is particularly true of MentorNet. These profiles are included to give a better overview of the organizations available to help women in computer; the profiles of the organizations specifically focused on computing, which appear in Chap. 8, are given in much greater detail.

© Springer International Publishing Switzerland 2016
W. Aspray, *Women and Underrepresented Minorities in Computing*,
History of Computing, DOI 10.1007/978-3-319-24811-0_6

6.1   Society of Women Engineers (SWE)

The first organization that had any significant impact on broadening participation for women in the computing disciplines was the Society of Women Engineers (SWE), founded in 1950. Between 1946 and 1949, local groups of women engineers had been meeting on the east coast – in Boston, New York City, Philadelphia, and Washington, DC. The most active of these groups was created by women engineers at Drexel Institute of Technology in Philadelphia and City College and Cooper Union in New York City. Women had been encouraged to take on engineering jobs during the Second World War to replace men who had gone off to serve in the military. Once the war ended, women were discouraged from working or studying engineering so as to protect engineering positions for male veterans.[1]

SWE's objectives were stated in its Certificate of Incorporation as a nonprofit organization (February 13, 1952 as cited in Homsher 2011):

> To inform the public of the availability of qualified women for engineering professions; to foster a favorable attitude in industry toward women engineers; to contribute to their professional advancement; to encourage young women with suitable aptitudes and interest to enter the engineering profession; and to guide them in their educational programs.

There were internal struggles in SWE during the 1950s over how to carry out these objectives.[2] In the 1950s there were tensions between whether the goal should be to use activism to attain professional equity with male engineers or to increase public recognition for the accomplishments of women engineers as professionals. For example, there was debate over whether SWE should provide its own awards inasmuch as the traditional engineering societies such as the Institute of Radio Engineers and the American Institute of Electrical Engineers had done little to recognize and promote women as professional engineers. During its early years, SWE formed a Professional Guidance and Education Committee that encouraged and advised high school girls on engineering careers. In the mid-1950s, SWE formed Junior Engineer and Scientist Summers Institutes for high school girls to gain further experience with science and engineering, as well as receive college and career advice.[3]

These tensions continued throughout the 1960s and 1970s, mostly in connection with determining what SWE's role should be, if any, in supporting national legislation in favor of the rights of women: equal pay laws, community property laws, special women's legislation, Social Security, and above all the Equal Rights Amendment. Some women pushed hard for SWE to actively support these pieces of legislation, but

---

[1] For an interesting example of women replacing male engineers during World War 2, see the story of the Curtiss-Wright Engineering Cadets as told by Meiksins et al. (2011).

[2] For a more general discussion of women in engineering and science in the United States, see for example Bix (2004), Hacker (1981), Oldenziel (1999), Rossiter (1982), and Zuckerman et al. (1991). On the history of women in computing, see for example Abbate (2012), Edwards (1990), Ensmenger (2010), Fritz (1996), Grier (2005), Light (1999), and Misa (2011).

[3] On SWE's outreach to high school girls in the 1950s, see Bix (2004, 2013).

others were concerned that to do so would jeopardize the organization's nonprofit tax status or its professional credibility. The leadership was more conservative than the rank and file on these issues. In the 1970s, SWE had a period of activism: passing a resolution in favor of the Equal Rights Amendment, helping in 1973 to form the Federation of Organizations for Professional Women, and voting in 1977 to support the National Organization of Women's boycott by not holding the SWE National Convention in any state that had not ratified the Equal Rights Amendment. In the late 1970s, however, SWE's participation in the women's rights movement waned as some of the other organizations in the women's rights movement began to speak out in opposition to technology as something that undermined women's position in society.[4]

Membership in SWE stood at 61 in its initial year of 1950. Because of the Cold War and in particular the space race stimulated by the launch of the Russian artificial satellite Sputnik in 1957, there was a large increase in federal support for the study of science and engineering as a national defense measure.[5] A small amount of this support went to scholarships and fellowships for women engineers. By 1961, when SWE took offices in the newly opened United Engineering Center in Manhattan with all of the other major American professional engineering societies, membership stood in the 700s. Both the Civil Rights Act of 1964 and the Title IX legislation of 1972 increased university interest in educating women and minorities in the engineering disciplines. The first student chapters of SWE were formed in the late 1950s – at CCNY, Drexel, Georgia Tech, MIT, Purdue, and the Universities of Colorado and Missouri.[6] Rapid membership growth occurred in the 1960s and 1970s, especially among student members, and by 1982 total membership had reached 13,000. Membership today is more than 27,000.[7]

It is interesting to consider the place of race and ethnicity within SWE. A few African American women joined SWE in the early 1950s. The 1957 National Convention was held in Houston, and one woman, Yvonne Clark, was not permitted to stay in the conference hotel because of her race. As a result, SWE set a policy not to hold any of its national conventions in the South, which it only lifted after passage of the Civil Rights Act in 1964. In the 1970s, when new engineering societies such as the Society for Advancement of Chicanos and Native Americans in Science (SACNAS) and National Society of Black Engineers (NSBE) were formed, SWE members who were African American, Native American, or Latina felt torn between loyalty to these race-centered organizations and SWE.[8]

Today SWE offers a variety of programs: ones to interest girls and young women in engineering such as bringing a working engineer to meet middle schools girls,

---

[4] For more on SWE's political activism, see Kata (2011).

[5] For a discussion of SWE and the Cold War, see Puaca (2008, 2014).

[6] Bix (2004).

[7] On membership and the influences that have shaped it, see the SWE web pages and also Daniels et al. (2011).

[8] For a discussion of race and sex in SWE, see Watford (2011). For a more general discussion of race, ethnicity, and gender in science and engineering, see Leggon (2006, 2010) and Leggon and Eller (2011).

scholarships, professional awards, professional development programs offered in the form of webinars, conferences, and a career center with a large job board. A recent survey indicates that most women join SWE primarily to meet other women interested in engineering or to find support for their own career.[9]

It is difficult to identify how many of the members of SWE were particularly interested in IT-related disciplines or to gain any overview that has statistical reliability. But we do have accounts of a few individual women involved in SWE who have been interested in IT-related education, research, or work – and we tell several of their stories here.

Gwen Hays studied engineering at the University of Pittsburgh – the only female in an engineering department with 1500 students – and received a bachelor's degree in electrical engineering. She worked at the National Security Agency, where she became interested in software design, and at Westinghouse on computer-aided design and radar development, before retiring from Westinghouse and becoming a llama farmer. She tells how, when her local chapter of SWE wanted to join the Engineering Center of Baltimore, it was resisted for some time because the men did not want women members. She also discussed how she used SWE to learn "more about society and how to work with society [in a way] that you wouldn't [learn] through the engineering channel." (Hays 2010)

Suzanne Jenniches earned a bachelor's degree in biology and taught high school biology until she switched 4 years later to being a computer test engineer – the only female doing this job at Westinghouse. She moved up through the technical and management ranks to become the vice president and general manager of the government systems division of Northrop Grumman. She held leadership positions in a number of professional organizations, including the presidency of SWE. Although she found that she was able to develop the "rhino skin" that she needed to survive in the workplace and had a supportive husband at home, she valued SWE for being "a very nurturing and caring environment, with people who can relate and understand," a place where you can "be yourself," a place to practice leadership skills, and a place to network with technical women. She used her presidency of SWE to attract more executive-level technical women from industry into SWE because she believed they are the people most able to make a difference in the workplace environment.[10]

Thelma Estrin, who earned her bachelors, masters, and doctoral degrees in electrical engineering at the University of Wisconsin in the late 1940s and early 1950s, was a leading scholar in the application of electronics and computers to biology. She has also had two daughters who have had highly successful careers in the computing field – one as an entrepreneur, the other as a college professor. She looked enviously on the opportunity for young women in SWE today to have a "community," a place to get support; and she would have probably become active in the organization had she been younger, but as she was coming up through the ranks, she did it without the support of organizations or other women. As she said, "I was the first woman engineer I ever knew." (Estrin 1992, 2002, 2006)

---

[9] See Daniels et al. (2011).

[10] The direct quotations are taken from Jenniches (2003). But also see Jenniches (2010).

In 1979 Paula Hawthorn received a PhD in computer science from the University of California at Berkeley with work on database systems. She worked for Hewlett Packard, Lawrence Berkeley Laboratories, and several startup firms. Her experience early in her career was that there were a lot of women in her field because "you didn't have to have a certain set of prerequisites to be a computer scientist or a computer programmer. You had to have a good mind... There wasn't a sense that you had to have a degree in computer science. So there were lots of women, who came from lots of other occupations." (Hawthorn 2002) However, the major field advisor at Berkeley when she was a doctoral student in the early 1970s urged her to drop out because she was a woman with children without her husband present. As she remembered the advisor saying: "You cannot be a serious student if you have children...to be a graduate student at Berkeley, you have to give up everything. This has to be your whole life. There is no time for anything else...If we had known that you had children, we would not have accepted you."

Throughout her years as a doctoral student at Berkeley, the number of women in the graduate computer science courses kept dwindling. Part of the reason was that the department had what they believed to be objective criteria for the awarding of financial aid, such as having a large number of undergraduate mathematics and physics courses – and this (unnecessary) set of criteria prevented many women from receiving the financial aid they need for attending graduate school. So, as discussed in Chap. 10, Hawthorn, fellow doctoral student Barbara Simons, and Women's Center staff member Sheila Humphreys developed a re-entry program to take in as doctoral students a group of students with nonstandard degrees. Hawthorn also formed women's groups at Berkeley and at Lawrence Berkeley Lab – and revived one at Hewlett Packard which had been shut down by management for fears of its becoming too militant – where she could relax, be among other women, and compare notes on how to handle certain situations that came up.

Interestingly, neither Hawthorn nor Simons had any interest in SWE. Their attitude is represented by the following quotation:

> It is true that the women that are the most successful are those who absolutely do not believe that they are discriminated against. Barbara and I used to call them the "My Daddy Was An Engineer" women. That was why we never wanted to join SWE, the Society of Women Engineers: there were so many engineers in SWE whose daddies were engineers, who felt that there was absolutely no issue with them being a woman in engineering, and that anyone who talked about anyone being discriminated against was just making it up! (Hawthorn 2002)

6.2   Association for Women in Science (AWIS)

In 1967 the physiologist G. Virginia Upton from the Veteran's Administration began to organize receptions at each annual meeting of the Federation of American Societies of Experimental Biology, so as to build a community of women researchers. At the 1971 meeting in Chicago, 27 women continued Upton's tradition but turned it into an organizational meeting to create the Association of [later, 'for'] Women in Science (AWIS). Elected as co-presidents were Judith Pool from the Stanford University Medical School, who handled fundraising, and Neena Schwartz from the University of Illinois College of Medicine, who handled program and volunteer development (Rossiter 2012).

The earliest activity of AWIS was to build a registry of women scientists to be considered for jobs, appointments, and awards; and to increase the number of women appointed to technical panels and study sections that evaluate NIH proposals – the largest funder in the biomedical sciences. When meetings of AWIS representatives with NIH officials did not result in significant changes, AWIS became the lead plaintiff in a suit against the NIH's parent organization, the Department of Health, Education and Welfare (*AWIS et al. v. Elliot Richardson*). The effect of their legal victory was an increase within a year's time in the percentage of women sitting on these technical panels and study sections from 2 to 20 %. The legal victory was also a powerful recruiting tool, enabling AWIS to rapidly attract a thousand members. The American Association for the Advancement for Science helped AWIS through its initial organizational growing pains.

During the 1970s, in addition to building the registry and establishing an office, AWIS formed an educational foundation to receive donations and offer fellowships and other grants. It began publishing a newsletter that focused on both current policy issues and career development. The organization established and expanded a local chapter system in the 1970s and 1980s, focused on both career development of women scientists and the encouragement of girls and women from the local community to enter science careers. In the 1990s, AWIS became actively involved in mentoring undergraduate and graduate students using an unusual community-mentoring model, which was developed with the support from the NSF and the Sloan Foundation, and which won a Presidential mentoring award. In the first 5 years, the program involved 6000 student protégés and 2500 mentors.[11] AWIS has also been involved in research on issues concerning the advancement of women in science, such as a study sponsored by the Sloan Foundation in the 1990s of the chilly academic climate for women scientists.[12]

AWIS's initial success in reforming the NIH peer review system led it to become actively involved in policy issues related to women and science. In the 1970s and 1980s AWIS used the legal system to ensure that affirmative action and equal opportunity employment laws were enforced to protect women scientists. The Association

---

[11] See Bird and Didion (1992), Didion (1995), and Fort (1995). For a more recent version of AWIS mentoring, see Fridkis-Hareli (2011), which describes the mentoring activities in AWIS's Massachusetts chapter. The AWIS process involves building mentoring circles of three to five peers and one to two mentors, all with similar interests and career goals, who meet in person monthly during the academic year.

[12] On the chilly academic climate study, see Didion et al. (1998). For a recent snapshot of AWIS's full range of activities, see its 2014 strategic plan at http://c.ymcdn.com/sites/awis.site-ym.com/resource/resmgr/Files/Strategic_Plan_FINAL_NOV1720.pdf. The plan includes 40 action areas organized under the headings: advocate for positive system transformation, help all women in STEM achieve success, and maximizing our impact by optimizing organizational capacity.

lobbied extensively for creation of the Commission on the Advancement of Women and Minorities in Science, Engineering, and Technology Development Act (CAWMSET), which was signed into law by President Clinton in 1998. More recently, AWIS lobbied Congress to strengthen the use of Title IX legislation to apply to science and engineering departments in higher education. The Association also publishes an electronic newsletter that keeps members apprised of relevant policy issues.[13] In 2014, AWIS entered a partnership with SWE to carry out public policy work together.

AWIS reaches more than 20,000 professionals each month through its members and chapters. 79 % of AWIS members hold an advanced degree and 66 % are at the middle or senior levels of their careers. About half of its audience is academic – the rest spread across industry, government, and non-profits.[14] While women in the computing fields have no doubt benefited generally from AWIS research and policy efforts on behalf of STEM women, and very likely some individual female computer scientists have benefitted from AWIS's mentoring activities, the emphasis of AWIS is on science – and particularly on the biological sciences – so AWIS has had a limited impact on helping women in the computing disciplines. One area is which AWIS has perhaps had its greatest impact on the computing field is in biocomputing. A number of biocomputing researchers have been involved with AWIS, including Hua Fan-Minogue from the Stanford Medical School, Sayanti Roy from Notre Dame, Estefania Elorriaga from Oregon State, and Hoda Abdel-Aty-Zohdy from Oakland University.

6.3 Women in Engineering ProActive Network (WEPAN)

With persistently low numbers of women in engineering education programs, a number of colleges and universities have formed women-in-engineering programs to support their women students and faculty as well as to recruit additional women.[15] The first such program was created in 1969 at Purdue University.[16]

---

[13] For more information on the history of AWIS, see "History of AWIS", https://awis.site-ym.com/?page=history, accessed 16 May 2015.

[14] The information in this paragraph comes from the AWIS fact sheet (http://c.ymcdn.com/sites/awis.site-ym.com/resource/resmgr/Fact_Sheets/AWIS_General_Fact_Sheet.pdf).

[15] In addition to the sources cited in the body of this section, this account relied heavily on the WEPAN website (wepan.org), in particular the pages entitled "The First Ten Years: 1990–2000" and an article reprinted their entitled "Yes, WEPAN" (Home Douglas 2009). It also relied on oral history interviews with co-founders Suzanne Brainard (2015), Jane Daniels (2015), and Susan Metz (2015).

[16] Daniels (2015) indicates it was the interest in diversity of Arthur Hansen, who was the president of Purdue University from 1971 to 1982, that enabled Purdue to be a pioneer in broadening engineering to include more women and more African Americans. In this interview Daniels gives a significant amount of detail about the women in engineering program at Purdue that goes beyond the scope of this study.

Jane Daniels,[17] who directed the program from 1978 until 2000, had answered numerous inquiries from other universities wanting to establish similar programs and knew that there was a need for a central place to provide information and guidance to women in engineering programs.[18] With NSF support, Daniels first carried out a survey of deans of engineering and SWE advisors at various colleges and universities across the nation to identify interested parties. Then she organized a meeting in Washington in 1990, attended by 200 people interested in women in engineering. The group rejected proposals to continue to meet as a special interest group at SWE or ASEE meetings. Instead, it decided to create a new national organization, the Women in Engineering ProActive Network (originally the Women in Engineering Program Administrators Network) for people who were advocating programs for recruiting and retaining women and girls in college engineering programs. The organization of the conference and the creation of the Network were carried out by Daniels with the full involvement of Susan Metz[19] and Suzanne

---

The program, she reports, "wasn't all altruistic." A downturn in student interest in majoring in engineering in the late 1960s was intentionally offset by drawing students from a wider pool that included women and African Americans. Funds associated with the federal Women's Education Equity Act of 1974 helped to strengthen Purdue's program for women in engineering, which provided funding for career outreach activities to high school girls and a course in the School of Technology, entitled Tools and Engines, in which female students could get hands-on experience with power tools and wiring circuits. However, most of the early funding of the program came from industry, especially from General Motors and IBM.

Daniels remembers that in the late 1970s the attitude was to "go find the women, bring them to Purdue, and fix them. Fix them so that they would do engineering just like our men always have, because, after all, Purdue has a wonderful reputation in engineering, so we don't want to change anything, we want to keep doing things the way we've always done it." Over time, effort was redirected to changing the system in various minor ways to make it a more inclusive community.

[17] Daniels (2015) sees lots of similarities between engineering and computer science: the similar, persistently low numbers of women in the two fields; "the environment is not, I don't see, as welcoming to women if they have more international students and faculty from countries that do not value women's educational rights and abilities"; and engineering and computer science are among the few STEM disciplines in which there is "meaningful employment and [the ability] to make important contributions to society" with only a bachelor's degree. (Daniels 2015)

[18] On women in engineering programs at colleges and universities, see Knight and Cunningham (2004).

[19] Stevens Institute of Technology, founded in 1870, was an all-male institution for 101 years. When the trustees voted to admit women in 1971, only a small number applied. Motivated by a concern about adequate enrollment and an optimism that women could be good engineers, Dr. Edward Friedman, Dean of the College at Stevens, suggested exploring external funding to develop pre-college programs to introduce women and their parents to engineering. Why would women consider going into a male-dominated field and why would their parents, teachers and guidance counselors endorse the idea of majoring in engineering if no one knew what engineers do all day? Metz was drawn into this initiative, received a grant from Exxon Company and subsequently, Stevens established the Office of Women's Programs, directed by Metz in 1980. National Science Foundation funding expanded the pre-college programming to include a series of four-week summer programs that attracted hundreds of high school women throughout the country to learn about careers in engineering and science. During that time, Metz began to do research on underrepresentation in STEM and the impact of pre-college programs. The American Society for Engineering Education (ASEE) Annual Conference typically offered a panel session on women in engineering,

Brainard,[20] who ran the women in engineering programs at Stevens Institute of Technology and the University of Washington, respectively.

These three organizers took turns in leading the organization over its first 10 years.[21] The target audience was people who worked with women engineering students at the undergraduate or graduate levels, e.g. directors of women in engineering programs, advisors, and interested faculty members.

By the year 2000, WEPAN had signed up more than 500 members and ran its programs out of regional centers based at the three universities of its founders. WEPAN established an annual conference, beginning in 1990, serving as an important community-building and networking opportunity for administrators and faculty who ran university-based women in engineering programs. This annual event has evolved and continues today under the title of WEPAN Change Leader Forum, catalyzing discussion and action on the impact of culture on engaging and retaining diverse communities of women in engineering – both in education and the workforce. (Private communication from Susan Metz, March 19, 2016) WEPAN also has partnered on a number of conferences, meetings, and workshops with other organizations including AAAS, the National Academy of Engineering, AWIS, SWE, the New York Academy of Sciences, NSF, the U.S. Department of Energy, and the National Association of Minority Program Administrators (NAMEPA).[22]

---

and it was at this event where she met Daniels, year after year, until they teamed up with Brainard and founded WEPAN. More recently, Metz has served as Executive Director for Diversity and Inclusion and Senior Research Associate, reporting to the president of Stevens. (Private communication from Susan Metz, March 19, 2016)

[20] Brainard served as the executive director of the Center for Workforce Development at the University of Washington until her retirement and held affiliate faculty positions in both women's studies and human-centered design and engineering. She has served as chair of the NSF Committee on Equal Opportunity in Science and Engineering and served on several National Academy studies on diversity in engineering.

[21] For details about the running of the organization, see Brainard (2015), Daniels (2015), and Metz (2015). Metz notes that it was difficult identifying someone who would run for President of WEPAN who was not one of the founders because there was no staff to support any officer position.: "The very first President after the Founders all rotated through that position – and I held it for 5 years – was Jan Rinehart who was at Texas A&M at that time. Transitioning from a Founder-run organization to other elected officers is no easy task, especially for that first person. Although Jan was reluctant, Dr. Karan Watson, Jan's supervisor at Texas A&M was very supportive and provided some release time. Jan was an outstanding leader, paving the way for other non-Founders to run for office. Eventually, WEPAN hired a full-time executive director and CEO, Diane Matt, who is still in that position today. (Private communication from Susan Metz, March 19, 2016)

[22] Brainard singled out the strong relationship that WEPAN had with the National Academy of Engineering during the years in which Bill Wulf was NAE president. She also pointed to close ties with Shirley Malcom and Yolanda George at AAAS. WEPAN had strong relations with the National Association of Multicultural Engineering Program Advocates (NAMEPA), an organization of educators and representatives from the public and private sectors to enhance recruitment and retention on underrepresented minorities in engineering careers. (See http://www.namepa.org/index.php?option=com_content&view=article&catid=19%3Adefault&id=62%3Ahistory&Itemid=76 for NAMEPA's history.) WEPAN had good, if not extensive working relations with AWIS from the beginning. However, there were some rough patches in WEPAN's early relations with SWE. When the two organizations eventually came to an understanding and agreement that SWE

Of particular importance were WEPAN's regional training seminars for educators wanting to create or strengthen women-in-engineering programs on their campuses. Between 1992 and 2001, with support primarily from AT&T, the Sloan Foundation and the Fund for Improvement of Post-Secondary Education (FIPSE), representatives from more than 150 institutions attended these seminars.[23] Topics at the seminars included acquiring resources, conducting pre-college outreach programs, developing retention and mentoring programs, implementing student needs assessments, evaluating initiatives, and encouraging industrial participation. Materials prepared for these training sessions found their way into published books funded by FIPSE entitled *Increasing Access for Women in Engineering* by Susan Metz and *Curriculum for Training Mentors and Mentees* by Suzanne Brainard.[24] Working together with Carol Muller, WEPAN became the incubator for the electronic mentoring program, MentorNet, for the period 1996–2001. (See the discussion of MentorNet later in this chapter.)

It is probably not coincidental that the number of women in engineering programs in the United States increased from 26 in 1991, to 66 in 1995, to more than 100 by 2004. However, over the last 10 years, the number of women in engineering programs has noticeably declined through underfunding and merger with minority programs.[25] Daniels notes that in the early years, most of these centers were managed by individuals with backgrounds in education or the social sciences, but there has been a trend in recent years for these programs to be managed by young people with engineering backgrounds. Daniels sees this as a good thing because then the

---

was primarily about professional development for women engineers and students and WEPAN was primarily about working with faculty and administrators to develop programs and initiatives to increase awareness about engineering, retain engineering students and understand and impact the culture of engineering, they became more collaborative and effective working together. (Brainard 2015; Daniels 2015; private communication from Susan Metz, March 19, 2016)

[23] These training sessions appear to have ended formally in 2001. The three founders extended this kind of work by entertaining a series of visitors at each of the home institutions and by making site visits to other college women in engineering programs.

[24] An effort was taken to move these materials online eventually because the Ford Motor Company was interested in having them available to its engineering staff members who were serving as mentors to college engineering students. Ford paid the full cost of this transfer of materials online.

[25] The Sloan Foundation funded a multi-institutional study of the impact of women's programs in the late 1990s, entitled Women's Experiences in College Engineering, which was not able to demonstrate benefit of such programs on recruitment and retention of women in engineering. (Private communication from Carol Muller, 24 February 2016)

Brainard (2015) tells the story of the evisceration of the center at the University of Washington in 2004, while Denice Denton was the dean of engineering. Denton was the first female dean of an engineering school at a major research university. On women's engineering centers generally, Brainard indicated it was much more difficult to convince university administrators to continue operation of a women in engineering center than a minorities in engineering center. One reason was that the women "typically had higher grades than the guys did; and they did very well or they dropped out because they didn't like the climate they were in," so there was not a large group of women engineers who were performing poorly academically to target as the need for the center. Another reason was that industry was generally more interested in increasing minority numbers than numbers of women in its engineering ranks. (Brainard 2015)

program director has first-hand experience of what it is like to be an engineer and perhaps can relate better to the students.[26] (Daniels 2015)

Beginning in 1993, each year the University of Washington surveyed its male and female undergraduate engineering students about perceived barriers to their education. Topics included the quality of teaching assistants, teaching, engineering labs, departmental assistance, and curriculum. Based upon this survey and with funding from the Engineering Information Foundation, WEPAN developed a national climate survey that it administered in 1998. More than 8000 undergraduate engineering students (57 % female) at 29 institutions responded to the survey. The survey results showed that female students reported less self-confidence in engineering and physics than male students; and that female students had less confidence in asking questions in class and lower comfort levels with lab equipment. This study received considerable attention from engineering schools. (Metz et al. 1999) The national survey was repeated in later years, when funding was available. It enabled individual schools to see how they measured up to national averages as well as to track their progress over time. The WEPAN staff used these surveys, from time to time, to assess the needs at a given institution and then suggest a course of programmatic changes the institution could implement.

Another important early activity was the creation of WEPAN's Knowledge Center. In the early days, program directors simply tried things out, to see if they would work. Program directors were hungry for research that would help them to shape and justify their programs. The Knowledge Center collected published research on relevant topics from the education and social science literatures. It also became a repository for data on women in engineering.

The Strategic Initiatives and Programs page on WEPAN's current website provides a snapshot of WEPAN's activities today, a time when none of the three founders are actively involved in the daily leadership of the organization. The annual conference has received added importance, in part because it is now the financial underpinning for the organization. WEPAN is currently supporting four strategic programs listed in Table 6.1, the last three of which are supported by the NSF. All four of them involve transforming engineering culture, although each does it in a slightly different way and sometimes with a slightly different audience.[27]

It is worth taking time out to discuss in more detail the last of these programs, the ENGAGEEngineering.org Project. It began with a 2.6 million dollar Extension Service grant in 2009 from NSF's Research on Gender in Science and Engineering Program directed by Jolene Jesse. The principal investigator was Susan Metz, Executive Director of Diversity and Inclusion at Stevens, with partners Diane Matt,

---

[26] "Not all see this as a good thing – however. These folks have less training in the social sciences and education to help them understand how to address institutional issues, and to appreciate the underlying causes and potential remedies for women's historical exclusion from engineering studies and professions. They are often not well-placed in terms of influence and status in university hierarchies." (Private communication from Carol Muller, 24 February 2016)

[27] Cultural change in an organization is notoriously slow and difficult to achieve. It may be too early to see many results of these cultural change programs.

Table 6.1 WEPAN's strategic initiatives (as of early 2015)

The *Advancing Culture in Engineering Initiative* engages educational and workplace leaders in changing the engineering environment in which women and underrepresented minorities learn and practice engineering. It is based on a four-step Gender Inclusive Organizations framework created at Simmons College:

Equip the Women by equalizing experience between men and women;

Create Equal Opportunity by eliminating structural and procedural barriers within organizations that impede women;

Value Difference by appreciating rather than eliminating the differences between men and women; and

Re-Vision Engineering Culture by addressing the assumptions, norms, and practices that lead to gender inequities within organizations.

Transforming Engineering Culture to Advance Inclusion and Diversity (TECAID) is a project working with five mechanical engineering departments to sustain interactive cultures in the various formal and informal aspects of engineering education – in the classrooms and labs, in faculty meetings, and in informal student interactions whether they are at work or play.

Engineering Inclusive Teach (EIT) is a faculty professional development project to implement best practices to help faculty members create inclusive engineering learning environments

Engaging Students in Engineering (ENGAGE) provides small grants to engineering faculty members to put into practice research-based classroom strategies to enhance student engagement and retention.

Source: WEPAN Strategic Initiatives and Programs (https://www.wepan.org/?page=528)

the Executive Director of WEPAN, and Patricia Campbell, the President of Campbell-Kibler Associates. Campbell is a long-time participant in work on gender and race in STEM.[28] The goal of the project was to increase retention of women in STEM through the dissemination of three research-based strategies.

The project identified three strategies "that have a rigorous body of research connecting each with retention of engineering students." (Metz 2015) The strategies include: increasing faculty-student interaction in and out of the classroom, using everyday examples that are "relevant to students to teach technical concepts", and assess and improve students' spatial skills ability among those who have weak skills.

> [T]he reason we chose these three strategies was that although using them improves the educational experience for everyone, they disproportionately impact women and underrepresented minorities. In addition, the strategies were not wholesale changes to the curricu-

---

[28] When discussing the NSF ADVANCE program, Metz (2015) observed how difficult it is to build a program, say, for the advancement of women faculty in engineering – although she could have been speaking of broadening participation programs more generally: "My women faculty don't want to hear that in front of their male colleagues. So you still tread a very thin line of supporting women in a way that they don't feel needy, that their male colleagues don't point fingers and say why aren't we getting these and we could use these too, or, yes, women are needy and they need the ADVANCE initiatives so that's good that you're doing that. Communicating what we're doing, how we're doing, being inclusive is really, really challenging without disenfranchising the women, making some women say 'I don't want any part of ADVANCE', like they did for women in engineering programs. Other students don't want to be part of anything that's just for women; so it's tricky."

lum. We did not want to deal with the bureaucracy and challenges of getting things through curriculum committees. We wanted faculties to be able to take these plug-and-play resources and use them in their classroom. This worked very well with the faculty-student interaction and everyday examples strategies. Spatial visualization, identified as a critical cognitive skill connected to persistence in STEM, is not as straightforward. Although an individual faculty member can implement a spatial visualization assessment and training program, the ideal approach is for a school to assess all incoming engineering students through a 20-minute Purdue Spatial Visualization Test: Rotations (PSVT:R) and then provide the NSF-supported and -tested training program developed by Dr. Sheryl Sorby to those students who fall below 70 % on the test. (Metz 2015)

ENGAGE initially used a train-the-trainer model by identifying a team of individuals in universities who, armed with the necessary information and resources, could implement the strategies on their campus – usually in three-person teams. The training occurred during a three-day workshop. The first year, the project included teams of faculty and administrators at ten high-profile, large schools because the organizers believed that success in these schools would give the program credibility and make it more desirable for other schools to participate. (Private communication from Susan Metz, 19 March 2016)

While the workshop went well – participants valued the information, planned to use it and rated the sessions highly – "the translation to other colleagues at the engineering schools was challenging." As Metz (2015, modified in a private communication, 19 March 2016) went on to explain:

The faculty and administrators who attended the workshops were excited about the three strategies and many implemented one or more of the strategies at their schools personally. But the idea was [to] get your colleagues together, share this information, and have them use the strategies in their classrooms. That was not happening. We realized that faculty are really not comfortable in this train-the-trainer domain. They are used to being experts in their discipline. These retention strategies are not [within] their area of expertise so they have some vulnerability. They were not sharing information with their colleagues that they [were] totally comfortable with…[29]

As a result, the project organizers changed the dissemination strategy to directly involve individual faculty members.[30] Instead of giving mini-grants ($10,000 –

---

[29] Metz (2015) explains that although the trainers were uncomfortable communicating research results that they were not entirely familiar with to their faculty colleagues, that none of the faculty were particularly concerned with the research basis behind these practices: "We really thought that faculty were interested in the research behind these strategies, that we had to convince them that these were evidence-based strategies, there's reason to use them; but faculty didn't want to know the details. What they said was, "if you're telling me this is research-based and these are all the references – I believe you! Just explain what I should do. Again, we shifted our emphasis in faculty's professional development to the implementation of these strategies."

[30] It might seem that it would be easier to disseminate these research-based practices more widely through the train-the-trainer program, but in fact ENGAGE was able to disseminate these practices more widely after they adopted the strategy of involving individual faculty members. The original promise in the grant proposal was to reach 33 educational institutions. After the change in strategy, the project "started involving many more schools in the process. … [W]e started doing a lot of virtual events, webinars, discussions at ASEE Conferences, which was a terrific opportunity to share the research, share the experiences of schools who were doing it." (Metz 2015)

$12,000) to engineering schools, even smaller mini-grants ($2000–$2500) were given to individual faculty members to implement and document in their class-rooms. Faculty at more than 70 institutions received mini-grants. These schools implemented one, two, or three of the ENGAGE strategies. (Private communication from Susan Metz, 19 March 2016)

> …if you are a faculty member and you are teaching hundreds of 1st and 2nd year engineer-ing students – using the ENGAGE retention strategies can make a difference. We tabled the professional development workshops and instead began developing and conducting webi-nars – each focused on one ENGAGE strategy. Engineering faculty was the target group and we partnered with ASEE and ASME primarily to get the word out. That started gaining some traction. As a result of the webinars and enhanced electronic communications, web-site downloads of everyday examples, lesson plans, papers, presentations, and resources for each strategy hit 233, 359 downloads in 2015. (Metz 2015)

One of the reasons that the ENGAGE project is so interesting for the purpose of this book is that, in the fourth year of the project, the principals decided to collabo-rate with the National Center for Women & Information Technology (NCWIT). Thus we have an opportunity to see the ways in which broadening participation in STEM and computing are similar and different.

Prior to the joint project, the ENGAGE project had not done any work with com-puter science and only limited work with computer engineering. Most of ENGAGE's effort had been with mechanical engineering. The principals in the ENGAGE proj-ect were familiar with the people at NCWIT, particularly with NCWIT's senior scientist Joanne Cohoon, because both groups had received grants from the NSF program on Research on Gender in Science and Engineering and had come to know one another through the annual principal investigator meetings. NCWIT also had a well-developed Extension Service model for bringing research-based practices into academic departments (see Chap. 8).

> …that's when we got together and said, okay, ENGAGE is focused more on retention and engineering. NCWIT is focused more on computer science and recruitment. If we got together, melded our strategies, and focused on those departments that had the lowest rep-resentation of women, let's see if NSF would fund that. … We focused on four departments: Computer Science, Computer Engineering, Mechanical Engineering, and Electrical Engineering – two of the largest engineering disciplines [with] the lowest representation of women. And we used the NCWIT consulting process too – each "client school" is assigned an Extension Service Consultant, someone who has received extensive training by NCWIT. NCWIT does a great job in data collection. They have a tracking tool, so [schools are] required to document enrollment and retention numbers. (Metz 2015)

While the ENGAGE team has statistics about the number of downloads of course materials they developed for making the classroom more engaging to various types of underrepresented students (which number in the hundreds of thousands), they did not have the funding through the NSF Extension Services grant to track the retention of students. The philosophy of this grant program was to disseminate research-based practices, not to repeat the research that confirms the best practices.[31] Evaluation and

---

[31] Metz (2015) made an interesting comparison between progress at broadening participation in the academic and industry sectors: "I think we're all heading in the right direction. But it's a slog;

use data supports the claim that ENGAGE created and disseminated resources that faculty use, but it is not definitive what impact these resources had on retention of a diverse student body. "We have to assume that it has because of the implementation approach." (Private communication from Susan Metz, 19 March 2016)

While both the original ENGAGE grant (through no-cost extensions) and the ENGAGE-NCWIT grant are still ongoing, there is a sense that both programs have had some success, but neither has been revolutionary:

> Neither one of these grants has really penetrated engineering schools holistically. There have been some [signs of progress], ... there are packets of faculty who embrace it, do it, and slowly it trickles out into their schools; but, boy, it's a slow process. The numbers ... I don't know ... That's the biggest frustration. The numbers are still challenging. (Metz 2015)

This slow process is problematic in that NSF typically expects to see results in a three-year time frame.[32] However, the ENGAGE program did not begin to show real progress until the fourth, fifth, sixth, and seventh years. Extension Services grants have the luxury of time – more than others. NSF also expects principal investigators to be able to sustain successful programs, typically without ongoing NSF funding after the initial grant runs out. When asked how ENGAGE plans to do this, Metz (2015, extensively revised in private communication of 19 March 2016) replied:

> This is what this last year is devoted to. Faculty who have used the strategies are committed to them. We need to keep pushing out the information to the community. ENGAGE launched a redesigned website that has a simplified, modernized interface that is mobile-adaptive, making the website more user-friendly and accessible for our visitors who are more frequently using mobile devices to access the website (21.3 % of website visits come from a mobile device, up from 2.6% in 2011). Popular, heavily viewed and downloaded resources and content are highlighted on the website, and redundant, infrequently used information was eliminated. Information on the website is organized by ENGAGE strategy, and then in three simple sections: "Why it Works" – key, compelling highlights of the evidence behind the strategy; "Learn More" – additional detail on the research evidence; and "Take Action/Resources" – a complete list of information and links for easy downloads of all resources and tools related to implementing/using the strategy. This page effectively functions as a toolkit for each strategy, and WEPAN is committed to keeping this going.

When asked whether working in a computer science environment was different from working in an engineering environment, Metz (2015, extensively revised in a private communication of 19 March 2016) replied:

> I don't think so. I think it's very similar and all disciplines are concerned about recruitment and retention and broadening diversity. However the concept behind the NCWIT-ENGAGE

---

academia is much slower than industry in embracing change, particularly in terms of the culture and climate in engineering. They don't have the profit incentives that industry has. McKinsey and Company (Women Matter) and many others have researched and documented the real value of diversity including impacting the bottom line. Women have so many career choices and culture matters. Why should they go into a culture where they have to struggle and continually prove themselves."

[32] This issue of expectations by NSF of short-term windows for progress and of operations becoming self-sustaining after one or two rounds of NSF funding is discussed at several places in Aspray (2016).

Extension Services grant was for the engineering departments (mechanical, electrical and computer) to work with the computer science department collaboratively. Sometimes this worked well and sometimes it did not work at all. However it is unclear at this point how critical the collaboration was in terms of impact. Since computer science is often in a different school from the engineering departments and since universities tend to be very siloed, faculty don't know each other, the culture is different, and logistically it is just more challenging to work together. In many cases, engineering and computer science worked to address the goals of the grant independently. NCWIT is exploring the value of collaborating across schools.

Returning to a discussion of WEPAN's current Strategic Initiatives and Programs, the WEPAN webpage identifies three programs in the area of dissemination of research and knowledge. One is an online repository (mentioned above) called the Women in STEM Knowledge Center, created in 2008 by WEPAN and the American Society for Engineering Education, with more than 2000 resources about gender underrepresentation in STEM. A second one is professional development webinars on topics such as academic coaching, salary negotiation, and stereotype threat, which are viewed by almost 3000 people each year. The third one is its national change leader forum.

WEPAN is also entering into strategic collaborations to advance diversity in engineering. These include two collaborations with ASEE – on diversity in corporate settings and diversity of the ASEE membership; and with the University of Colorado BOLD Center on strategies for success of underrepresented engineering students.

WEPAN's mission specifically is to serve women in engineering, not women in computing. With the exception of the ENGAGE project, WEPAN has not focused on computing. But strategic partnerships have always been a key component of all WEPAN's initiatives and the combination of NCWIT and WEPAN constituencies could be very powerful. The centers at Washington, Purdue, and Stevens all have strong ties to the computing community. Computer science resides in the engineering school at Washington and at Stevens, so the women in computer science were important clients of Brainard's and Metz's programs. Both Washington and Purdue had Virtual Development Center projects associated with the Anita Borg Institute (see Chap. 8), and Brainard was actively involved in that work. Brainard's program has also been actively involved in the Grace Hopper Celebration (see Chap. 8) for many years, and has provided the official external evaluation of the NCWIT programs.

6.4 MentorNet

In 1987, 2 years after completing her doctoral degree at Stanford in education administration and policy analysis, Carol Muller returned to her undergraduate institution, Dartmouth College, as an assistant dean in the engineering school. She felt as though she had stepped back in a "time warp" and that Dartmouth's engineering school had not yet internalized the social changes of the 1950s, 1960s, and 1970s. Though her newly created job was not at all focused on women, or even on students, she was curious to determine why there were so few women in engineering at Dartmouth. In 1989, she contacted Karen Wetterhahn, a Dartmouth professor of chemistry newly appointed to the role of Associate Dean of the Sciences at Dartmouth, and together they founded the Dartmouth Women in Science Project, launched in 1990 with support from the president's office and individual faculty members' research funds. Soon, that effort was joined with more major funding from the Sloan Foundation and the National Science Foundation, which enabled the two founders to hire a director and continue the expansion of the Women in Science Project (WISP).[33] WISP sponsored paid research internships for first-year female undergraduate students, working in labs with a faculty member; lectures by visiting women scientists to offer role models; a newsletter; and site visits to industrial research and engineering organizations.[34] (Muller 2014; also see Muller et al. 1996; Cunningham et al. 1996)

An e-mentoring program piloted as part of the Women in Science Project eventually led to the creation of MentorNet. Muller had arranged for WISP students to make a field trip to the IBM facility in Burlington, VT. This was a long enough trip that, even though a number of students had signed up, the students were too busy with their own studies and lab sessions to follow through on the six-hour outing. When it became apparent that just one student would attend, the visit had to be rescheduled. Muller, who was driving the students to these industrial site visits, could see these visits were having a great impact on the students' interest and confidence in science and engineering fields; and she "began to think about [whether] is there a way we could get some of that benefit by connecting students with working professionals more readily, especially in a field like engineering, and also in the sciences." (Muller 2014, slight corrections by interviewee 24 February 2016)

Because John Kemeny had made computing an integral part of the Dartmouth curriculum and everyday life while he was president in the 1970s, Dartmouth had become an early adopter of email. The scientific community generally was also sooner to adopt email than the population at large, and Muller had the idea of an e-mentoring network for women in science and engineering. Muller wrote a four-page proposal that led to IBM funding for a small pilot project at Dartmouth in 1993 (Muller 2014).

---

[33] The Sloan Foundation funding, awarded in 1992, was used in part to fund faculty development retreats to engage faculty in learning about improved and exemplary teaching and mentoring processes in the STEM fields. (Muller 2014)

[34] The Women in Science Project was initiated by both Muller and a new Associate Dean responsible for the sciences in the College of Arts and Sciences, the chemist Karen Wetterhahn. The Dartmouth president was receptive to the idea because he was looking for ways to recruit more women to Dartmouth after a journalistic article about the fraternity system had made it sound as though Dartmouth was a bad place for women to attend college. The faculty was receptive because a group of psychologists at Yale, Brown, and Harvard had recently prepared a study that reported, after taking into consideration every factor they could think of such as courses taken in high school, there was still a large unexplained gap in this rigorous, data-driven scientific study concerning women's low enrollment in science, mathematics, and engineering at these schools (Muller 2014).

A rigorous evaluation of the pilot project showed the promise of e-mentoring, and both the Sloan and Intel Foundations provided funds in 1996 to plan for a national e-mentoring program using email. Sloan Foundation program officer Ted Greenwood pointed out that there were already some experiments underway with what we would now call e-mentoring, and the foundation arranged for a meeting in 1996 at the Boston airport of all the stakeholders in Muller's project, plus some others who were developing other e-mentoring programs. Participants in this meeting included, for example, Dorothy Bennett, who was running a tele-mentoring for young women in computer science at the Bank Street School of Education in New York City; and Lee Sproull from Boston University, whose book with Sara Kiesler, *Connections: New Ways of Working in the Networked Organization* (Sproull and Kiesler 1991), was in consonance with Muller's idea.[35] (Muller 2014)

Significant funds were provided in 1997 by the Intel and AT&T Foundations to implement the new e-mentoring program, by this time known as MentorNet.[36] Muller had relocated to Silicon Valley for family reasons and opened MentorNet's offices in 1997 at San Jose State University, where the Dean of Engineering, Don Kirk, was interested and offered office space. Muller was on the board of directors of WEPAN, and for its first 3 years, MentorNet was operated as a program of WEPAN so as to avoid added costs and complications of setting up a new non-profit organization (Muller 2014).

The plan was "to create a very large scale infrastructure using technology, and automating as much as possible, but with a lot of intelligence behind it so that multiple college students from campuses anywhere could connect, find the appropriate professional to serve as their mentor for them, and take off on a mentoring relationship."[37] (Muller 2014) To avoid having to take special steps to protect minors as well as get parental permission, MentorNet focused on college instead of younger students. Some people wanted MentorNet to allow only women to mentor women, but Muller disagreed:

> Some people had strong feelings that only women should mentor women and that just didn't seem right to me, so I made the point that if we relied on all the women, they are only 10% of the workforce and furthermore, we weren't really going to change things if we never gave men the opportunity to learn anything from the students they mentored about what [are] the experiences or life of the women. (Muller 2014)

---

[35] Another early e-mentoring program, which Muller was aware of, was one run out of Hewlett Packard by David Niels. His program branched into mentoring in Africa. It also mentored high school students. (Muller 2014)

[36] The term 'e-mentoring' was coined, at least in this context, in 1993 by Amy Mueller (no relation), a Dartmouth graduate who worked at AT&T and served on the advisory committee for the Dartmouth Women in Science Project. (Muller 2014)

[37] As a later MentorNet CEO, Mary Fernandez, explained the MentorNet process: "it was a very early version of eHarmony except for [being for] women in STEM fields. So you fill out a profile, you are algorithmically matched with another person and then our program is a guided mentoring program. So we guide and mentor and the protégé with discussion topics that are relevant to the mentees, level of their education and some of their personal experiences and that has evolved considerably over time." (Fernandez 2014)

From the beginning, MentorNet was not only about "fixing" the students who were receiving mentoring (known in MentorNet as 'proteges'), but also about teaching the mentors to understand what these women students were facing, in the hopes that they could fix the system in which these individuals studied and worked.[38] Muller called it a "two-way learning relationship." (Muller 2014)

> [I]t did seem to me that this was an opportunity that could enable us to have a much bigger impact than just being the band aid or "fix or equip the women", whatever you want to say, because of the interactions students would have with their mentors. We were careful in our evaluations to look at what the mentors were learning as well as what the students were learning. Another thing we were really trying to do… through our coaching curriculum and our training was to impart… more information about some of the causes and perpetuation of the imbalance by gender and/or other diversities in the field. So that even when mentors who signed up because they thought perhaps they could help 'clueless' people they might have their eyes opened about experiences to help them see that these individuals were not lacking. (Muller 2014)

At first, MentorNet was about providing e-mentoring experiences primarily for college women. Both prospective mentors and protégés completed online applications, indicating not only demographic characteristics, areas of study/work, and interests, but also expressing their preferences for the individual with whom each would be matched. This information fed a bi-directional matching algorithm that considered the needs and preferences of both mentors and protégés.

> The initial program connected undergraduate and graduate students with professionals working in industry or government, providing different coaching and training for mentors and for the protégés, depending upon the students' educational levels. Soon, we were hearing from those who wanted to find additional mentors for academic careers, beyond their advisors. We slowly began to realize they were expressing the value of having *external* mentors. With a grant from NSF's ADVANCE program, we rebuilt our systems once again to accommodate a mentoring option for students and for early career academics (both post-docs and pre-tenure faculty members), linking them with faculty members as their mentors. At about the same time this new capability was deployed, we had sufficient mentors, protégés, and experience to move our matching to an "on demand" year-round system, rather than linking the timing to an academic calendar.

---

[38] Mary Fernandez, a regular mentor for MentorNet and today the CEO, explained the impact that a mentoring relationship could have on the mentor: "I think what the mentor is faced with is often having to understand their own choices in their career, in their professional and personal lives, as a way of helping the mentee or the protégé understand those choices for themselves. So there is a high degree of introspection that the mentor goes through. I think successful mentors do this quite naturally. And it's fascinating and … the upshot of this is that what we find – and this has been reported over and over and over again – the mentor feels a great sense of personal fulfillment, they feel an increased connection to their profession, their profession brings more meaning to them… It's the way of you understanding why is it that that you are doing what you are doing, what joy does it bring you and what frustration does it bring you. So almost everyone who I have spoken to, especially what we call our master mentors, the mentors who have been with us for years and years and years, … they all report that they feel they have grown and sometimes grown more than the mentee or protégé, which is very interesting because we focus of course on the needs of the protégé." (Fernandez 2014)

[T]hen five years into it…we rebranded the whole thing very deliberately as the e-mentoring network for diversity in engineering and science. And so we really deliberately changed MentorNet's identity, went through all the materials and really turned it around. We recognized we wanted to partner with, and could have stronger relationships with, the whole community of underserved people in engineering and science, all of whom could benefit from MentorNet. (Muller 2014)

One of the early challenges for MentorNet was to get the colleges these students attended to consider a cooperative arrangement in which all would help support the system financially:

[W]e were trying to create a collaborative among many different colleges and universities with the primary champions of the idea in those colleges and universities often being people who didn't have a lot of influence. They tended to be [, for example,] the directors of women and engineering programs. And so that was challenging all by itself. They were used to thinking of the other colleges and universities as competitors, not collaborators. (Muller 2014)

Another issue was overcoming the disbelief that mentoring could be effective over email. As Muller remembers:

[U]sing email by itself was new for a lot of people, using the web was new. But using electronic communications for mentoring, I gave numerous talks and wrote papers earnestly justifying in great detail the ways in which people actually could build relationships with others they had never met face to face. At the time [mid to late 1990s], many people just felt that building productive relationships via email alone was pretty unlikely. (Muller 2014)

Muller argued that there were differences between e-mentoring and face-to-face mentoring, and that each had its own strengths and weaknesses.[39] In these pre-Skype and Facebook days, the advantages of face-to-face mentoring included the ability to read facial expressions and body language that "people find quite rewarding and stimulating and humanizing," which could not be duplicated in a written exchange. However, e-mentoring had its own advantages.[40] Email is user-friendly and widely available. Being asynchronous, it is useful for communicating across time zones or between people who have different work and study schedules. When one takes time to compose an email message, one can be more thoughtful and deliberate. Moreover, the email communication provides a lasting record that the protégé can return to

---

[39] Muller (2014) points out that, today, there are opportunities to get some of the advantages of face-to-face mentoring while interacting remotely online through the use of videoconferencing, and that some recent mentoring programs blend online and in-person contact between the mentor and the protégé. Also see Muller (2002).

[40] Muller (2014) points to the scholarship on mentoring and its use in workplace settings by Kathy Kram, Belle Rose Ragins, Stacy Blake-Beard, and Lois Zachary as being particularly insightful. She also praised Rhodes (2002) work in evaluating youth mentoring programs. MentorNet also benefited considerably from the active scholarship on mentoring and writing undertaken by its first program manager, Peg Boyle Single, who led authorship of a number of papers and studies based on the MentorNet work.

read many times.[41] (Muller 2002) Muller also noted some more subtle advantages of using email:

> [T]he early opportunities for people to mentor across race and even across gender without having visual cues to remind them or even in some cases let them know that the person they are mentoring was quite different from them, contributed to building stronger relationships and better understanding across differences... [P]articularly where status differences are concerned, e-mentoring flattens hierarchy as any electronic communications does and enables discourse on a more level field for the mentor and the protégé. Email conversations often work better for people who are a little more shy or reserved and for those who need to take some time whether because English isn't their first language or for other reasons want more time to compose their thoughts... [T]he process of writing down, which of course is what you do in an email, your concerns, your questions enables you to take a step farther in solving your own problems and in identifying what it is that you are really trying to achieve. (Muller 2014)

MentorNet faced funding challenges. Muller knew that the initial foundation support would come to an end and was not likely to be renewed. Clearly, the business model had to rely upon support from corporations and others interested in building a diverse STEM workforce. But what was the best way of selling this program to industry? The organization had "terrible ups and downs in founding cycles, the worst of it [was]...in about 2001, when the Internet bubble had popped." (Muller 2014) But there were other funding challenges. When the stock market crashed after 9–11, or when there was a natural disaster somewhere in the world, it was hard for MentorNet to attract foundation funding. Corporations, similarly, were reluctant to invest in the workforce through MentorNet when the stock market was weak. In one particularly trying time, MentorNet had to lay off its entire R&D team.

MentorNet also faced technological challenges. The technology involved building an interface, an automated algorithmic process for matching mentors and protégés, as well as providing an online facility for training, coaching, and program communications "while we were working with first hundreds and then thousands of mentoring pairs and doing it in an economically feasible way." (Muller 2014) While the technology does not seem remarkable in our current world of social networking, MentorNet's software platforms were built mostly by a single employee (Stephanie Fox) linking email, homegrown databases and a proprietary complex, dynamic matching algorithm, a number of off-the-shelf Microsoft applications, and building what is now called a "customer relationship management" system, at a time well before the existence of most of the social networking sites that we know today.[42] Each iteration of matching and related evaluation helped the team to improve the processes for more and more successful mentoring relationships.

---

[41] Muller has authored or co-authored more than 40 papers on e-mentoring. In addition to the one cited here in the text, we mention only two more recent ones that concern populations that are double minorities, in these cases women of color: Muller et al. (2012), Blake-Beard et al. (2011).

[42] "We did apply for a patent, but the actual write-up, done on a *pro bono* basis by a law student, ... [W]hile the patent was still pending, both Stephanie and I left MentorNet, so the follow-up was left to David Porush, and I gather he wasn't able to follow through successfully." (Private communication from Carol Muller, 24 February 2016)

Muller learned a great deal about mentoring through her experiences at MentorNet. She noted that 'mentoring' is a word that encompasses many different possible activities. She pointed to Sheryl Sandberg's riff on Dr. Seuss in her chapter entitled "Are You My Mentor?" in her book *Lean In*. (Sandberg 2013) Muller also observed that it is important for there to be clear, shared understanding of the expectations in a mentoring relationship so that there is not dissatisfaction on the part of the mentor or the protégé. The staying power of mentors is known to be particularly important with at-risk youth who have sometimes been let down by adults in their lives in the past and who may be particularly disheartened by the failure of a mentoring relationship. Muller expressed a certain amount of agreement with Sheila Wellington (2001), the former president of the non-profit organization Catalyst, which advocates for better workplace environments for women, who argues that people can learn by themselves many of the things that they rely on mentors for. Muller also pointed out the value of what the late Margaret Ashida, an executive at IBM and STEMx who was an early leader in broadening participation in computing, had called 'penalty free mentoring', where a student could have a mentor who was someone other than their advisor, "where you could ask dumb questions and not [be] worried that somebody was going to think they were dumb; or you could confide that you are thinking of dropping out or starting a family or other things that students often rightly might have assumed their advisors would take in a negative way or maybe would reduce their chances for success in their academic environment." (Muller 2014)

In 2008, after 12 years of founding and leading the organization – a much longer period of time than she had ever anticipated staying – Muller decided to leave MentorNet to pursue new learning and other interests. The board selected David Porush as the next chief executive. He was founder of the program in electronic media at Rensselaer Polytechnic Institute; Executive Director of Learning Environments for all the campuses of the State University of New York, where he had been a leader in developing online degrees; and co-founder and chairman of the social networking organization SpongeFish.

Porush made wholesale changes in the organization. He received grants from the Sloan and Bechtel Foundations to do strategic planning and platform re-engineering for the aging technology used by MentorNet.[43] He laid off the remaining staff and outsourced technology development overseas. In 2008, just after Muller left, MentorNet received NSF funding for a grant proposal she had written to extend the organization's mentoring to the geosciences (Porush 2010). In 2011 MentorNet was opened to all students in science and engineering, not just those from colleges and universities that were formal partners of MentorNet.

---

[43] The original MentorNet technology had been cutting-edge when first developed by Stephanie Fox, and it was robust enough to handle the scale growth of MentorNet over its first decade; but the original technology did not reflect the rapid advances in networking technologies that occurred during MentorNet's first decade. (private communication from Carol Muller, 24 February 2016)

Additional major changes came in 2013, when MentorNet chose Mary Fernandez as CEO.[44] Fernandez was committed to mentoring because of her own personal experiences. She had benefited from a mentor throughout her doctoral student years in computer science at Princeton University, provided through a program funded by AT&T for women and minority STEM students. During a summer internship at AT&T Bell Labs and while still a Princeton graduate student, she had a very positive mentoring experience with her mentor, the famous Bell Labs computer scientist Brian Kernighan.

And I had – as many graduate students do, … some bumps in the road along the way. I wasn't quite sure if I was going to make it through and I had difficulty with my topic area. The person I intended to study with left [Princeton] not long after I arrived. Just one thing after another. Through the crisis of confidence, and Brian really helped me overall with those bumps although he would tell you that he didn't do anything at all. But I was really affected by having this incredible resource available to me, this objective person… who is really [standing] in my corner. (Fernandez 2014)

Upon graduation, Fernandez joined AT&T Bell Labs (soon renamed as AT&T Labs), first as a research scientist and later as a research manager. AT&T was a sponsor of MentorNet, and it was natural that Fernandez would work through MentorNet to give back to the community. Over the years, she served as a mentor to 17 students through MentorNet.

Fernandez's primary contribution to the organization so far has been to modernize the technology on which MentorNet operates.[45] As she explained about the durable technology that had powered MentorNet during Muller's tenure:

Carol in many ways, in fact … is quite the visionary because MentorNet predates open social networks, Facebook, LinkedIn; it predates pervasive mobile communication; it predates everything that we do today and … the interaction the digital natives are so accustomed to – the people who are maybe … 25 and younger. The way they interact in the world, their virtual and physical world that are really in some sense one. … [T]hey are very accustomed to this way of communicating and interacting, whereas [for] those of us who are digital immigrants, there's still to a certain degree some foreignness to it. (Fernandez 2014)

---

[44] In Fernandez (2013), the new CEO of MentorNet discusses her 15 years of experience as a mentor.

[45] Fernandez notes that the new technology created during her tenure at MentorNet could be used for other purposes than women and minorities in science and engineering in the United States. She has been contacted both by organizations that want to broaden STEM participation in countries outside the United States and by U.S. organizations interested in unrelated issues. While she believes that the MentorNet platform could work for these organizations, Fernandez is not pursuing these opportunities at this time – for scale reasons. However, she is also cautious in her response to these other organizations, cautioning them that they will not be successful simply by applying this platform, that "the hard part is figuring out programmatically how to serve the needs of that target community and that is where subject matter expertise around the needs of your target community are absolutely critical, right. So I think it's the case and there has been huge amount, there is huge body of research, social science research around effective mentoring." (Fernandez 2014)

But the original technology "was not robust enough to handle the scale and growth of MentorNet and did not reflect the rapid advances in social networking technologies." (private communication from Mary Fernandez, 27 March 2016)

In creating the architecture for the new MentorNet technology, Fernandez drew on her experience with Software As A Service at AT&T, borrowed from LinkedIn "the paradigm of being presented with people who might be of interest to you and engaging in a protocol for inviting them and for starting a relationship," and copied Facebook's ability for online communication in order to facilitate chat. While MentorNet has a custom interface, the goal was to make it familiar to young people who were experienced with the popular social networking sites.[46]

One of the advantages of the new technology is that it enables the MentorNet staff to continually add or change the questions that are being asked of mentors and protégés. This enables scalable, large data collection and the use of statistical analysis and machine learning techniques to aid in formal evaluation, e.g. to determine which attributes of a mentor-protégé relationship are good predictors of positive mentoring outcomes. These attributes can be more finely structured than in MentorNet's earlier years so as to do targeted mentoring (or targeted research on mentoring) in the same way that advertisers do targeted advertising or customized content.

I like to say that gender is too blunt an instrument or too blunt an attribute for understanding any individual, just like race is in some sense too blunt a dimension ... let's say you have a Latina woman who is first generation in her family to attend college. She lives in the Southwest with a lower socioeconomic status. She is attending a two-year community college with the hope of transferring to a four-year university. Because of the circumstances of her family she does not have a lot of exposure or very limited exposure to the career opportunities that might be available to her if she were to get a two-year degree or a four-year degree ... That's a complex person. She has many different characteristics.

Now I will take another Latina woman who lives on the Upper East Side of New York City. Her mother is an anesthesiologist and her dad works at Goldman Sachs. She is interested in getting a degree in computer science because it seems like a really hot topic. She went to a private school. ... Now if you were to just focus on women in the same bucket because they happen to be women and they happen to be Latina, Hispanic women, you're not comparing apples and apples here, certainly not with respect to the likelihood that [this] person is going to succeed in [her] academic trajectory.

So the vision for our program over time is that we have the ability to understand who that person is really from a multidimensional standpoint – from many different perspectives – and to ask them both based on the knowledge that we have about people with those various characteristics as well as engaging them in the question about what are your biggest challenges. (Fernandez 2014)

LinkedIn was MentonNet's primary sponsor of the new mentoring platform and program.[47] A short-term goal is to connect the data collected by MentorNet with the

---

[46] Fernandez also says that, on the organizational level, MentorNet is following the model set out by Michael Wu in *The Science of Social* (Wu 2012). In particular, MentorNet is following Wu where the "question is really a matter of how, in business, you would call your 'go to market strategy', how are we going to establish the strategic relationship that allow our acquisition gear to be very efficient [for example, to reach a particular Hispanic population]." (Fernandez 2014)

[47] NIHGMS (NIH General Medical Sciences) is MentonNet's largest overall sponsor.

data collected by LinkedIn to make the mentoring process more effective. There are commonly more protégés than mentors, and sometimes there is a pool of protégés who have not yet been matched. If one does data analysis on the unmatched pool of protégés, one can determine the attributes of the mentor types most in demand. One can then do direct mentor recruitment with some specificity on the LinkedIn platform. (Fernandez 2014)

In addition to changing the platform, there have been some important changes in the mentoring process – all based in social science research results from the past 15 years. One is that they encourage a new mentor and protégé to connect with one another by videoconference (e.g., Skype or Google Hangout) in the first week after they are matched together – even if they plan to carry out all further interaction through some asynchronous means of communication such as email. This is done because research has shown that "if you see a person's face and hear their voice early on … this [is] … like an imprinting… [Y]ou feel a stronger sense of connection and also you can often discover quickly that maybe you are [they are] … not necessarily going to be good match." (Fernandez 2014) Similarly, Fernandez has introduced formal training for mentors in how to mentor effectively based on recent social science literature on organizational development. Working with an outside consulting firm that worked for corporations, the obligatory training for mentors now teaches them about the method of Socratic questioning and the power of storytelling as a mentoring tool.[48] "So we guide the mentor and the mentee with discussion topics that are relevant to the mentees' level of education and personal experiences. The topics have evolved considerably over time." (Private communication from Mary Fernandez, 27 March 2016)

Fernandez has noted a sizable increase in interest for mentors in the computing disciplines. This is tracking in consonance with the Bureau of Labor projections that 70% of new STEM jobs over the next decade will be IT jobs.

In 2014, MentorNet became a division of Great Minds in STEM (GMiS), a national non-profit operating out of Los Angeles founded in 1988 to provide STEM awareness programs in underserved communities. Under this arrangement, Fernandez continues on as the President of MentorNet and reports to the CEO of GMiS (N.A. 2014).

---

[48] Fernandez has observed that in a previous generation, people were generally homophilic: mentors wanted to mentor protégés like themselves – women mentoring women, Hispanics mentoring Hispanics, computer scientists mentoring computer scientists. "The younger generations are not self identifying as strongly with respect to their ethnic or cultural heritage. They are more fluid in their own identity which is fascinating. And I saw this at AT&T before I left in fact. Specifically, in our employee resource groups. Employee resource groups traditionally have been founded around racial identify, Hispanic, Latino, Asian, South Asian, Pacific Islander et cetera. And what a lot of corporations are finding is that their younger employees don't self identify in that way such that it's important with respect to their professional development. So that actually kind of changes the way that people cluster. And so it's interesting for us because our mentors are of one generation and our protégés are of a different one. So I think that will be a ongoing, definitely an ongoing exploration." (Fernandez 2014)

References

Abbate, Janet. 2012. *Recoding gender: Women's changing participation in computing*. Cambridge, MA: MIT Press.

Aspray, William. 2016. *Participation in computing: The National Science Foundation's expansionary programs*. London: Springer.

Bird, S.J., and C.J. Didion. 1992. Retaining women science students: A mentoring project of the association for women in science. *Initiatives* 55(3): 3–12.

Bix, Amy. 2004. From 'Engineeresses' to 'good engineers': A history of women's U.S. engineering education. *NWSA Journal* 16(1): 27–49.

Bix, Amy Sue. 2013. *Girls coming to tech! A history of American engineering education for women*. Cambridge, MA: MIT Press.

Blake-Beard, S., M.L. Bayne, F.J. Crosby, and C.B. Muller. 2011. Matching by race and gender in mentoring relationships: Keeping our eyes on the prize. *Journal of Social Issues* 67: 622–643.

Brainard, Suzanne. 2015. Oral history interview by William Aspray. Charles Babbage Institute Oral History Collection, June 1.

Cunningham, Christine M., Mary L. Pavone, and Carol B. Muller. 1996. Factors influencing women's pursuit of a college science major or science career: An evaluation of the Women in Science Project (WISP). In *Proceedings of the women in engineering conference: Capitalizing on today's challenges*, ed. S.S. Metz, 289–294. Hoboken: Stevens Institute of Technology.

Daniels, Jane. 2015. Oral history interview by William Aspray. Charles Babbage Institute Oral History Collection, June 11.

Daniels, Jane, Sabina Bajrovic, and Nicole M. DiFabio. 2011. Why women and men joined SWE over the last 60 years. *Journal of the Society of Women Engineers* 5: 50–59.

Didion, Catherine Jay. 1995. Mentoring women in science. *Educational Horizons* 73(3): 141–144.

Didion, C.J., M.A. Fox, and M.E. Jones. 1998. *Cultivating academic careers: AWIS project on academic climate*. Washington, DC: Association for Women in Science.

Edwards, P.N. 1990. The army and the microworld. Computers and the politics of gender identity. *Signs: Journal of Women in Culture and Society* 16(1): 102–127.

Ensmenger, N. 2010. *The computer boys take over: Computers, programmers, and the politics of technical expertise*. Cambridge, MA: MIT Press.

Estrin, Thelma. 1992. Oral history interview by Frederik Nebeker. IEEE History Center, August 24–25.

Estrin, Thelma. 2002. Oral history interview by Janet Abbate. IEEE History Center, July 19.

Estrin, Thelma. 2006. Oral history interview by Deborah Rice. Society of Women Engineers, March 16.

Fernandez, Mary. 2013. A path between: Mentoring the next generation of computing professionals. *Computing Research News* 25(9). http://cra.org/crn/2013/10/a_path_between_mentoring_the_next_generation_of_computing_professional/. Accessed 20 Nov 2015.

Fernandez, Mary. 2014. Oral history interview by William Aspray. Charles Babbage Institute Oral History Collection, September 24.

Fort, D.C. 1995. *A hand Up: Women mentoring women in science*, 2nd ed. Washington: Association for Women in Science.

Fridkis-Hareli, Masha. 2011. A mentoring program for women scientists meets a pressing need. *Nature Biotechnology* 29: 287–288.

Fritz, W.B. 1996. The women of ENIAC. *Annals of the History of Computing, IEEE* 18(3): 13–28.

Grier, David Alan. 2005. *When computers were human*. Princeton: Princeton University Press.

Hacker, Sally. 1981. The culture of engineering: Woman, workplace, and machine. *Women Studies Quarterly* 4: 341–353.

Hawthorn, Paula. 2002. Oral history interview by Janet Abbate. IEEE History Center, July 5. http://ethw.org/Oral-History:Paula_Hawthorn. Accessed 20 Nov 2015.

Hays, Gwen. 2010. Oral history interview by Sheldon Hochheiser. IEEE History Center, February 17.

Home-Douglas, Pierre. 2009. Yes, WEPAN. *Prism Magazine*, March: 40–43.

Homsher, Betsy. 2011. Priceless treasures: The society of women engineer's archives as a source for American History. *Journal of the Society of Women Engineers* 5: 13–21.

Jenniches, Suzanne. 2003. Oral history interview by Lauren Kata. Society of Women Engineers, May 29.

Jenniches, F. Suzanne. 2010. Oral history interview by Frederik Nebeker. IEEE History Center, April 13.

Kata, Lauren. 2011. The boundaries of women's rights: Activism and aspiration in the society of women engineers, 1946–1980. *Journal of the Society of Women Engineers* 5: 36–49.

Leggon, Cheryl B. 2006. Women in science: Racial and ethnic differences and the differences they make. *Journal of Technology Transfer* 32: 321–329.

Leggon, Cheryl B. 2010. Diversifying STEM faculties: The intersection of race, ethnicity, and gender. *American Behavioral Scientist* 53: 1013–1028.

Leggon, Cheryl B., and Troy Eller. 2011. Women in engineering: The illusion of inclusion. *Journal of the Society of Women Engineers* 5: 83–92.

Light, Jennifer. 1999. When *computers were hum*an. *Technology and Culture* 40(3): 455–483.

Meiksins, Peter, Anne M. Perusek, and Tanya Zanish-Belcher. 2011. Curtiss-Wright engineering cadets: 21st century questions and issues. *Journal of the Society of Women Engineers* 5: 22–35.

Metz, Susan. 2015. Oral history interview by William Aspray. Charles Babbage Institute Oral History Collection, September 25.

Metz, Susan Staffin, Suzanne Brainard, and Gerald Gillmore. 1999. National WEPAN climate study exploring the environment for undergraduate engineering students. *Proceedings 1999 international symposium of technology and society, women and technology: Historical societal, and professional perspectives*, 61–72.

Misa, Thomas J. 2011. *Gender codes: Why women are leaving computing*. Hoboken: Wiley.

Muller, Carol B. 2002. MentorNet: Large scale e-mentoring for women in science, engineering, and technology (SET) fields. *Proceedings of the 2002 eTEE conference, E-Technologies in engineering education*. August 11–16. Davos, Switzerland.

Muller, Carol. 2014. Oral history interview by William Aspray. Charles Babbage Institute Oral History Collection, November 17.

Muller, Carol B., Pavone, Mary L., and Wetterhahn, K. E. 1996. Toward parity: A model campus project for support and systemic change. *Society of women engineers annual convention technical paper*.

Muller, C.B., S. Blake-Beard, S.J. Barsion, and C.M. Wotipka. 2012. Learning from the experiences of woman of color in MentorNet's one-on-one program. *Journal of Women and Minorities in Science and Engineering* 18(4): 317–338.

N.A. 2014. 'MentorNet joins the great minds in STEM family' announced at 26th annual HENAAC conference. 24–7 Press Release. http://www.24-7pressrelease.com/press-release/mentornet-joins-the-great-minds-in-stem-family-announced-at-26th-annual-henaac-conference-394465.php. Accessed 11 Dec 2014.

Oldenziel, Ruth. 1999. *Making technology masculine: Men, women, and modern machines in America*. Amsterdam: Amsterdam University Press.

Porush, David. 2010. The e-mentoring network for diversity in science and engineering. In *Cases on online tutoring, mentoring, and educational services: Practices and applications*, ed. Gary A. Berg, 12–22. Hershey: Information Science Reference.

Puaca, Laura Micheletti. 2008. Cold War women, professional guidance, national defense, and the society of women engineers, 1950–1960. In *The educational work of women's organizations, 1890–1960*, ed. Anne Meis Knupfer and Christine Woyshner, 57–77. New York: Palgrave Macmillan.

Puaca, Laura Micheletti. 2014. *Searching for scientific womanpower: Technocratic feminism and the politics of national security, 1940–1980*. Chapel Hill: University of North Carolina Press.

Rhodes, Jean. 2002. *Stand by Me: The risks and rewards of mentoring today's youth*. Cambridge, MA: Harvard University Press.

Rossiter, Margaret. 1982. *Women Scientists in America: Struggles and Strategies to 1940*. Baltimore: Johns Hopkins.

Rossiter, Margaret. 2012. *Women scientists in America: Forging a new world since 1972*. Baltimore: Johns Hopkins University Press.

Sandberg, Sheryl. 2013. *Lean in: Women, work, and the will to lead*. New York: Knopf.

Sproull, Lee, and Sara Kiesler. 1991. *Connections: New ways of working in the networked organization*. Cambridge, MA: MIT Press.

Watford, Bevlee. 2011. On the intersection of race and gender. *Journal of the Society of Women Engineers* 5: 76–82.

Wellington, Sheila. 2001. *Be your own mentor: Strategies from top women on the secrets of success*. New York: Random House.

Wu, Michael. 2012. *The science of social: Beyond hype, likes, and followers*. London: Lithium Technologies.

Zuckerman, Harriet, Jonathan R. Cole, and John T. Bruer (eds.). 1991. *The outer circle: Women in the scientific community*. New York: Norton.

Chapter 7
Organizations That Help Underrepresented Minorities to Build STEM Careers

Abstract This chapter discusses organizations that help underrepresented minorities to build science and technology careers. Many of these organizations were founded in the 1970s in response to the Civil Rights movement. The first section discusses three organizations principally serving African Americans: the National Society of Black Engineers (NSBE), the National Action Council for Minorities in Engineering (NACME), and the National Consortium for Minorities in Engineering and Sciences. The second section discusses three organizations principally serving Hispanics: the Society for the Advancement of Chicanos and Native Americans (SACNAS), Latinos in Science and Engineering (MAES), and the Society of Hispanic Professional Engineers (SHPE). The third section discusses two organizations principally serving American Indians: the American Indian Higher Education Consortium (AIHEC) and the American Indian Science and Engineering Society (AISES).

This chapter discusses organizations that help underrepresented minorities to build STEM careers. Many of these organizations were founded in the 1970s in response to the Civil Rights movement. The first section discusses three organizations principally serving African Americans: the National Society of Black Engineers (NSBE), the National Action Council for Minorities in Engineering (NACME), and the National Consortium for Minorities in Engineering and Sciences. The second section discusses three organizations principally serving Hispanics: the Society for the Advancement of Chicanos and Native Americans (SACNAS), Latinos in Science and Engineering (MAES), and the Society of Hispanic Professional Engineers (SHPE). The third section discusses two organizations principally serving American Indians: the American Indian Higher Education Consortium (AIHEC) and the American Indian Science and Engineering Society (AISES).

© Springer International Publishing Switzerland 2016
W. Aspray, *Women and Underrepresented Minorities in Computing*,
History of Computing, DOI 10.1007/978-3-319-24811-0_7

7.1   Organizations Principally Serving African Americans

Three organizations were created in the 1970s to support African Americans study-ing science and technology disciplines. The National Society of Black Engineers promotes the study of African American engineering students at both the under-graduate and graduate level. The National Action Council for Minorities in Engineering principally uses fellowship programs to enable African American and other minority students to attend college to study engineering and also to encourage colleges and universities to improve their records of recruiting and retaining minor-ity engineering students at the undergraduate level. The National GEM Consortium focuses on graduate education of minority students in science and engineering.

7.1.1   National Society of Black Engineers (NSBE)

The National Society of Black Engineers is a large nonprofit organization devoted to increasing the number of Black engineers, enhancing their educational and career experiences, and enabling them to more effectively contribute to society. On their website (NSBE 2014), they list their objectives:

- Strive to increase the number of minority students studying engineering at both the undergraduate and graduate levels
- Encourage members to seek advanced degrees in engineering or related fields and to obtain professional engineering registrations
- Promote public awareness of engineering and the opportunities for Blacks and other minorities in that profession
- Function as a representative body on issues and developments that affect the careers of Black Engineers

The organization achieves these objectives through programs at the local, regional, and national levels including "tutorial programs, group study sessions, high school/junior high outreach programs, technical seminars and workshops, a national com-munications network (NSBENET), two national magazines (*NSBE Magazine* and *NSBE Bridge*), an internal newsletter, a professional newsletter (*Career Engineer*, a supplement in *NSBE Magazine*), resume books, career fairs, awards, banquets and an annual national convention "(NSBE 2014).

NSBE had its origins at Purdue University. Purdue graduated its first African American engineer in 1894 (in civil engineering), and its first African American graduate in electrical engineering in 1914. However, by 1965 the African American student population was still small – only 129 of more than 20,000 students. During the late 1960s, Purdue experienced nonviolent campus protests regarding the treat-ment of African Americans on campus and in society (as did many other college campuses). In 1969 a Black Cultural Center was opened on campus, in 1970 an interdisciplinary Afro-American Studies program was started, and in 1971 students gained the opportunity to major or minor in Black Studies (Purdue Libraries 2010). That same year two Purdue undergraduates, Edward Barnette and Fred Cooper, approached the dean of engineering about establishing a student organization to improve recruitment and retention of African American engineering students – in light of the fact that in the late 1960s only 20% of the African American freshman who matriculated in the engineering program graduated with an engineering degree. The dean was sympathetic, and the sole Black engineering faculty member, Arthur Bond, agreed to serve as the faculty advisor to this student group, which was known as the Black Society of Engineers.

The situation for African American engineering students at Purdue began to improve in 1974. That year, the Engineering Department started the Minority Engineering Program and the Black Society of Engineers began to grow. In 1975 a letter was sent to the president or dean of all 288 U.S. colleges and universities with accredited engineering programs, asking that they identify Black student leaders, organizations, and faculty who might have an interest in belonging to a national society serving the same function as Purdue's Black Society of Engineers. Approximately 80 schools responded – a number of which already had programs similar to those at Purdue – and 48 students, representing 32 schools, came together at a meeting at Purdue in 1975 to form the National Society of Black Engineers.

Over the years, NSBE steadily grew. By 1979 there were 88 chapters and by 1990 that number had reached 180. Today, there are almost 400 chapters and more than 30,000 members (including chapters and members outside the United States.) That first national meeting at Purdue grew into the annual national conference, with attendance that at times has reached 8000 people. The organization opened a head-quarters office in Alexandria, Virginia and hired its first full-time executive director in 1987, but much of the organization continues to be run by elected student officers. While the founders were all male students, female students began to actively participate and several have served as the national chair of the organization: Virginia Booth (1978–1980), Carolyn Cooper (1980–1981), Donna Johnson (1985–1987), Regenia Sanders (1997–1998), Chancee Lundy (2004–2006), and Stacyann Russell (2009–2010) (NSBE 2010).

NSBE plays an important role for African American undergraduate engineering students at both HBCUs and minority institutions across the entire country. While it has an Information Technology Think Tank to help NSBE members to enhance their IT skills and while it has occasionally published articles about opportunities in computer science (see, e.g. Barger and Addison (2008)), the society has had few programs or events focused specifically at its members interested in IT careers. One exception is the NSBE Hackathon, which was held at the University of Memphis in 2015. (http://www.memphis.edu/cs/news_and_events/news/2015_nsbe_hackathon. php) Another is having employers from the IT industry (along with employers from many other STEM areas) attend the 2015 annual NSBE/SHPE career fair.

7.1.2 *National Action Council for Minorities in Engineering (NACME)*

In 1974, a group of corporate executives, with encouragement from minority and science policy leaders and the academic community, formed and funded the National Advisory Council for Minorities in Engineering (NACME) in response to the call made in the Sloan report described in Chap. 1. Six years later, three other organizations with similar interests were merged into the organization, and the name was changed to its present one: National Action Council for Minorities in Engineering. Those three other organizations were the Committee on Minorities in Engineering, the Minority Engineering Education Effort, and the National Scholarship Fund for Minority Engineering Students.[1] While NACME also carries out K-12 programs, research activities, and engineering public policy advocacy and education, its emphasis is on scholarships for minority students pursuing engineering education and careers. Over its first 40 years, NACME has supplied financial support to more than 24,000 students at 160 colleges and universities, at a cost of more than $142 million.

Because few universities were actively recruiting minority engineering students, NACME funded the universities directly, with the understanding that these funds from the Incentive Grant Program would be used to recruit minority engineering students to their campus. Some 150 colleges and universities received funding from this NACME program.

NACME realized over time, however, that one weakness with this scholarship program was that it was focused solely on recruitment, while retention was lagging considerably at many of the schools they funded. So, NACME changed its funding model. Instead of sending the funds directly to the schools and have them provide funding to their students, a national competition was held in which individual students would apply directly for funding. Approximately 500 students were supported in this way. These students were typically high academic achievers with high college retention rates. NACME decided that, while this approach was good for a small number of academic achievers, it did not achieve NACME's goal of increased national enrollment or graduation of minority engineering students.

The third iteration of the scholarship program, the NACME Scholars Block Grant Program, returned to making awards to the schools rather than to the individual students. However, this time schools were required to pay attention to retention as well as recruitment. They had to record and report data, so as to inform NACME about both recruitment and retention. For a school to remain in good standing with NACME, 80% of the scholars had to persist to graduation (as did 90% of transfer students). Grants were made for 5-year renewable periods, but

---

[1] This is taken from the 40th anniversary program for NACME, supplied by Brit Byrnes, Marketing and Communications Manager, NACME (private communication with the author, 19 September 2014).

schools that did not achieve these percentages were placed on probation, renewed for shorter terms, or eliminated from the program.

Faculty and administrators were free to determine the particular means they thought would work best for retention in their schools; NACME simply monitored the percentage of students who made it through to graduation. Grants were not made to community colleges, but the 4-year colleges and universities were encouraged to enter into formal articulation agreements with local community colleges to feed minority students into their engineering programs. In the competition to receive grants in the first place, colleges were judged on four criteria[2]:

- *Recruitment*: Institutional leadership committed to recruiting and admitting promising students from high schools in underserved communities and 2-year colleges
- *Admissions*: Published admissions policies and procedures that reflect a holistic approach that goes well beyond SAT/ACT scores and high school GPA in evaluating student potential to complete the baccalaureate degree in engineering
- *Pre-matriculation enrichment programs*: Summer programs designed to enrich intellectual exchange and socialize students for participation in the life of the university and the engineering community
- *Community building*: A campus community of faculty, students, and administrators designed to increase student engagement and provide institutional support for the academic success of all students

NACME is primarily interested in engineering, but to the extent that computer science is taught within engineering schools, it has also fallen within the scope of NACME. Currently, 3 % of NACME scholars major in computer science or information systems technology, and 11 % major in computer engineering (Goode 2015). As part of its policy efforts, NACME has shown concern about the low placement rate of African Americans and Hispanics in Silicon Valley firms (Roach 2014).

7.1.3 National Consortium for Graduate Degrees for Minorities in Engineering and Sciences (GEM)

The National Consortium for Graduate Degrees for Minorities in Engineering and Sciences (GEM), which is "a network of leading corporations, government laboratories, top universities, and top research institutions that enables qualified students from underrepresented communities to pursue graduate education in applied science and engineering", was created in 1976 at the University of Notre Dame (GEM 2015). In 1972, J. Stanford Smith, a senior vice president at General Electric, had called for a ten-fold increase in minority engineering graduates within 10 years. In

---

[2] These criteria are taken verbatim from the Rating Sheet for NACME Block Grant Proposals (private communication to the author, September 19, 2014, from Aileen Walter, NACME's Vice President for Scholarships and University Relations).

1973, the National Academies held the conference on minority engineering described in Chap. 1. In 1974, Ted Harbarth, the affirmative action officer at Johns Hopkins Applied Physics Laboratory and later president of GEM, wrote a proposal for a national consortium; and later that year, 40 representatives from research centers, universities, and advocacy organizations met at Notre Dame to discuss methods for advancing minority participation in engineering. Several people from that meeting revised Harbarth's proposal, and it was distributed by the president of Notre Dame, Father Theodore Hesburgh, to 53 organizations.[3]

In 1976, the first six fellowships - for master's study in engineering - were awarded. Over time, the fellowship program continued to grow: 106 fellowships awarded in 1985, doctoral fellowships in both science and engineering added in 1990, and 223 fellowships (both masters and doctorate) awarded in 1998. Recently, more than 300 fellowships have been awarded each year, totaling more than 4000 fellowships since 1976. Throughout the program's history – up until today - more than 3000 of their fellowship students have received a master's degree in engineering or a doctorate in science or engineering.

GEM has attracted a distinguished set of presidents. They include John A. White, dean of the College of Engineering at Georgia Tech and later Chancellor of the University of Arkansas; Charles Vest, president of MIT; Kurt Landgraf, CEO of Educational Testing Service; Ronald Goldsberry, an African American chemist who held senior executive positions at Occidental Petroleum and Ford Motor Company; Juan Andrade, president of the U.S. Hispanic Leadership Institute; Alfred Grasso, CEO of MITRE Corporation; and Eric D. Evans, Director of MIT Lincoln Laboratories.[4]

GEM added the Faculty Bridge Symposium (now called the Future Faculty Professionals Symposium) in 1996. It brings together graduate students and junior faculty members from underrepresented groups each year for a 3-day symposium with senior faculty members, managers, and researchers. One aspect of the symposium is to provide advice to these young people about choices to be made during the remainder of their graduate education and about post-graduate careers – particularly for those who are planning to become faculty members. It is also intended to build lasting mentoring relationships and a peer network. Topics include "effective mentoring, conflict resolution, dissertation and grant writing, multiculturalism in the workplace, research opportunities, and successfully managing career transitions" (GEM 2015, "Future Faculty and Professionals").

In 2006, GEM created GRAD (Getting Ready for Advanced Degree) Lab, which is a 1-day event held at different locations over the course of the year, mostly on college campuses but sometimes at meetings of other minority organizations such

---

[3] This and the next paragraph are based on the GEM History page on the GEM organizational web pages (GEM 2015).

[4] Ted Habarth was the first president. Edward W. Seberger was president beginning in 1987. Michael L. Vaughan, the senior assistant dean of the University of Delaware College of Engineering, was interim president of GEM in 2009. GEM's offices were located at the University of Notre Dame until 2007, when they moved to the Washington, DC area.

as SHPE (described later in this Chapter). GRAD Lab reaches almost a thousand underrepresented STEM undergraduates each year and encourages them to pursue a graduate degree in science or mathematics. The lab brings in a range of people, from current graduate students to senior researchers and managers, to talk to undergraduates about why it is important to attend graduate school (both for the value that research brings to society, and for how it improves the careers and lives of graduates), how to prepare for graduate school, and what daily life is like for an intern or a researcher. The labs encourage and advise students to apply for a GEM Fellowship and admission to graduate school (GEM 2015, "Getting Ready for Advanced Degrees (GRAD) Lab").

For masters students, the GEM fellowship couples the student with a mentor during the paid summer internships. The master's fellowship pays – for up to 2 years – $4000 per semester in living stipend plus full tuition and fees (as well as two paid summer internships) if the student matriculates at one of GEM's University Members. For doctoral students, the fellowship program provides a $16,000 per year stipend, tuition, and fees at a GEM University Member institution, as well as an internship for at least one summer.[5] A senior scientist or engineer mentors each summer intern. The students awarded these fellowships typically have high educational attainment, with an average GPA that exceeds 3.5 (on a 4-point scale). GEM carefully matches the skills of students with GEM employer members, and over 80 % of the students accept job offers from these employers (GEM 2015, "About the GEM Fellowship Program").

GEM broadly supports science and engineering – its web pages identify 33 academic disciplines it supports. These disciplines include several that are closely associated with computing: computer science, electrical engineering, information systems, and operations research. A number of the GEM employer members indicate that computer scientists or computer engineers are among the top three majors that they recruit through GEM: Adobe, Aerospace Corp., Brookhaven National Laboratory, Caterpillar, Cisco, ExxonMobil, Fermi National Accelerator Laboratory, Georgia Tech Research Institute, IBM, Intel, Johns Hopkins University Applied Physics Laboratory, Lawrence Livermore National Security LLC, Lexmark International, MITRE, Motorola Solutions, National Renewable Energy Laboratory, Northrop Grumman, Oak Ridge National Laboratory, Qualcomm, Raytheon, SpaceX, and United Technologies Research Center (GEM 2015, "Employer Members").

---

[5] There are over 100 member universities. These include some Minority Serving Institutions (e.g. Howard, New Mexico State, Prairie View A&M, Tuskegee, Puerto Rico at Mayaguez, Texas at El Paso) and many majority institutions, including elite institutions (e.g. Cal Tech, Carnegie Mellon, Cornell, Harvard, MIT, Princeton, Rice, Stanford, Berkeley, and Yale) (GEM 2015, "University Members").

7.2 Organizations Principally Serving Hispanics

In 1973 and 1974, three organizations were formed to advance the opportunities for Hispanic scientists and engineers: the Society for Advancement of Hispanics/ Chicanos and Native Americans in Science (SACNAS), Latinos in Science and Engineering (MAES), and the Society of Hispanic Professional Engineers (SHPE). SACNAS is primarily oriented towards graduate education and research careers, whereas MAES and SHPE are primarily focused on professional degrees and engineering careers.

7.2.1 Society for Advancement of Hispanics/Chicanos and Native Americans in Science (SACNAS)

SACNAS is a scientific society with the goal of advancing the success of Hispanic and Native American scientists in their postsecondary education and their careers. It was formed in 1973, incorporated in 1986, and approved for tax-exempt status the following year.[6] According to the SACNAS website, the organization has three specific goals:

- To increase the number of Hispanics/Chicanos and Native Americans with advanced degrees in science and the motivation to be leaders.
- To increase the number of Hispanics/Chicanos and Native Americans in science research, leadership, and teaching careers at all levels.
- To increase governmental commitment to advancing Hispanics/Chicanos and Native Americans in science resulting in increased resources, elimination of barriers, and greater equity (SACNAS 2014).

In 1972, with funding from the National Institutes of Health, a professor of biochemistry at the University of New Mexico Medical School, Alonzo Atencio,

---

[6]There are issues concerning SACNAS's name. The SACNAS website addresses these issues as follows: "The name under which the organization was founded and incorporated was Society for Advancement of Chicanos & Native Americans in Science, which remains the legal name today. In 1973, the nomenclature for US born individuals of Mexican heritage was "Mexican American." The term "Chicano" was adopted by various "Mexican Americans," including many SACNAS founders, who self-identified as members of a social-political movement—the "Chicanismo" initiative of the early 1970s. Over the years, as times, demographics, and language have evolved, SACNAS has in practice inserted the word "Hispanic" into its name, in order to reflect a broader and more inclusive ethnic demographic within underrepresented minorities. "Hispanic" and "Latino" are sometimes used interchangeably in reference to the same populations. SACNAS chose the word "Hispanic" because that is the designation used by the United States Census Bureau." (http://bio.sacnas.org/uploads/Marketing/SACNAS_History.pdf, accessed 4 December 2014)

convened a meeting of Hispanic and Native American scientists in Albuquerque.[7] At that meeting, it was decided to form a Federation of Chicano and Native American Scientists.[8] At a follow-up meeting later that year, the SACNAS name was chosen. In 1973, SACNAS held its first annual meeting in Atlantic City, NJ, co-located with the meetings of the Federation of American Societies for Experimental Biology. Approximately 50 people attended. When the newly created board of directors met that year, it was decided to create a pre-doctoral Graduate Fellowship Program to recruit Hispanics and American Indians into doctoral study[9] (Gonzales 2014).

SACNAS has grown steadily over the past 40 years. The annual national conference, which first attracted 50 people, now has attendance near 4000. There are currently some 25,000 members, including some 7000 dues-paying members.

One of the most important events of the year is the annual conference. About two-thirds of the attendees are undergraduates, but the meeting also hosts graduate students, postdocs, professors, and industry representatives and exhibitors. Attendance is typically about 60 % Hispanic, with the other 40 % evenly divided between American Indian, African American, and Caucasian participants. It is an intensive 4-day event that includes scientific paper and poster sessions, a graduate student symposium, professional development workshops, and cultural events. The NIH has been the principal funder in recent years, so that there has been increasing emphasis in the program on the biological sciences. The Central Intelligence Agency provides computer support and equipment for the conference. Universities located near the conference site are also typically sponsors and host visits by conference attendees to their campuses as a way to showcase their institution in order to recruit both students and faculty.

There is a philosophy within SACNAS that, while scientific advancement is the goal, culture and communication are keys to the success of Hispanics and American

---

[7] Attendees at that initial meeting, known as the Founders, were: Alonzo Atencio, Ciriaco Gonzales, Ruben Duran, Arthur Diaz, Ricardo Griego, Don Ahshapanek, Vicente Llamas, Bill Rivera, Robert Pozos, Reynaldo Morales, Jose Martinez, Richard Tapia, Fred Young Begay, Zenaido Camacho, Sigfredo Maestas, Orlando Cuellar, and Eugene Cota-Robles – all of them doctorates. The SACNAS website describes these Founders: "Many of them were the first people in their communities to receive PhDs in science, the first Hispanics/Chicanos and Native Americans to be hired in their departments, and the first mentors for a new generation of Hispanic/Chicano and Native American scientists. Over the years, SACNAS and its founders have flourished. Founders are now leaders at federal scientific agencies, tenured full professors, and university deans." (Gonzales 2014).

[8] A creation myth is told on the SACNAS website to explain the small beginnings: "Legend says that SACNAS was founded in an elevator at an American Association for the Advancement of Science (AAAS) meeting in the early 1970s. At that time, there were only a handful of Native American and Chicano scientists in the U.S., and most of them had converged to attend the AAAS meeting. After attending a networking event, they all got into the elevator together. One looked around and joked, "If this elevator crashes, it will wipe out the entire population of Chicano and Native American scientists!" (http://sacnas.org/about/our-history, accessed 4 December 2014).

[9] The board of directors at that time included two American Indians (Don Ahshapanek and Fred Young Begay) and seven Hispanics: Eugene Cota-Robles (president), Ruben Duran, Alonzo Atencio, Orlando Cuellar, Ciriaco Gonzales, Arthur Diaz, and Richard Tapia.

Indians and are not to be neglected. Thus there is a premium on face-to-face meeting and on celebrating cultural heritage at the annual conference and other events throughout the year.

SACNAS hosts a variety of other activities, including regional meetings, in addition to the annual meeting.[10] Professional and (88) student chapters sponsor activities throughout the year. The professional chapters run mentoring programs and networking events. In 2002, SACNAS won the National Science Board Public Service Award, with special recognition awarded to its mentoring activities. In 2004 the organization won the Presidential Award for Excellence in Science, Mathematics and Engineering Mentoring. Scholarships, internships, and fellowship are regularly awarded to students; and over the past decade SACNAS has provided scholarships to 4700 students to attend the annual meeting. To advance professional development, SACNAS offers an online listing of jobs, awards, and career prospects. It also has built up resources for students in the form of articles, videos, and tools to aid career planning and professional development.

SACNAS holds an annual Summer Leadership Institute, started in 2009 with the help of the husband-and-wife team of Robert Barnhill and Marigold Linton, both of whom have held senior offices within SACNAS. Through a highly selective process, 10 postdocs, 10 early career, and 10 mid-career faculty members and professionals are chosen.[11] The goal is to help these 30 individuals develop their careers by working on skills that will enhance their abilities in research or science administration. The institute is an intensive, week-long course, usually held in Washington, DC, and taught by psychologists who specialize in leadership. Barnhill and Linton, together with two other former SACNAS presidents, serve as mentors. One of the common topics is how to recognize and deal with ethnic and other kinds of bias. Part of the process is a bonding exercise, and each year's cohort of 30 stays in touch through a listserv. The alumni from these leadership institutes meet each year at the annual meeting (Barnhill 2015). While it is perhaps too early to do a formal evaluation of the impact of the Leadership Institute, Barnhill believes that it has made a difference in the participants' careers: "some of the first [graduates] have done really well since then and have indeed become deans and become better known in their scientific specialties. That itself is hard to measure, as you know, but there've been some really outstanding successes" (Barnhill 2015).

SACNAS also engages in science policy development and advocacy.[12] The basic principle behind its policy work is that it is important to have everyone with the talent

---

[10] On SACNAS's programs to interest young people in science and engineering, see Horwedel (2006).

[11] When asked what SACNAS looks for in selecting people to participate in the Leadership Institute, Barnhill replied: "What difference it would make if they were to know a lot more about leadership? What they are doing and what they want to do and whether this is a reasonable thing where leadership would help them achieve that? That's tough to judge but anyway, we do the best we can" (Barnhill 2015).

[12] SACNAS has operated a small office in Washington, DC for about a decade. It is housed at the national headquarters of the American Chemical Society. ACS is the largest professional society focused on an individual science or engineering discipline, and it has been a generous colleague in

to do so participating in the scientific enterprise. Barnhill argues that that one reason that Hispanics and American Indians are under-served in the science community is the lack of awareness in the federal government because these two groups are geographically the most remote from the District of Columbia, with large populations of both groups located in the American Southwest. So one goal of the SACNAS policy program is to increase involvement of its 25,000 members in federal science policy. This means promoting SACNAS members to serve on NIH panels and NSF panels and boards, and for these representatives to bring back to their universities or companies knowledge of how Washington works. Barnhill points to the competitive funding process in federal agencies: "some of this is not ability ... so much as it is presentation on proposals. If you've never seen [a funding proposal] before, you're pretty unlikely to write one that's going to be a winner in today's competitive climate with only about 15 or 20 % of proposals being funded at these agencies" (Barnhill 2015).

Another goal is to have funds set aside so that the tribal colleges can compete for funds. They do not currently have the strength to compete successfully against major research universities such as Michigan or Harvard. "They don't have to have complete set aside to maintain a competitive environment but they have to have kind of different leagues [from the Harvards and Michigans], if I may use that sports analogy" (Barnhill 2015).[13]

A third goal is to promote infrastructure programs that enable the Hispanic-serving institutions and tribal colleges to participate more fully in national science initiatives. This includes various NSF programs including the Tribal Colleges and Universities program (TCUP); other Minority-Serving Institution programs; the education, opportunity and training aspects of the national computer alliances (EOT-PACI); and the Rural System Initiatives.

While there are many distinguished scientists among the early leaders of SACNAS, the one who stands out to the computing community is Richard Tapia. He is a member of the National Academy and a former member of the National Science Board. He was also one of the Founders of SACNAS. He attended the initial organizational meeting, served on its first board of directors, and was appointed vice president in 1975. He has spoken several times at the annual conference but he has not been actively involved in the organization in recent years[14] (Barnhill 2015).

---

helping out smaller STEM-related organizations interested in having a presence in Washington. For example, ACS also houses the office of the Council of Scientific Society Presidents.

[13] Barnhill (2015) points here to the NIH's Ginther Report (Ginther et al. 2011), which shows that African-Americans and Hispanics receive awards at a much lower rate than others. Barnhill laces his discussion with the thoughts on disruptive innovation of Harvard Business School professor Clayton Christiansen.

Barnhill notes that the situation is somewhat different for Hispanics than for American Indians because some of the Hispanic Serving Institutions, notably the University of Texas at El Paso, are significantly stronger research institutions than the tribal colleges, but even UTEP cannot compete with the likes of Michigan or Harvard.

[14] By one account, Tapia approached SACNAS in the 1990s about getting them to sponsor a conference solely on computation open to all people of color and not just to Hispanics and American Indians. This idea never reached fruition (Gates 2014).

SACNAS is of course interested in all of the sciences, but it has a long tradition of interest in the computational sciences because of the background of Richard Tapia and some other early leaders in the organization. Some 15 years ago, mathematicians began to hold a day-long mathematics institute prior to the regular SACNAS annual conference, and then there would be sessions throughout the regular conference designed to be of interest to the mathematicians. Over time, those sessions have grown to be well attended. Following on the model of the mathematicians, the computer scientists are now working to have a workshop before the conference and special tracks during the regular SACNAS meeting.[15]

7.2.2   Latinos in Science and Engineering (MAES)

In 1974 Robert Von Hatten, an aerospace electronics engineer with TRW Defense Space Systems in Redondo Beach, California, formed the Society for Mexican American Engineers and Scientists, today called Latinos in Science and Engineering but still using the acronym MAES. Von Hatten was stimulated to form MAES through his volunteer work for several years in programs intended to reduce the number of high school dropouts. The organization's official mission is "to promote, cultivate, and honor excellence in education and leadership among Latino engineers and scientists" (MAES 2014). The organization is today headquartered in Houston.[16]

MAES sponsors a number of activities and events targeted at middle and high school students. Its Science Extravaganzas are 1-day events organized by its student and professional chapters around the country. These events are intended to involve fun, hands-on activities for children – teaching them about science and engineering, and introducing them to potential role models including college students and professional scientists and engineers. The goal is to encourage students to complete high school and attend college. Parents are also a target audience, with information about college financial aid provided at the Extravaganzas. MAES also supports the annual 2-day Texas Science and Engineering Festival, which has a similar purpose to the Extravaganzas.

MAES sponsors programs targeted at college students. The annual Symposium, for which there is grant support covering the registration and travel costs for needy students, includes various activities of interest to students such as research poster session open to high school, undergraduate, and graduate students as well as to STEM professionals; and a College Decathlon, in which various student chapters from around the country compete against one another in ten events. The National Leadership Conference workshop teaches communication, leadership, and

---

[15] See the discussion in Aspray (2016) in connection with the account of the Computing Alliance of Hispanic Serving Institutions.

[16] This account has been written primarily by pulling material from the MAES web pages (MAES 2014).

teamwork skills to STEM professionals. MAES provides competitive scholarships to attend college. It has organized student chapters on 40 2- and 4-year campuses across the country, and these chapters carry out ongoing activities during the academic year. Students receive the twice-a-year *MAES Magazine*, which covers topics of interest to the Latino community about STEM education, careers, and research.

MAES serves its professional members through the annual symposium, the magazine, and an awards program, as well as through the activities of local professional chapters. The organization covers the entire gamut of engineering. However, in regions where there is a strong computing presence, this is reflected in the MAES membership and activities. For example, the Austin, TX chapter of MAES, founded in 1998, has a large number of professional members from the semiconductor, computer manufacturing, and software development industries (http://www.maes-texas.org/welcome.html, accessed 7 January 2016).

7.2.3 Society of Hispanic Professional Engineers (SHPE)

In 1974 a small group of engineers who worked for the city government of Los Angeles founded the Society of Hispanic Professional Engineers. It was intended to be a "national organization of professional engineers to serve as role models in the Hispanic community." (http://national.shpe.org/index.php/about) To build its membership, the founders sent letters to the 500 Hispanic engineers licensed in California, leading to 50 new professional memberships.[17]

In the 1970s, "when other minority organizations were aggressively addressing societal issues, SHPE conservatively focused on professional issues" (SHPE 2014b, p. 28). The primary goal was networking, and one of the first activities was to form student chapters at local universities: first at the University of Southern California and California State University at Los Angeles, next at California State Polytechnic University at Pomona and California State University at Northridge. The student chapters steadily grew: 15 by 1980, 97 by 1990, 125 by 1995, 205 by 2004, and more than 250 today. The numbers would have been even higher in the earlier years, but the organizers did not have the resources to establish chapters at community colleges until the year 2000.

Professional chapters were organized in 1981 in Dallas and Chicago. The number grew over time – to 32 in 1990, 41 in 2004, and more than 75 today. Corporate chapters came later, with the first formed at Lockheed Martin Aeronautics in Ft. Worth, Texas. Today there are approximately 7000 student members and 1800 professional members. While there are only a few corporate chapters, between 40 and 50 companies support SHPE's activities in a typical year.

---

[17] This account of SHPE is drawn closely from its website (SHPE 2014a) and the organization's 40th anniversary issue of its magazine (SHPE 2014b).

Another early activity was the formation in 1976 of the SHPE Foundation to raise and award scholarships for students pursuing careers in science and engineering. In 2014, the Foundation awarded $269,000 in scholarships.

In 1978 SHPE held its first seminar – a single-day event in Los Angeles attended by a few hundred students and 50 companies. This evolved into SHPE's annual conference. Until 1992, all of these conferences were held in Los Angeles. Indeed it was not until 1990 that outsiders viewed SHPE as a national organization. By 1992 there were 2000 registrants at the annual conference, and it included a major job fair with on-site interviewing. Today, the meeting lasts 4 days and typically attracts 4500 registrants and 300 exhibitors.

In 1980 SHPE received a grant from the U.S. Department of Education to conduct a program for high school students. It involved visits by working engineers to high schools, tutoring sessions for students who expressed an interest in pursuing a science or engineering career, and shadow-an-engineer days. This project evolved into SHPE's Advancing Careers in Engineering program under which the central organization channels small grants to local chapters for various K-12 programs. A more recent addition to SHPE's K-12 activities is the Pre-College Symposium, which is co-located with the annual conference and typically offers hands-on STEM activities, a financial aid workshop, and a college fair to some 1000 local middle school and high school kids.

Between 1984 and 1994, SHPE carried out eight international technological exchanges, mostly with organizations in Mexico. This exchange program included several technology and trade conferences held along the U.S.-Mexico border, as well as an international student conference in Costa Rica in 1990. Today, there are active SHPE chapters in Mexico, Colombia, and the Dominican Republic.

With the rapid expansion of student chapters in the 1980s, SHPE wanted to ensure that the student leadership of these chapters was strong. In 1986 it organized the National Student Leadership Development Conference in Chicago, attended by 67 student chapter presidents. This became an annual event, changing its name in 2003 to the National Institute for Leadership Advancement, with a goal not only to make the student chapters stronger but also to teach leadership skills that these student leaders could employ when they progressed to the workplace. In 2005, SHPE added the Executive LeaderSHPE Institute for engineers with at least 10 years of work experience; and the following year it added a Management Growth Training institute for mid-level managers.

The 2014 annual conference, held in Detroit, provides a good snapshot of the range of SHPE activities. These included professional programs (continuing education and certification programs for people at the beginning, mid career, and executive levels), technical seminars (including some that focus on the likely job impacts of current science and engineering trends), an engineering research symposium, a distinguished lecture series, a corporate readiness program (joint with Johnson & Johnson, to help SHPE student leaders transition effectively into the professional workplace), a Latinas in STEM track (involving both networking and role models), a dean's summit (on best practices for recruitment, retention, and student college success), a pre-college symposium (to encourage high school completion and performance, and to steer students toward the choice of a STEM major in college), several competitions (best technical paper, extreme engineering 24-h student teams, and an Academic Olympiad with student teams competing in a quiz format), corporate tours, and a large career fair.

One student indicated the value of SHPE to him:

> … the Society for Hispanic Professional Engineers…[and] organizations like that help students have a support network to help you get through the tough times, the tough tests or tough class or professor, and who [to] take and don't take, and when to take this class versus that class. Those types of networking and support groups are very, very important, because you can also have study groups inside your classes, but sometimes you're more comfortable if someone's Latino…(Quoted anonymously in Taningco 2008)

SHPE is interested in all engineering areas and is not focused specifically on computer engineering. Nevertheless, computer engineering is an interest of SHPE. At its 2015 annual conference, the keynote speaker for the graduate luncheon lecture was Jose Martinez, a computer scientist from Cornell University. In 2015 SHPE partnered with GitHub, the popular platform for building software, on a new initiative called SHPECodes. This initiative is intended to recruit both new student memberships and attendance at the annual SHPE conference among students majoring in computing-related disciplines. (Eligible under this promotion are "Majors, Minors, and Certifications in: Computer Science, Computer Programming, Software Development, Electrical & Computer Engineering, Information Technology, Information Systems, Video Game Design, Web Application Development, Network Administration, Database Administration" (http://shpe-codes.org/faq/, accessed 7 January 2015)).

7.3 Organizations Principally Serving American Indians

The final section of this chapter discusses two organizations created in the 1970s to support American Indian students interested in science and technology.[18] The American Indian Higher Education Consortium (AIHEC) represents the interests of

---

[18] One organization that we have not discussed in this chapter is the National Institute for Native Leadership in Higher Education. Here is a synopsis of its history:

> NINLHE was founded in 1993 and established as a 501(c)(3) corporation by Dr. Colleen Larimore and Dr. Jim Larimore, the former directors of the Native American programs at Dartmouth College and Stanford University. With funding from the Intel Foundation, NINLHE began as a strategic, self-help coalition of a dozen directors from the most successful Native student retention programs in the country. Though the directors faced nearly identical challenges on respective home campuses, at that time, each worked in isolation from one another. They realized that by working together they could do much to help themselves, and by extension, all of the students. By coming together through an organization as NINLHE, it offered the opportunity to expand the impact to effect much needed change in Native education practice and policy at the institutional and national levels as well. NINLHE's success as a national professional higher education organization since its incep-

the tribal colleges and universities. Particular interests of AIHEC include agriculture, science, and information technology – all of which are viewed as critical to the advancement of American Indians in the modern world. The American Engineering Science and Engineering Society promotes interest and opportunities in science and engineering from pre-college through professional career levels.

7.3.1   American Indian Higher Education Consortium (AIHEC)

The first tribal colleges were founded in the late 1960s, and by 1972 there were six of them.[19] They were all seriously underfunded (mostly through short-term, discretionary federal grants), and they had not so far paid much attention to each other's efforts and predicaments. This is somewhat understandable, given that each of these schools represented a different tribe, and many of the tribes had longstanding rivalries. In 1972 David Risling (a Hoopa involved in the founding of D-Q University), Gerald One Feather (an Oglala Lakota who helped found Oglala Lakota College), Helen Scheierbeck (a Lumbee who worked for the U.S. Office of Education), and Pat Locke (a Hunkpapa Lakota who worked for the Western Interstate Commission on Higher Education in Denver) convened a meeting of the tribal college presidents in Washington, DC.[20] At that meeting the group formed the American Indian Higher Education Consortium with the purpose of bonding together to lobby for federal support of all tribal colleges and universities. AIHEC continues to this day to be the lobbying arm of the American Indian higher education community – not only for funding but also for educational policy. (AIHEC 2014)

AIHEC succeeded in its initial lobbying efforts when Congress passed and President Carter signed into law the Tribally Controlled Community College Assistance Act in 1978, which authorized federal support of $4000 (later increased) for each American Indian student enrolled in the tribal college.[21] At the time, many of the tribal colleges were in a precarious financial position; and this legislation

---

tion has been achieved through the support of several philanthropic organizations. These include the W.K. Kellogg Foundation, the Educational Foundation of America, the David & Lucile Packard Foundation, the GE Foundation, the Daniels Fund, and the Lumina Foundation for Education. NINLHE partnered in the past with several sponsors that included: the University of Northern Colorado, the University of New Mexico, Dartmouth College, the University of Washington, Stanford University, Buffalo State College, and the Mohegan Tribe. (http://www.unco.edu/ninlhe/history.html)

[19] This history of AIHEC is drawn primarily from Boyer (1998).

[20] The tribal colleges then in existence included Navajo Community College, Hehaka Sapa College of D-Q University, Oglala Sioux Community College, Sinte Gleska College, Turtle Mountain Community College, and Standing Rock Community College.

[21] See Pease-Pretty on Top (2003) for an excellent account of the politics of passage of the Tribally Controlled Community College Assistance Act. Passage was anything but straightforward. Also see Gipp (2009) for an account of the early years of AIHEC by one of the participants.

helped to stabilize the tribal college movement. The passage of the bill was complicated because the fight over this legislation occurred at the same time as the political
battle over Indian self-determination, and many who were fighting for self-
determination did not support legislation that routed funding directly from the federal government to the tribal colleges without offering any control to the tribal
governments.[22] In its early years, AIHEC also played an important role in advising
tribes that wanted to create new colleges. It has continued to provide advice to tribal
college administrators on various administrative and technical infrastructure issues.

While the organization thrived in the 1970s, a hostile political environment in
Washington in the 1980s, during the Reagan Administration, meant that AIHEC had
cash-flow problems and had to scale back operations. This was a time when the
actual Congressional appropriations were much lower – sometimes less than a
third – of the amount authorized by the Tribally Controlled Community College
Assistance Act.

During the 1990s, public and Congressional support for American Indians grew.
AIHEC was able to rebuild its staff and redouble its lobbying efforts. Two major political victories of this decade were to get Congress to pass legislation[23] in 1994 recogniz-
ing the tribal colleges as land-grant institutions - making them eligible for new
funding – and convince President Clinton to sign an Executive Order in 1996 directing
federal agencies to strengthen their relations with the tribal colleges.[24] The following
year the Carnegie Foundation published an upbeat report about the tribal colleges,
Native American Colleges: Progress & Prospects (Boyer 1997a; also see Boyer
1997b). Enrollment at the tribal colleges and universities, which had totaled less than
2500 students in 1982, steadily grew to where the number exceeded 30,000 by 2003.

Today, AIHEC continues its work in public policy, advocacy, research, and programmatic activities for the 37 tribal colleges and universities that are its members.
The colleges represented by AIHEC have more than 75 campuses, located across
15 states, with students drawn from more than 230 tribes.

AIHEC collects, organizes, retains, and makes available to its various constituencies a wide collection of materials (project documents, contact lists, Internet
links, calendars, databases, etc.) through password-protected online portals. One of
the most important portals is AIMS, the American Indian Measures for Success,
created by AIHEC in 2004 with support from the Lumina Foundation. It provides
qualitative and quantitative metrics for the success of the tribal colleges in carrying
out their missions. Other portals include one for chief academic officers, one that
provides resources for culturally sensitive Indigenous evaluation, one that provides

---

[22] The most vocal opponent on these grounds of undermining tribal self-determination was Patricia
Locke from the Western Interstate Commission of Higher Education. Another opponent was
Congressman Gerald Ford from Michigan, who was opposed to segregated higher education and
who believed it was not cost effective to expend funds on the education of a small minority group.

[23] This was part of the Elementary and Secondary Education Reauthorization Act passed in October
1994.

[24] This Executive Order also led to the creation of the President's Board of Advisors on Tribal
Colleges and Universities, as well as to the White House Initiative on Tribal Colleges and
Universities.

resources for faculty and staff in the STEM disciplines, one intended to increase student success rates at minority-serving institutions that is supported by the Walmart Foundation and run jointly with the Hispanic Association of Colleges and Universities (HACU) and the National Association for Equal Opportunity in Higher Education (NAFEO, which serves the Historically Black Colleges and Universities),[25] and a virtual library that contains resources about Native Americans.

AIHEC continues to broker STEM educational activities among the tribal colleges and universities. For example, in 2010 it partnered with Northwest Indian College, Dine College, Tohoma O'odham Community College, Haskell Indian Nation University, and the College of the Menominee Nation to create an introductory curriculum on climate change. The purpose behind this curriculum was to interest students in STEM disciplines early in their college careers. The courses are using the infrastructure of Facebook and YouTube, which many of the students are already familiar with (Pember 2010).

We have already discussed in Chap. 5 the AIHEC Virtual Library Project as well as AIHEC's efforts to address the digital divide and enhance the computing infrastructure in tribal colleges and universities. Up until now, information technology has been more of a tool in the tribal colleges than a principal object of study in its own right.

7.3.2   American Indian Science and Engineering Society (AISES)

In 1977, Los Alamos National Laboratory scientist Arnold Anderson – a Mohawk – and five other American Indian scientists formed the American Indian Science and Engineering Society. AISES's goal is to increase the number of American Indians and Alaska Natives in STEM education and careers.[26] AISES has wide representation across the American Indian community. More than 200 tribal nations are represented in AISES. There are 3000 members, and 184 college and university chapters and 13 professional chapters have been formed. 160 affiliated schools are teaching more than 55,000 K-12 students. AISES has received substantial support from both government and industry, including both high-tech companies such as Google and Intel, and regional companies centered in Indian Country such as Burlington Northern Santa Fe.[27]

---

[25] In 1999 AIHEC, HACU, and NAFEO formed the Alliance for Equity in Higher Education. It represents more than 350 minority-serving institutions.

[26] The other founding members were Al Qoyawayma, Carol Gardipe, George Thomas, Jerry Elliot, and Jim Shorty.

[27] This account of AISES has been written primarily from material on the AISES website. One can piece together parts of AISES's program from its house magazine, *Winds of Change*. Also see Begay-Campbell (2002). For a different take on how AISES, AIHEC, and other organizations enrich STEM education on an American Indian campus, see Oglala Lakota College (2013).

In 1978 AISES introduced its premier event, a national conference, which has become an annual event. It brings together high school juniors and seniors, undergraduate and graduate students, high school and college teachers, and representatives from government and industry. Attendance at a recent conferences reached 1600 from across the United States. The conference includes social and professional networking, mentoring, research presentations, award ceremonies, professional development workshops, various focused meetings, campus tours, and a large career fair.

At the K-12 level, AISES supports STEM education through "teacher training, regional science bowls, science fairs, leadership development, mentorship, scholarships, internships and other programming designed to support students and their families." (AISES 2014) The purpose is to enhance the quality of pre-college education and events in the STEM fields for American Indians, and to encourage high school students to pursue a college education and continue their interest in the STEM fields.

At the college level, AISES programs are focused on providing access to STEM programs for American Indian students, and to enhance the success of these students. Activities include "college chapters, regional and national conferences, leadership development, mentorship, scholarship, internships and career resources." (AISES 2014) For example, a number of the college chapters of AISES, including College of the Menominee Nation in Wisconsin, and Navajo Technical University and Southwestern Indian Polytechnic University in New Mexico, have actively participated in rocket design competitions sponsored by NASA. (Davis 2014)

In 2014 AISES received a 5-year grant from the NSF to support a program entitled Lighting the Pathway to Faculty Careers for Natives in STEM. It is a national effort to provide American Indian students with "motivating encouragement, practical skills and intellectual experiences that will help compel them to stay in their chosen STEM field so that they will earn the necessary academic credentials to land tenure track positions at U.S. colleges and universities" (NSF 2014). Participants study one of biology, computer science, chemistry, engineering, mathematics, geology, or astronomy at the bachelor, graduate, or postdoctoral level. Students receive individualized mentoring from an American Indian who holds a Ph.D. The program includes a monthly virtual seminar in which the students learn about how to succeed in their academic coursework, apply to graduate school, write a resume, balance work and life, and integrate American Indian culture into their research. Each student is given an opportunity to participate in research and present research results in a conference. The students all attend the annual AISES conference to network with American Indian faculty and work on developing other skills.

At the professional level, AISES focuses on supporting its members who are pursuing STEM careers. This support is provided at the early career, mid-career, and executive levels through professional chapters, regional conferences, awards programs, professional development workshops, employment services, and chances for these professionals to mentor students (AISES 2014).

The computing disciplines are just one of many science and engineering areas of interest to AISES. The National Science Foundation grant mentioned above identifies computer science as one area to be pursued. Fellowship support from Intel awarded through AISES supports students studying in areas of interest to Intel (computer science, computer engineering, and electrical engineering, as well as chemical engineering, and material science), while a Google fellowship program administered through AISES supports students in the areas of computer science, computer engineering, and management information systems.

References

American Indian Higher Education Consortium. 2014. *Organizational website*. http://www.aihec. org. Accessed 5 Nov 2014.

American Indian Science and Engineering Society. 2014. *Organization website*. http://www.aises. org.

Barger, Theresa Sullivan, and Eric Addison. 2008. Blacks in Computer Science: The Secrets of their Success. *Career Engineer*, National Society of Black Engineers, November/December: 46: 48–50

Barnhill, Robert. 2015. *Oral history interview by William Aspray*. Charles Babbage Institute Oral History Collection, January 16.

Boyer, Paul. 1997a. *Native American colleges: Progress and prospects*. The Carnegie Foundation for the Advancement of Teaching. San Francisco: Jossey-Bass.

Boyer, Paul. 1997b. First survey of tribal college students reveals attitudes. *Tribal College: Journal of American Indian Higher Education* 9(2): 36–41.

Boyer, Paul. 1998. Many colleges, one vision: A history of the American Indian higher education consortium. *Tribal College: Journal of American Indian Higher Education* 9(4): 16–22.

Davis, Tom. 2014. *Oral history interview conducted by William Aspray*. Charles Babbage Institute Oral History Collection.

Ginther, Donna K., Walter T. Schaffer, Joshua Schnell, Beth Masimore, Faye Liu, Laurel L. Haak, and Raynard Kington. 2011. Race, ethnicity, and NIH research awards. *Science* 333(6045): 1015–1019.

Gipp, David M. 2009. The story of AIHEC. In *Tradition and culture in the millennium: Tribal colleges and universities*, ed. Linda Sue Warner and Gerald E. Gipp, 7–16. Charlotte: Information Age Publishing.

Gonzales, Ciriaco. 2014. *SACNAS – The beginning*. http://sacnas.org/about/our-history. Accessed 4 Dec 2014

Goode, Robin White. 2015. Diversifying the field of engineering. *NACME in the News*, May 1. http://www.nacme.org/news/articles/129-diversifying-the-field-of-engineering. Accessed 7 Jan 2016.

MAES. 2014. *MAES: Latinos in science and engineering*. http://mymaes.org. Accessed 29 Oct 2014.

National GEM Consortium (GEM). 2015. *Organization website*. www.gemfellowship.org. Accessed 13 Jan 2015.

National Science Foundation. 2014. *Education and workforce program*. http://www.nsf.gov/funding/pgm_summ.jsp?pims_id=13396&org=CNS. Accessed 20 Oct 2014.

National Society of Black Engineers. 2010. *The history of NSBE*. http://www.nsbe.org/NSBE/media/Files/Official%20Downloads/NSBEHistory61610.pdf. Accessed 16 June 2016.

National Society of Black Engineers. 2014. *NSBE mission and objectives*. http://www.nsbe.org/About-Us/NSBE-Mission-Objectives.aspx#.VD6u8b4ea4g. Accessed 15 Oct 2014.

Pease-Pretty On Top, Janine. 2003. Events leading to the passage of the Tribally Controlled Community College Assistance Act of 1978. *Journal of American Indian Education* 42(1): 6–21.

Pember, Mary Annette. 2010. Initiative grants access to STEM curriculum: AIHEC partnership helps build capacity to develop and sustain STEM education and research programs. *Diverse Issues in Higher Education*, March 18: 7.

Purdue Libraries, Archives and Special Collections. 2010. *Or the fire next time: A timeline of African American history at Purdue*. http://collections.lib.purdue.edu/orthefirenexttime/ Or%20the%20Fire%20Next%20Time.pdf. Accessed 4 Apr 2016.

Roach, Ronald. 2014. Experts: Minority talent available for Silicon Valley. *NACME in the News*, October 20. http://www.nacme.org/news/articles/106-experts-minority-talent-available-for-silicon-valley-2. Accessed 7 Jan 2015.

Society for Advancement of Hispanics/Chicanos and Native Americans in Science (SACNAS). 2014. *Organization website*. http://www.sacnas.org. Accessed 4 Dec 2014.

Society of Hispanic Professional Engineers. 2014a. *History and mission*. http://national.shpe.org/ index.php/history-mission. Accessed 3 Sept 2014.

Society of Hispanic Professional Engineers. 2014b. *SHPE: The Official Magazine of the Society of Hispanic Professional Engineers*, Spring. Celebrating 40 Years of Changing Lives Anniversary Issue. http://www.nxtbook.com/nxtbooks/shpe/spring14/#/0. Accessed 15 Dec 2014.

Taningco, Maria Teresa V. 2008. *Latinos in STEM professions: Understanding challenges and opportunities for next steps*. Los Angeles: Tomas Rivera Policy Institute, http://files.eric.ed. gov/fulltext/ED502064.pdf. Accessed 16 June 2016.

Chapter 8
Organizations That Help Women to Build Computing Careers

Abstract This chapter discusses organizations with the mission of increasing the numbers of, and support provided to, women in computing-related education and work. Four organizations are profiled in the chronological order in which they appeared. The Anita Borg Institute (ABI), created in 1987, is strongest in its work on behalf of professional women in computing. The Computing Research Association Committee on the Status of Women in Computing Research (CRA-W), created in 1991, focuses on women graduate students in computing disciplines as well as women computing researchers in both universities and industrial research labs. In 1993, the computing professional organization ACM established its Committee on Women in Computing. Its scope includes undergraduate education and IT workers generally, not just computing researchers. The National Center for Women & Information Technology (NCWIT), founded in 2004, is the largest of these four organizations and has the widest scope, with interest in female students, academics, industrial workers, and entrepreneurs.

This chapter discusses organizations with the mission of increasing the numbers of and support provided to women in computing-related education and work. We will cover four organizations in the chronological order in which they appeared. The Anita Borg Institute (ABI), named after the late computer scientist Anita Borg who started these activities in 1987, is strongest in its work on behalf of professional women in computing. The Computing Research Association Committee on the Status of Women in Computing Research (CRA-W), created in 1991, focuses on women graduate students in computing disciplines as well as women computing researchers in both universities and industrial research labs. In 1993, the computing professional organization ACM created its Committee on Women in Computing. Its scope is broader than that of CRA-W, with more attention to undergraduate education and to IT workers more generally, not just to computing researchers. The National Center for Women & Information Technology (NCWIT), founded in 2004, is the largest of these four organizations. It has the widest scope of the four organizations, with interest in women students, academics, industrial workers, and entrepreneurs. All of these organizations continue to this day, and there are various types of cooperation and collaboration between them, so that there are many parallels and even some overlap between these accounts.

© Springer International Publishing Switzerland 2016
W. Aspray, *Women and Underrepresented Minorities in Computing*,
History of Computing, DOI 10.1007/978-3-319-24811-0_8

8.1   Anita Borg Institute (ABI)

In 1981 Anita Borg (1949–2003) completed her doctoral dissertation at the Courant Institute at New York University and joined Auragen, a start-up company building fault-tolerant operating systems.[1] In 1987, while in Austin, TX attending the ACM Symposium on Operating Systems, the leading operating systems conference, she took the first step at creating what became Systers, an electronic mailing list for computer scientists.[2] As Anita explained these origins in an oral history conducted by historian Janet Abbate (Borg 2001):

> I always look at the list of attendees to see how many women are there. Of four hundred attendees, there were only about thirty women. I ran into a friend of mine in the bathroom. We began talking about why there were so few women. Each time someone came in, she joined the conversation. It's a little bathroom with only two stalls, and we wound up with about eight of the women from the conference crammed in there, talking about it! We said, "You know, we should meet somewhere else. Why don't we try to get all the women at the conference to come to dinner together?" …We got, I think, all but two of the women, and it was great. …everything from graduate students, to me in the middle, to these senior women. It was so extraordinary that I collected their email addresses (for those who had [one] – not everybody had email at that point, or some had email, but it was just inside their companies) and set up a mailing list and came up with this funny little name.

At first, Systers was only open to women conducting research in the area of operating systems. Anita's close friend and later the CEO of the Anita Borg Institute, Telle Whitney, was originally turned down when she applied to join Systers because her research was in a different area of computer science. However, at the encouragement of Barbara Simons, another Ph.D. in computer science who eventually served as president of the ACM, Systers was eventually opened up to all technical women in computing. In Borg's mind, the golden years for Systers were when it numbered a few hundred people because it was then the right size to serve as a community. "Before Systers existed, there was no community of women in computing. It didn't exist. We all existed as individuals: we had a few women that we knew, but there was no community. There was no notion of how many women were out there, doing what" (Borg 2001).

At the time that Systers was formed, the technology was not capable of carrying out the tasks that Borg wanted it to perform. For example, most mailing systems assumed that all users were reading their email on the same hardware (a reasonable assumption for mail systems inside a corporation, but not true for of those who wanted to join Systers). So, Anita took on the task of building a mail system (Mecca) that would serve her needs. She wanted to build a system that not only handled

---

[1] This section is written primarily from the oral histories conducted by Janet Abbate with Anita Borg (2001) and Telle Whitney (2002), the oral history conducted by this author with Telle Whitney (2014), and Abbate (2012).

[2] While Systers was the first electronic mailing list for women in computing, it is not the only one. Others include ResearcHers organized by Susan Landau in 2004, PhdjobhuntHers and JrProfessHers organized by Rachele Pottinger in 2006, and ProfessHers created by Rebecca Wright in 2008 (Soffa 2011).

communication but also enabled individuals to have an online profile, much like a social networking site does today.

Over time, Systers grew and grew. As of 2014, it had approximately 4000 sub-scribers. At first it was primarily made up of faculty members and students from the academic sector, but over time a number of women working in industry also joined. The sense of an intimate community was lost as the numbers increased – even with technological improvements that enabled the formation of subgroups with like interests – and some of the original women on the list chose to become inactive on Systers.[3] New features were added, e.g. the Systers' Pass It On program, which provided grants to established technical women in computer science so that they could help women who were just beginning their technical careers.

The Systers website gives a sense of the purpose of the mailing list:

> Systers continues to serve this purpose by providing women a private space to seek advice from their peers, and discuss the challenges they share as women technologists. Many Systers members credit the list for helping them make good career decisions, and steering them through difficult professional situations. (Systers http://anitaborg.org/get-involved/systers/, accessed 6 October 2014)

Anita Borg apparently had to defend making Systers available open to women. In an article entitled "Why Systers Excludes Men" she argued that:

> Systers is not analogous to a private all-male club. It is different because women in com-puter science are a small minority of the community. It is different because systers is not interested in secrecy or in keeping useful information from the rest of the community. Useful messages regularly are made public after checking with the contributors. *It is unlikely that an underempowered minority will keep inaccessible information from the large empowered majority that has every means of communication available to it.* (Borg 1993, emphasis in original)

Borg goes on to list the reasons for keeping Systers female-only: women need (1) a place to find each other, (2) female role models and mentors, (3) a place to discuss their issues, and (4) to discover their own voice (Borg 1993; numbering added; drawn from the topical sentences throughout the text)

While some of the Systers discussion concerned women-specific issues, often involving career or workplace advice, other topics – in the TechTalk section – were about purely technical topics. Abbate has argued (2012, pp. 167–168):

> The FAQ for the main systers list also notes, "We allow people to ask for help on purely technical topics…because many women feel uncomfortable asking certain types of ques-tions on lists dominated by men." These comments reveal that what is usually regarded as "purely technical" discussion is actually gendered: the aggressive tone in which many online technical debates are conducted is felt as masculine and alienating by some women who would otherwise be interested in the technical content. Rather than concluding that it is impossible to be both feminine and technical or that women must accept and adopt the aggressive style as inherent in the nature of technical discourse, TechTalk challenges the masculine norm by offering a less confrontational mode as an equally valid alternative.

---

[3] There were, for example, ResearcHers for women in research careers and systers-entrepreneurs for women interested in entrepreneurship (Jeffries 2006).

Perhaps the best-known activity with which Anita Borg was involved is the Grace Hopper Celebration of Women in Computing, co-founded with Telle Whitney. The first of these conferences was organized in 1994 (Borg 1994). The conference had multiple origins. One strand involves the vision of Nico Habermann, the founding dean of the College of Computer Science at Carnegie Mellon University, who had agreed to join the National Science Foundation in 1991 as head (the official title is Assistant Director) of the computing directorate. He was keenly interested in enhancing opportunities for women in computer science, and at his initiative there had been a series of conversations during 1992 in Washington, DC about how to do so (Borg et al. 1994). Unexpectedly, Habermann died of a heart attack in August 1993.

One activity that came out of these discussions was the Windows of Opportunity Symposium for Female Students in Computing, held in May 1993. The symposium, sponsored by a grant from the National Science Foundation to Computing Research Association and hosted by Professor Dianne Martin at George Washington University, brought together 208 students from more than 100 universities to network and learn about careers and research and funding opportunities (Martin 1993). In his keynote address, Habermann challenged the computing community to increase the percentage of female graduate students in the computing disciplines to 45% by the year 2000. Anita Borg was one of the keynote speakers at the symposium.

Among Anita Borg's many activities, she was actively involved in CRA-W (described below), which had been formed in 1991 and was both studying and acting on issues related to women in computing research. She was also actively involved as a volunteer in ACM. She used both of these connections when she looked for funding to hold the first Hopper conference.

In her interview with Abbate, Borg mentions Marlene McDaniel as a direct stimulus to create the Hopper conference. McDaniel was working for a company that organized large trade shows, and she approached Anita about founding a conference of women in computing as a way to make money. Anita tracked her thinking about a possible conference:

> The whole idea of making money off all these women didn't appeal to me, but the idea of a conference was pretty interesting. I thought, First of all, wouldn't it be fun to get them all together and meet them? It would make such an even better experience for the community to get together. Then the question was, what kind of a conference could we have where somebody's boss would pay for them to go – where you could use money that wasn't out of your own pocket to go to this conference? And we thought: a technical conference. Wouldn't that be extraordinary, for women to get together and hear what everybody else is doing? So there were all these possibilities. One was, what would a technical conference that was almost all female be like?
>
> Would it be the same? I don't think so! I think it would be wildly different.
>
> So we planned, as close as we could, a top-notch technical conference.... I didn't know exactly what we were going to get...Some of our principles were: This was not going to be a half-assed conference. I've been to lots of get-togethers about women, and it's always on the cheap. This was not going to be on the cheap. This was going to be at a good place. It was going to be as high-class as SOSP or any other really good computer science conference: at a nice hotel, really well done. We figured we could do that if we raised fifty or a hundred thousand dollars. We raised two hundred and fifty thousand dollars! We had room

for four hundred people: we squeezed in four hundred fifty, and we had to turn away a hundred people. This was in Washington. It was extraordinary; it was absolutely extraordinary, and I am wildly, wildly proud of that, because a lot of people said it wouldn't happen. (Borg 2001)

As Valerie Barr, a professor of computer science who has attended every Hopper Conference, remembers the 1994 event:

[It] really was an amazing event. And part of what I absolutely loved about it was because it was only about 450 people. Which was amazing for everyone who was there, because who knew that there would be that many? We were all in this one auditorium the whole time, and basically it was one wonderful technical talk after another by all of these leading women computer scientists when most of us had no idea that there were even enough leading women computer scientists to give that many research talks. So it was really a great event. There were these fabulous talks, and then birds-of-a-feather sessions. And really just an incredible experience for the people who attended.[4] (Barr 2014)

At first, the conference was held every 3 years. In the early 2000s it started to be held every other year. Since 2006 it has been held annually. Long before the 2014 conference, for example, all 8000 attendee spaces had been sold out.

The conference always includes high-quality technical tracks. But it also includes student, academic, industry, and career tracks as well as a career fair. Other organizations piggyback on the conference, such as CRA-W holding career mentoring workshops and the Computer Science Teacher's Association holding computing teacher's workshops.[5] Borg's institute itself has piggybacked a Senior Women's Summit on the conference.

---

[4] Others who attended concurred with Barr. For example, Leah Jamieson (2015) remembers: "It really took my breath away to walk into that room and be in a room with 450 women." Camp (2015) remembers: "I was just blown away. I was so inspired by the senior women that were in the room. I had never had a senior female professor. I don't think I had a single one up to that point, at least no one's coming to mind. Many of them instantly became major role models for me. I got to meet Anita Borg. I remember when I met her, I was so nervous meeting her and she was just, at the end of our conversation, she gave me a hug. I'm glad I grabbed that confidence to go up and introduce myself, because that was a wonderful conversation and I ended up helping her with the next Grace Hopper in 1997, I was the Scholarship Chair for student scholarship."

[5] Originally, Anita Borg was a member of CRA-W and CRA-W was a co-sponsor of the conference. As Jamieson (2015) explains CRA-W's attitudes about the conference: "We are not getting along well. We're all in this space, but I think there was a sense of this was clearly an important gathering, an important forum… I think on the CRA-W side, I think it was more a stretch about what the connection with research was. My recollection is that CRA-W was not thrilled about being relegated to a no-role or very minor role and did watch, over time, as actually some of the collaborations came. For example, CRA-W would start to host some of the mentoring workshops and locate them with Hopper, which I think was the *rapprochemont* thing that happens. … Here's something that CRA-W does and does well. Divorcing it from Hopper was actually probably missing an opportunity, but it felt like a phenomenon of being in a crowded space and a whole bunch of organizations trying to claim some of that space." In the end, CRA-W signed over all rights and revenues concerning the Hopper conference to Borg's organization. In fact, the author (at the time serving as the executive director of CRA) of this book negotiated and signed this agreement with Borg.

The conference struggled with growing pains. Borg's institute experienced serious financial problems in 2004 (although it has been financially stable since 2006), and ACM stepped in to secure the finances; and in later years ACM continued to provide substantial funding for students to attend the conference.[6] Computing Research Association was also an initial sponsor, but like Borg's institute it did not have the financial wherewithal to underwrite the conference, and it was better for the sake of efficiency to have fewer organizers (Weyuker 2014). The early conferences were run entirely by volunteers, but as the conference grew in attendance it was necessary and prudent to bring in professional conference staff.[7] While some applaud the increase in the numbers of women attending the conference, others believe that this growth has come at a cost. The sense of intimacy that was so special at the first few conferences is lost when there are thousands in attendance. The organizers have had to work hard to maintain their preferred attendance ratio of half students in the face of companies wanting to sign up hundreds of their female employees. Some of the original members have decried the growing importance of the career fair, arguing that "the conference is becoming less valuable to faculty members because it is expensive and does not count in their departments as a research conference, and question whether … for the companies, [sending large numbers of employees has] becomes an easy way to say 'See, we really are trying to do something about diversity'"; also an easy way to recruit new women so as to bolster their diversity numbers without commitment to changing internal practices or paying women equal wages to men. If faculty attendance is dwindling over time, as Barr suspects, this means there is less opportunity to get out the message about the values of an academic career (Barr 2014).

As much as she loves the Hopper Conference, Barr has nagging doubts about its role in solving the problems facing women in computing:

> I was saying this to one of my ACM-W people the other day, we were talking about Hopper and the fact that it was 8000 and I predicted that their goal for next year would be 10,000. And then I asked one of the ABI VPs what she thinks for next year and she said, "Oh I think we'll get 10,000." I know that they're planning five years out in terms of venues. I think what will happen is that, and okay, this is the cynic in me again, I think that something like Hopper will be very big for the next few years, and there will be this incredible hiring frenzy and the companies that are, at the moment, only 17 % women on their technical staff

---

[6]As White (2015) tells this story: "What I do remember is how we pulled together and cemented the current relationship we have with Hopper, which dealt with some dark days for Anita Borg's Institute. She was extremely ill, dying. The Institute was in a precarious position. We had discussions of this on the ACM executive committee, and we decided to do two things. One was to provide ABI some funding so that they could survive, and two, we wanted as part of that to have a relationship with the Hopper conference going forward.

"We wanted to see the Hopper conference grow because it would be the main funding stream for ABI to stay alive and do its good work. If anything ever happened to the Anita Borg Institute or the conference was not being supported by the right groups, ACM would have the right to step in and sponsor the Hopper conference going forward. That was all understood and agreed to, and it was at that time ACM became the presenter of the Hopper conference, which has continued."

[7]For a few years after 2004, ACM also provided conference management services for the Hopper Conference.

will inch up toward 30 %, maybe 35 %. And then everybody will say, "Well, things are so much better." And something like Hopper will begin to shrink in size, because the hiring frenzy won't be necessary anymore. In five years we're going to look around and those of us who are sitting here now saying, "But we're still earning 86 cents on the dollar, and there isn't really meritocracy, etc, etc…" will be the ones who will still be going to Hopper in five years, in ten years, and saying, "We still have a problem. Just because the numbers are better doesn't mean the attitudes have changed." I would love to be proven wrong. But I certainly think that that's a plausible scenario. Sexism is really entrenched, as is racism. And I think it's going to take more than just a hiring frenzy to take care of it. (Barr 2014)

Since 1986 Anita Borg had been working in the California research labs of Digital Equipment Corporation. She felt that she had hit a glass ceiling in her career there, and in 1997 she used her contacts with John White and Mark Weiser, two of the principal members of Xerox PARC management, to convince them to allow her to relocate there and direct her own Institute for Women and Technology. The Institute subsumed responsibility for both Systers and the Hopper conference.

Already, Anita had been thinking about other projects associated with women and information technology. She had submitted an unsuccessful six-million-dollar proposal to NSF to support the Diversity Collection, a web-based database that collected information about every program in the United States intended to get women and minorities more actively involved in the STEM disciplines. She had received funding from the Kellogg Foundation to hold a meeting of women to read and discuss Pamela McCorduck and Nancy Ramsey's book, *The Futures of Women* (1997), which Anita claimed changed her life.

Borg was particularly interested in learning about how women use technology and how it impacts them, as well as to give women greater input in the design of technology. Beginning in 1999 with projects at MIT, Purdue, Santa Clara, and Texas A&M and eventually expanding to nine universities, this interest was carried out in the Institute through a program called Virtual Development Centers. The Institute worked with the universities and the local communities to put in place a model undergraduate course intended to engage both technical and non-technical students in the design of technologies that would serve the local community. Each university developed its own course. Some were associated with the introductory computer science or introductory engineering course, while others were upper-level courses for majors. Some of the schools had both male and female students involved, while other schools had women only.[8] Projects were carried out by student teams during

---

[8] For example, at Santa Clara University there were projects involving the use of RFID tags to keep track of children in day care, a self-cleaning house, and a project management and scheduling system for use by a family. (http://www.scu.edu/engineering/vdc/and linked pages). At Smith College, the VDC was an introductory engineering course, called Teach Our Youth (TOY) Tech, in which the Smith students used engineering principles to design fun, hands-on learning experiences for students at a public school in order to teach them about science and engineering. At the University of Arizona, the students helped a medical clinic to keep better track of patients. After consulting with the people at the clinic, the students recommended a helpful, but non-technical solution that involved rearranging the office so that workers had better access to the files they needed. At Purdue, an all-women class developed applications for a laptop they thought would be appealing to young girls. The University of Colorado at Boulder students developed websites for

the school year, and then the students were brought together in Palo Alto, CA at the end of the school year to present their results (Barr 2014; Jamieson 2015).

Anita was diagnosed with brain cancer and died in 2003 at age 54. She had two aspirations for her Institute that she was not able to realize. One was to broaden the scope from information technology to all technology; the other was to make the Institute international in its programmatic reach. Even today, nothing has been accomplished in broadening the Institute to all technology. Modest progress has been made on the international front, notably including the formation of the Grace Hopper Celebration in India in 2010.

In January 2002, Anita's friend Telle Whitney came in to help out her friend with the Institute on a temporary basis, first as an advisor to assist with fundraising, then as the interim president. Telle, who holds a PhD in computer science from Cal Tech and who had worked in several chip startup firms, had senior management experience as the vice president of the Canadian firm PMC-Sierra. She ended up staying on at the Institute as the CEO after Anita's death. This was surprising given that Telle saw herself more as a technologist than as a nonprofit executive focusing on women. Anita herself had never been interested in management, and the organization – renamed the Anita Borg Institute in 2003 – had to work hard to become economically viable. The VDC program, which Anita had held a strong personal interest in, was expensive to operate, hard to scale up, received diminishing interest from the academic partners, and was challenging to find sustaining funds for after the original HP Foundation grant ran out.[9] The VDC program was cancelled in 2005.[10]

Systers and the Hopper conference remain the hallmarks of the Institute. But ABI also has other programs. The TechLeaders program was intended to teach leadership skills to technical women and build up a community of technical women. Each TechLeaders workshop is intended to focus on a single leadership skill, as the titles of the initial workshops indicate: Leadership for Cultural Change, Skills and the Art of Leadership, Effective Technical Leadership Styles, Developing and Running Effective Organizations and Institutions, Developing and Turning Your Vision into Action, and Combining Theatre and Voice with Leadership (Goral and

---

local nonprofits. For a general discussion of VDC, see Barr (2014). Jamieson (2015) discusses both the VDC program generally and the highly successful EPICS program at Purdue. On EPICS, also see Jamieson (2001).

[9] In addition to the HP Foundation, the VDC program received funding from Sun and Microsoft.

[10] Some people, such as Valerie Barr, who was the chair of the advisory committee of the Institute while Borg was alive and for a short while afterward, thought the VDC idea was "brilliant" and were sorry that there were not funds to carry it forward, especially after these nine schools had given successful examples of VDC courses (Barr 2014). After Anita's death, there was discussion about ABI's strongest organizational capabilities, especially in the face of financial realities. Rick Rashid from Microsoft and Justin Rattner from Intel were on the ABI board and encouraged the organization to focus on high-tech careers rather than on technical education. Even the long-standing academic members (Fran Berman, Leah Jaimeson, and Bill Wulf) thought this work focus was appropriate so long as ABI did not become completely divorced from students and education (Jamieson 2015).

Harris 2006). ABI runs a Women of Vision awards banquet each year to honor women who are making significant contributions to technology. This is a main event on the Silicon Valley calendar. ABI also runs an annual Technical Executive Forum in which senior leaders – both male and female – come together to grapple with the organizational change and how to increase the participation of women in the tech field. The Forum prepares white papers, distilling social science research – mostly from outside researchers – into a form readily consumed by its corporate audience. ABI also holds Hackathons (Tech for Good), networking receptions, and a job service.

8.2 Computing Research Association Committee on the Status of Women in Computing Research (CRA-W)

In 1989 Ed Lazowska, a professor at the University of Washington, and Ken Sevcik, a professor at the University of Toronto, encouraged Maria Klawe, then a computer science professor and senior administrator at the University of British Columbia (and today the president of Harvey Mudd College), to run for a seat on the board of directors of Computing Research Association. CRA is a nonprofit organization that represents the research-oriented computer science and computer engineering departments and industrial research labs in the United States and Canada. It is important to the life of computing research in North America, and although it had been formed in 1972, there had not been a woman elected to the board during its first 17 years. To break this pattern, Klawe agreed to run and was elected in 1990.

At the time of Klawe's election, Peter Freeman from Georgia Tech was vice chairman of the board. He introduced Klawe to Nancy Leveson, a former colleague who was then a computer engineering professor at the University of Washington. She had strong views about what CRA should be doing to engage women in computing.[11] Over a lunch, the three of them made plans to form a committee for this purpose. As Klawe remembers:

> Nancy and I had some very strong ideas about how we were going to organize this committee. ... [B]ecause this was the Computing Research Association, it had to be entirely composed of strong researchers. This was not going to be a service committee. [If t]his was going to ... get the respect of CRA, it had to have strong researchers. Moreover, it was going to do things... so every person who was going to be on it had to have an active

---

[11] Leveson had already been interested in issues related to women in computing. In the late 1980s, she served on the advisory committee for the Experimental Integrated Activities Division within the computing directorate at NSF, chaired by Rick Adrion from the University of Massachusetts. Leveson and Ruzena Bajcsy, a professor from the University of Pennsylvania and later the head of the NSF computing directorate, requested the formation of a group to look into women in computing issues., and a subcommittee was formed at NSF. The first grant from NSF to what became CRA-W was written by Leveson and submitted through CRA; it was funded by the Experimental Integrated Activities Division.

project. You couldn't be on it just to have it on your resume. You had to lead, or co-lead, an active project. We also decided we'd have co-chairs so that it would be a reasonable workload to chair it. We decided that we would allow men to be on it, because there were men who really cared about women. And Joe O'Rourke from Smith [College] was the first male on the CRA-W. We also said that we would be co-chairs for the first three years, and that we would rotate the chair positions so that a lot of people [had a chance to lead] … so that we got new ideas coming in. (Klawe 2005)

In 1991 the Computing Research Association on the Status of Women held its first meeting, with Maria Klawe and Nancy Leveson as its first chairs. In an article that reviewed the first 2 years of the committee, Klawe and Leveson (1993) stated that there were already numerous studies and reports about the problems faced by women, and CRA-W's intention was to be action- rather than study-oriented. At its first meeting, CRA-W established an operating procedure that it continues today: each member of the committee must direct a project that addresses a significant problem (with inactive members regularly replaced), and these projects are typically carried out by groups of individuals, including many people who are not members of the committee.[12]

CRA-W received NSF project support from its first year.[13] The first NSF grant paid for the creation of a database – a project led by Joan Feigenbaum of AT&T Bell Laboratories – of female researchers in computing, which by 1993 contained more than 600 names (Feigenbaum 1992). Another early project, begun in 1991 and led by Fran Berman of University of California at San Diego, was the creation of a regular column (called The Pipeline) in *Computing Research News*, addressing issues affecting women in computing research. Eventually, 47 articles were published in this column on such topics as family-friendly leave policies for academics, childcare at professional conferences, and mentoring best practices.[14] CRA-W's second NSF grant, awarded in 1993, supported the CRA Distributed Mentor Project. In this project, which continues today under the name Distributed Research Experiences for Undergraduates (DREU), female computer science undergraduates were matched with a female computer science faculty member and given support to spend a summer working on research with that faculty member. The purpose was to overcome the lack of female role models and encourage female students to pursue graduate education and a computing research career. This project was directed in the beginning by Joe O'Rourke from Smith College.

---

[12] The material about the early years of CRA-W is drawn primarily from Klawe and Leveson (1993). The CRA board meets twice a year, and CRA-W provides a 6-month update on its activities as a report to the board at each meeting. The board briefing books, containing these CRA-W reports, are available from CRA from 2000 to the present. These reports provide finer-grain materials similar to those covered in the text here.

[13] Jamieson (2015) discusses the critical importance of NSF to the CRA-W program, and the critical role that Caroline Wardle and later Jan Cuny played as NSF program officers in supporting CRA-W's efforts.

[14] Most issues of *Computing Research News* are available on the CRA website. The first 3 years are missing, as are issues 4.1 and 6.2. So it is possible to construct a relatively complete history of CRA-W through these articles.

Other early CRA-W activities included a report prepared by Mary Vernon of the University of Wisconsin on strategies and experiences of female computer science faculty members on work-life issues; a booklet compiled by Sandra Baylor of IBM on graduate fellowships available to women in computing; another booklet targeted at high school women, entitled *Women in Computer Science* (50,000 distributed by 1998), presenting 18 brief biographies meant to inspire young women to pursue a career in this field; and efforts by Maria Klawe to establish and find funding for awards for outstanding male and female undergraduates in computing (Lopez et al. 1996; Irwin and Berman 1996; Francioni 1998). Based on a successful Academic Career Workshop that had been organized by Cindy Brown of Northeastern University at the 1993 Federated Computing Research Conference, CRA-W planned an academic careers workshop for 1994 organized by Jan Cuny of the University of Oregon and Dianne Martin of George Washington University.

CRA-W has organized consistently high-quality programs over the years, as attested by its winning the Presidential Award for Excellence in Science, Math, and Engineering Mentoring in 2003 and the National Science Foundation's Public Service Award in 2005 (CRA 2005). There are several characteristics of CRA-W's work and its membership that have helped to enable its success. It is focused on increasing the numbers of women computing researchers and the situations of and opportunities for them. Computing researchers have traditionally occupied a small number of occupations – primarily as faculty members in research-intensive universities or as researchers or research managers in government or industrial computing research laboratories. The skills and knowledge required for these occupations are well understood, stable, and vary little even if one moves from, say, a faculty position to a position in an industrial research laboratory. Thus, computing researchers represent a small and homogenous sector of the information workforce. The (primarily) women who have been active in CRA-W are extremely accomplished, and they often come from prestigious universities, which helps to open doors to funding. The goals of CRA-W align well with both the research and workforce goals of NSF, and NSF has been a generous funder of CRA-W programs. CRA-W also has had the deep pockets of successful IT companies to draw upon.[15] This is not to say that the programs created by CRA-W are not well designed and well implemented, but being so closely aligned with the goals of NSF and major research-oriented computing companies has made it easier to attract not only financial resources but also high-quality volunteer talent and places for internship experiences and job opportunities.[16]

---

[15] In its first 20 years, in addition to support from its parent organization (CRA), CRA-W received support from companies including Microsoft, Google, Sun, IBM, Intel, GM Canada, and Lucent Technologies; government agencies including NSERC and the National Science Foundation; professional societies including ACM and USENIX; and private foundations including the Henry Luce Foundation (Soffa 2011).

[16] Many of the people associated with CRA-W have remarked both privately and on tape about how committed the members of CRA-W have been and how well they have worked together. Jamieson (2015) is typical: "CRA-W just works unbelievably well. I do think the key part of that was, as I said, that everybody had a project. You could not be there just to be there. You had to work. The

Because of the narrow set of occupations that CRA-W targets, its programs have primarily fallen in a narrow range of activities that involve engagement of college students in computing research and recruitment of these students to graduate school and computing research careers; retention of students in graduate school to complete the doctorate; and advising and supporting women in their research careers as faculty members at research universities or in professional positions in industrial computing research labs – primarily at the beginning stages, but increasingly also at the mid-career and senior stages of their careers. As the title of CRA-W's column in *Computing Research News* suggests, CRA-W strongly adheres to a pipeline model of formal education to get women into the computing research workforce.[17]

As computing spreads through society, the task of CRA-W is becoming more complicated and CRA-W's traditional approaches may be challenged. This is partly because students in an increasingly wide range of academic disciplines (e.g. various other science and engineering disciplines, economics, and even digital humanities) are studying computing, so CRA-W has to target more than just computer science and computer engineering undergraduates. It is also because doctorally trained researchers are increasingly being placed in jobs that were not part of the traditional CRA-W target community, e.g. in faculty positions in departments other than computer science or computer engineering, or in a wide array of industrial sectors – not just the IT sector – and increasingly in companies that do not have their own research laboratories.

While CRA-W adds programs on an ongoing basis to fill in the gaps to meet the goals described above, many of its programs have been successful and had long staying power. Across these many years, the program goals have been unwavering. So instead of discussing the programs as they appeared chronologically, we organize the discussion by programmatic goal.

Three CRA-W programs are targeted at undergraduates: DREU, CREU, and the Distinguished Lecture Series. The Distributed Mentor Program (DMP) began in 1994 and changed its name to Distributed Research Experiences for Undergraduates (DREU) in 2009. This is a recruitment program. A number of the undergraduate women who could succeed in a graduate computer science program and in a computing research career are enrolled at colleges where the amount of computing research is limited. DREU enables these undergraduate students to gain experience

---

other thing that we would talk about actually fairly explicitly was that you'd get in discussions about leadership style and women in leadership styles and men in leadership styles was that just played out in really interesting ways in CRA-W. We would kid about the fact that at a CRA-W board meeting everybody would be talking. Things got done. There was very little worrying about who was actually managing the conversation. It was just friendly and collegial and constructive and a lot of energy and a lot of respect among all the people who were there. For me, it was just a pretty incredible model – both leadership but also effective organization."

[17] There has been talk from time to time within CRA-W of working on alternative pathways into computing research careers, such as transitional ramp-up programs into doctoral study that help women who hold undergraduate degrees in some field other than computer science and engineering get the computing background necessary for doctoral study in computer science; however, there has never been a successful CRA-W program on non-traditional pathways.

with computing research. The program matches highly qualified undergraduate women to a faculty mentor at another university and pays the expenses for the student to spend a summer working on a research project with the faculty mentor at the mentor's home institution. This gives the students a chance for daily interaction not only with the mentor, but also with graduate students and other faculty. Between 1994 and 2011, 545 students and 123 faculty members were funded under this program (Soffa 2011).

In 2008, CRA-W added the Collaborative Research Experience for Undergraduates (CREU) program. Whereas, under DREU, individual students went off to another university for a summer research experience, under CREU a team of undergraduate students work together with a faculty member at their home institution in a structured environment on a research project throughout the school year (and possibly the following summer). This arrangement enables undergraduates to gain a research experience without having to relocate, but it does require certain features to be in place in order to succeed: enough interested students to form a research cohort, a research-active faculty member who is willing to serve as the mentor, and adequate research facilities on campus. Between 1998 and 2011, 480 students had participated in CREU (Soffa 2011; Danyluk 2013).

In 2000, CRA-W added the Distinguished Lecture Series as another tool to recruit undergraduates into graduate study and a computing research career. Under this program, prominent female computer scientists visit an undergraduate campus – often one in which computing research activity is limited. The distinguished lecturer not only presents a technical talk but also participates in a variety of recruitment activities such as panel discussions, luncheons, and small group meetings where the visitor communicates to the students the nature and excitement of graduate study and a computing research career. More than 50 of these distinguished lecture visits were made between 2000 and 2011. (http://cra-w.org/ArticleDetails/tabid/77/ArticleID/53/Distinguished-Lecture-Series.aspx)

Four other CRA-W programs are targeted at graduate students: Career Mentoring Workshops, Discipline Specific Workshops, Distinguished Lecture Series, and the Grad Cohort. The CRA-W website describes the purpose of the Career Mentoring Workshops:

> Women often find themselves a minority in their own departments or research unit, typically have few female colleagues and role models, and may be concerned about their potential for success. CRA-W-sponsored Career Mentoring Workshops (CMW) bring junior researchers and educators together with women already established in their fields. The established professionals provide practical information, advice, and support to their younger colleagues. Workshops are generally co-located with major professional meetings, providing many attendees with the opportunity to attend technical talks and make contacts in their research areas. (http://cra-w.org/ArticleDetails/tabid/77/ArticleID/50/Career-Mentoring-Workshop-CMW.aspx)

The workshops, which began in 1993, were targeted in a one-size-fits-all approach at all women in computing research. More recently, there have been three different tracks of workshops:

The CMW-R track targets female junior faculty in research universities and senior PhD students interested in research faculty positions. The CMW-E track targets female junior faculty in teaching colleges and senior PhD students interested in undergraduate education. The CMW-L track serves female researchers in industry and government research labs, and senior graduate students interested in research lab positions. (http://cra-w.org/ArticleDetails/tabid/77/ArticleID/50/Career-Mentoring-Workshop-CMW.aspx)

The total number of people who have participated in these workshops is not available, but between 1993 and 2002, 352 students participated (Soffa 2011).

Since 2006, CRA-W has sponsored 36 discipline-specific workshops in various areas of computer science and engineering, such as data mining or machine learning.

The goal of these discipline specific workshops is to increase participation of members of underrepresented groups within a specific research area by providing career mentoring advice and discipline specific overviews of past accomplishments and future research directions. Specifically, the workshop should focus on helping young researchers at the graduate or post-graduate level become interested in and knowledgeable about the research and research paradigms of a specific discipline. (http://www.cra-w.org/ArticleDetails/tabid/77/ArticleID/52/Discipline-Specific-Workshops-DSW.aspx)

The Graduate Cohort program, which began in 2004 and is held annually, is a retention program for women in their first year of graduate study. The students meet for 2 days in both formal and informal settings with some 20 senior women computing professionals, where they learn about graduate school survival skills and the rewards of a research career, build peer networks, and identify potential mentors. By 2011, 1844 students had participated in this program (Soffa 2011).

Two programs are targeted at academic faculty: the Advanced Career Mentoring Workshop and the Borg Early Career Award. The workshops, which began in 2004, are directed at mid-career women who are focused on a teaching track, whether they are at a primarily teaching institution or a research university. Topics include "collaborations, mentoring up and down, managing up and down, promotion to full professor, promotion to other positions such as Dean, effective leadership, leading new initiatives, [and] negotiating." (http://www.cra-w.org/ArticleDetails/tabid/77/ArticleID/231/Advanced-CMW-E-formerly-known-as-CAPP-E.aspx) Named in honor of Anita Borg, the early career award is given to women in academia, government, or industry early in their career (no more than 8 years past the PhD) who have made significant research contributions but who have also contributed to their profession, especially to the advancement of women. (http://www.cra-w.org/ArticleDetails/tabid/77/ArticleID/47/Borg-Early-Career-Award-BECA.aspx) It has been awarded annually since 2004.

Five CRA-W programs are targeted at researchers in government and industry labs: the Career Mentoring Workshop, an Advanced Career Mentoring Workshop, Travel Support, the Distinguished Lecture Series, and the Borg Early Career Award. We have discussed the Career Mentoring Workshop, Distinguished Lecture Series, and the Borg Early Career Award above. The Advanced Career Mentoring Workshop for laboratories is similar to the one described above for mid-career academics but is intended instead for mid-career government and industrial research lab researchers, e.g. people who might hold the title of senior staff engineer or principal member of the technical staff. Topics might include advice on how to reach the top of the technical ladder within their lab, lead initiatives, or manage up and down. (http://cra-w.org/ArticleDetails/tabid/77/ArticleID/71/Default.aspx?IsPreview=true) Travel support provides funds to women in industrial and government labs to attend workshops and conferences – funds that might be in short supply in their organizations during tight financial times. The travel fund program made 34 awards between its founding in 2007 and 2011 (Soffa 2011).

Over the years, CRA-W has received significant support from NSF for its programs. Around 2010, NSF began asking CRA-W for additional hard scientific evidence of the effectiveness of the CRA-W projects NSF was supporting. As Camp (2015) explains: "Basically, what CRA-W needed to do was to show that our programs had an impact, [that] our participants had impact from our programs that non-participants who looked identical to our participants did not. That was the goal." Towards this goal, in 2010, the CRA-W/CDC Alliance received a grant to establish the data buddies project. 35 departments of computer science were randomly selected and signed up to collect data from students to use as a comparative baseline for measuring the success of various broadening participation interventions. Two years later, the Alliance received another grant to establish the CRA Center for Evaluating the Research Pipeline (CERP). The responses of students who had participated in CRA programs could be compared with those who had not with respect to career goals, confidence levels, and other issues – thus providing a means for evaluating the effectiveness of CRA programs (Stout 2013).

A second grant from the NSF BPC program was awarded to the CRA-W/CDC Alliance in 2012 to hire a professional social scientist specializing in the study of underrepresentation in STEM disciplines. The data buddies program has grown from 35 to 70 departments, and CERP is now helping to evaluate programs run by other organizations, not only those associated with CRA-W. CERP is able to prepare customized reports comparing the data reported by the client department's students to the data reported by students in similar departments. It also can provide evaluations of programs, and it has done so for various undergraduate, graduate student, and postdoc programs run by CRA-W, CDC, and DIMACS.[18]

A third grant, from NSF's Education and Human Resources Directorate, was awarded to CERP in 2014. This funding is being used to analyze CERP data in basic research on persistence among different undergraduate groups in their computing education and careers. It is too soon to be able to judge the success of this activity (Stout 2015).

---

[18] For more detail, see the CERP organization website at http://cra.org/cerp/. Also see CRA 2013a, b.

8.3 ACM's Women in Computing Committee (ACM-W)

In 1990, the computing professional organization Association for Computing Machinery (ACM) formed a committee to address issues of underrepresentation of women and minorities in the computing field.[19] The committee included Anita Borg, Sheila Humphreys (for many years the diversity director in computer science and electrical engineering at University of California at Berkeley, discussed in Chap. 10), and Elaine Weyuker (a professor of computer science at New York University), and was chaired by Shari Pfleeger (a well-known software engineer who had worked at RAND, MITRE, and other companies). ACM CEO John White (2015) remembers that the CRA effort on women came together more quickly than ACM's did, and there was "a bit of a struggle to get the initial vision for starting some activities [within ACM] off the ground." Weyuker, who was also a member of CRA-W in its early days, was appointed as the liaison between these two organizations. The ACM committee met a couple of times, made little progress, and was disbanded no later than 1992 (Weyuker 2014).

In 1993, ACM formed a new committee called the Committee on the Status of Women in Computing and later renamed as the ACM Committee on Women in Computing (ACM-W). There are no records of ACM-W's work in its first 3 years. Amy Pearl was the first chair. The size of the committee, the program, and the budget were all small at first (Gurer 2015). White (2015) remembers that ACM "struggled to get good people involved and to come up with powerful things to do" during these years.

In 1996 Anita Borg asked her friend Denise Gurer, a computer scientist at SRI International who had been helping out with the scholarship program for the Grace Hopper Conference, to take over as chair of ACM-W since Pearl wanted to move on to other things. Gurer invited Tracy Camp, a young computer science professor at the University of Alabama who she had met through the Grace Hopper Conference, to join the committee in 1996; and Camp served as Gurer's co-chair from 1997.

The first written records available for ACM-W, an annual report for FY1997 (July 1996–June 1997), coincides with Gurer's first year as chair. That report provides a mission statement for the organization:

> engage in activities and projects that aim to improve the working and learning environments for women in computing. This includes promoting activities that result in more equal representation of women in CS such as mentoring or role modeling, monitoring the status of women in industrial and academic computing through the gathering of statistics, providing historical information about women's accomplishments and roles in CS, and serving as a

[19] In her oral history Weyuker (2014) recalls that this committee was formed in the late 1970s or early 1980s; however she also says that it was at about the time when CRA-W was formed. Since CRA-W was formed in 1990, this author believes that Weyuker was off by a decade and the Pfleeger committee was formed around 1990. This was conformed by John White (2015), who pointed out that one issue they ACM faced in forming its women's committee was to determine who to appoint since they did not want too much overlap with CRA-W, which was forming at about the same time; and that there was a similar question about having overlap and inefficiencies between ACM and CRA's efforts on women.

repository of information about programs, documents and policies of concern to women. (ACM-W 1997)

Gurer remembers the senior staff of ACM – in particular the two most senior staff members, John White and Patricia Ryan – as unfailingly supportive but also conservative.[20] They wanted the committee to demonstrate its ability to be effective and use funds wisely, so the budget ACM provided started from a small base but grew in incremental steps. Gurer saw her task as building up the number of projects that ACM-W was undertaking. She remembers three projects as being of particular importance during her tenure with ACM-W. One was writing proposals to NSF to fund scholarships for students to attend the Grace Hopper Conference. The second was to engage some historical activities – a panel at the Hopper conference of women who had made a difference in computer history, and a special issue of the SIGCSE Bulletin, *Inroads*, on women in computing (Camp 2002). The third was the pipeline data project described below, on which Gurer worked with Camp on collecting statistics.

Perhaps the project that was most closely associated with the work of ACM-W and that received the widest attention within the computing community was Tracy Camp's project on monitoring the status of women in computing. This involved the publication of a much-cited article, "The Incredible Shrinking Pipeline" (Camp 1997), which popularized the pipeline model of participation in computing.[21] Camp analyzed National Center for Education statistics and showed how there had been a substantial drop in the number of women receiving computer science degrees over time. Camp pointed out that, while she was not the first person to talk about this numbers problem, the fact that she had a snazzy title and presented her results in the widely read *Communications of the ACM* meant that the topic received a lot of attention in the wider computer science community.[22] For example, she heard the director of the National Science Foundation use her phrase "the incredible shrinking

---

[20] ACM CEO John White confirmed Gurer's comment as a "fair assessment". "I believe that the [early years] of unspectacular performance of ACM-W contributed to there being, well, we're not going to dump money in ACM-W's lap. We want to see them come together with some good ideas. Then we, the EC [Executive Committee], will look at funding them" (White 2015).

White was strongly interested in engaging more women in computing. As a faculty member in the early 1980s at the University of Connecticut, he remembered that there had been large numbers of women in advanced undergraduate computer science courses such as compiler theory and advanced topics in software engineering; and he was concerned when the number of women in computer science courses and majors fell off in the 1990s. As manager of the computer science lab at Xerox PARC, he brokered the conversation between the director, John Seely Brown, and Anita Borg to house her research institute on women and computing there. When White was vice president or president of ACM, Barbara Simons encouraged him to have ACM become active in women and computing issues, and this was one of the origins of ACM-W. White had convinced Borg to become a member of the ACM Council while he was president, so with Borg and Simons, there were strong voices in support of women at high levels within ACM from the early 1990s (White 2015).

[21] Gurer (2015) indicates that she worked with Camp on some of this data analysis.

[22] Camp came up with the name of her paper when she was talking with her husband, who mentioned the 1981 film comedy with Lily Tomlin, *The Incredible Shrinking Woman*.

pipeline." Recently, there have been more than 500 papers that cited Camp's paper (Camp 2015).

We can command a more complete picture of the early activities of ACM-W by considering the 1996–1997 annual report. Anita Borg used Systers as her initial ACM-W project, but this online discussion group had been created outside of ACM-W and had a life of its own; the ACM-W participation seemed to make little difference to the success of Systers. Kathy Kleiman was directing an ENIAC programmers oral history project, which was funded by the Kapor Family Foundation; and it also had a life of its own. Denise Gurer and Teri Perl ran a Pathways project, which provided Internet-based mentoring for young women, who were matched with women computer scientists. This project was co-sponsored by the Math-Science Network but it never took off in the way that MentorNet did (see the separate discussion of MentorNet in Chap. 6). Susanne Hupfer ran the Ada Project, which was intended to be a clearinghouse of information about women in computing, but it never became well established. Judi Clark ran a project to create a CD-ROM of role models for women in computing.

For the decade 1993–2003, the committee carried out a set of loosely connected activities. Typically each member of the committee had responsibility for her own project. There was no effort to marshall the committee to do something of greater scale.[23] Elaine Weyuker, a later chair of ACM-W, called these early efforts "nominal projects", and criticized them for receiving funding "year-after-year…with no accounting." (Weyuker 2014) There were no procedures in place to keep each member working toward the promised goal, and the level of effort by the members varied considerably (Camp 2015).

New projects were added over time: Tracy Camp organized a teacher enhancement program for high school teachers of computer science, which today she does not remember as making a significant difference (ACM-W 1998; Camp 2015). An Ambassador Project was initiated to recruit representatives from around the world to report to ACM-W on women in computing issues in their home country (ACM-W 1999) An Ada Lovelace documentary project was added (ACM-W 2000). ACM-W joined a Coalition for Women in Computing, together with CRA-W, IEEE Computer Society, Usenix, IWT, WEPAN, NSF, AAAI, and SIAM (ACM-W 2002). There seems to have been little tangible development out of this coalition.

The activities of ACM-W began to take on more significance when ACM President Maria Klawe recruited Elaine Weyuker, at that time a distinguished software engineer at AT&T Labs, to be co-chair of ACM-W.[24] She joined Ursula Martin,

---

[23] It was only later that ACM-W decided to build student chapters to take advantage of the membership structure of ACM. This effort has been only moderately successful. In reflecting on bringing ACM's organizational capabilities to issues concerning women in computing, John White (2015) remarked that of the strong organizations with a mission related to women in computing (he mentioned CRA-W, ABI, and NCWIT) only ACM is an organization an individual can join as a member and contribute to these efforts, and "I just never felt like we've [ACM] been able to leverage [that advantage] well enough."

[24] Klawe had strong support from incoming ACM President David Patterson for the Weyuker appointment. Patterson indicated as part of the planning for his term in office: "I want to find

a distinguished theoretical computer scientist who was then at the University of Cambridge Computing Laboratory, in leading this effort.[25] Within a year, Weyuker was the sole chair, a position she held until 2012, when she handed off the chair duties to Valerie Barr, a computer science professor at Union College.

One of the major new activities of ACM-W was to provide scholarships to women to attend a research conference. In 2012–2013 this program – organized by Weyuker and Barr – provided 33 scholarships to 136 applicants (Barr 2014). Weyuker, Barr, and Anita Borg were close friends from their early days together at the Courant Institute at New York University, where Barr was a beginning doctoral student, Borg an advanced doctoral student, and Weyuker an assistant professor.

> So [with] Valerie, I started a set of scholarships for women to attend a research conference. The idea was as a recruiting-retention tool that this is not aimed at the fourth-year PhD at MIT or Stanford, this was meant for the undergraduate student at Backwater College or any place in the world to tell them, you know, there is a world of research out there and there is something more than they can aspire to or for a terminal masters students to encourage them to stay on and get the PhD ... [B]ecause I had been – I come from a working class background, my parents did not know about research. ... And so, I wanted to give women who might not have the resources that [women at research universities] had some kind of glimpse that there was something ... that was really wonderful that they could aspire to. (Weyuker 2014)

Weyuker actively raised funds for the scholarships from major US-based computer companies, including Microsoft, Google, and Sun, and convinced the Indian consulting firm Wipro to make a major donation. She also convinced many of the ACM technical societies (SIGs) to waive their high conference registration fees for these scholarship students. As the quotation above suggests, the program was targeted primarily at women students from smaller schools in the United States and third-world countries where there were fewer resources than at the major research universities (Weyuker 2014).

Barr thinks of the scholarship program in terms of the pipeline metaphor: "encourage undergrads to think about graduate school, masters students to think about [the] PhD, and help [retain] early PhD students [through to completion]." (Barr 2014) While there has been no formal evaluation of the impacts of this program, both Weyuker and Barr are big supporters.

A second major new activity of ACM-W was the Athena Award – established in 2006 by Weyuker and Susan Landau, a leading cybersecurity and privacy scholar. The award recognizes a woman for her technical contributions to computing, and

---

someone to lead ACM-W, who has probably been active in CRA-W and some of the more visible and more successful women's activities" (as remembered by White (2015)). White also remembers that it was hard to attract strong talent to serve not only as chair but as members of ACM-W because of the slow start that ACM-W had, in contrast to the strong starts by ABI and CRA-W; if a strong computer scientist had limited hours to contribute to volunteer work, she or he would want to make sure those hours were contributed to an organization that would get something important done.

[25] Part of Martin's responsibility at Cambridge was to direct the Women@CL project, which was sponsored by Microsoft and Intel to promote activities for women in computing.

the award winner presents a keynote lecture at one of the major ACM technical conferences. Weyuker explained the rationale for the Athena Award:

> I have been on the ACM Fellows selection committee and had chaired it for a year. And as we all know, ... there were very few women being selected. And the problem was that they were not being nominated. And then, I started looking at ACM Awards, so I go to the ACM Award banquet and the award winners would all be men and if a woman won an award, it was always for service, not for technical [contributions].... I was on many program committees, it was very common that I would be the only woman on the program committee, and when I would suggest another woman, nobody would ever say "oh, no, ... she is not competent" or if I would ... suggest a woman keynote speaker – it would never be, "oh, no, she is not competent," it [would] tend to be – "oh, geez, that is a good idea, I never thought of her." So basically, my feeling was that women were invisible. And it was not so much that people were biased against them, but that we were not even ... on anybody's radar ... so what tends to happen when you try to think of who should be the keynote speaker, you think of, who have I heard before who did a great job. Well, if you never heard them before, they do not come to mind. And so, my ideas for the ATHENA Award was that it would be for outstanding scholarship alone that is no commitment, no requirement that they have done great service or anything, purely for scholarship... (Weyuker 2014)

In order to get buy-in from the research arm of ACM (the SIGs) as well as to reinforce the memory of the leading women researchers, Weyuker and Landau organized the award so that the SIGs made the nominations and provide the venue for the award lecture.

> ...rather than having [an] individual nominate a person, instead have the SIGs do it; and that's what is required. ... [T]he candidate has to be nominated by their SIG Executive; and the reason is I felt that if they go through the exercise of picking who are the outstanding women in [their] field maybe they can remember them long enough when they came to nominating people for other awards or having them be program chair or be the keynote speaker... (Weyuker 2014)

A third new activity of ACM-W involved regional meetings, beginning in 2006. These meetings were intended to produce a similar effect to the Hopper Conference, but on a smaller, regional scale in a number of different locales around the country. They were intended for women who could not afford the time or cost of attending the Hopper Conference. Gloria Townsend, a computer science professor at Depauw University in Indiana, was the creator. She organized several successful regional conferences, and when these were successful the project was scaled up to 12 regions of the country for 3 years with a grant in 2010 that NSF awarded to ACM-W, ABI, and NCWIT. These conferences typically lasted no more than 24 h, and cost no more than $50 (including lodging and meals). Most people could drive to them, helping to keep costs low. Typical attendance was between 60 and 250 participants. The grant funds have now been expended, so the regional conferences program is once again an activity solely sponsored by ACM-W, using ACM internal funding.

To take greater advantage of ACM's individual membership structure, Barr has pushed to build up student chapters and professional chapters for ACM-W. So far these efforts have met with only limited success.

ACM has faced a common problem faced by many professional STEM organizations with affiliated women's groups. Although long-time ACM executive director

John White has given generously of his time and from his budget for diversity issues, and whereas recent ACM presidents including Alain Chesnais, Vint Cerf, and Alex Wolf have also been highly supportive of ACM-W activities, until recently there has been a residual attitude that the problem of underrepresentation of women is one for ACM-W rather than for the entire organization to fix.[26] Barr sees this attitude changing in good ways over the past 2 years:

> The first time I went to an ACM Council meeting I thought, "Oh my God, if I wasn't in the room for that conversation, that would've just been awful." Now I don't have to be in the room. I hear afterward. ... And what I hear these days is that the executive committee has amazing conversations, and there's a level of bringing up issues that have to do with diversity that happen even though I'm not in the room. So that's a great thing, because that means that it's really becoming part of the way other people are thinking about what ACM does and how it does it. ... It doesn't mean it's always a bed of roses, but it's definitely progress. [The attitude expressed by one of the ACM Presidents that] most of the work ACM does in this area is done by ACM-W, and I've already spoken with [him] and said, "No. That's not the view. We just, we can't have that view anymore. The problem with women in computing, the problem of minorities in computing is everybody's problem, and everybody has to work to fix it.... I'm fine if you want to say that ACM-W provides leadership. If you want to think of us as your conscience, I'm okay with that. But there is no way you can let the other 107,900 people off the hook. Everybody has to start to do this work. And every SIG, every SIG chair, everybody's got to be held accountable for the dearth of women and minorities and the changes that have to happen. And our job is just to keep reminding you of that." (Barr 2014; also see White (2015) who talks about the leadership Barr has offered in changing ACM culture on this issue.)

8.4 National Center for Women and Information Technology (NCWIT)

The National Center for Women & IT is a national nonprofit, founded in 2004 and headquartered in Boulder, Colorado, that aims to increase the meaningful participation of women and girls in computing. Over time, NCWIT has grown so that it is today the largest network of computing-related organizations in the world, involving more than 700 organizations. For example, its K-12 Alliance of member organizations can now reach 100% of the girls in the United States through their distribution channels. (Sanders 2014; update by Ruthe Farmer, private communication, 1 April 2016)

After 25 years at Bell Labs, where she had been a Bell Labs Fellow and manager of a large, international engineering laboratory, Lucy Sanders retired and eventually joined the ATLAS Institute at the University of Colorado, Boulder, as the executive in residence. ATLAS is an interdisciplinary research institute examining issues of information and communication technologies in society, at the time managed by Robert (Bobby) Schnabel, the university CIO and vice provost as well as a professor

---

[26] Barr (2014), Weyuker (2014), and Camp (2015) all mention the strong support they received from the ACM senior staff.

of computer science. Caroline Wardle, the NSF program officer who was responsible for the IT Workforce program in the computing directorate at NSF (see Aspray (2016) for details), was concerned that the funding NSF was putting into diversity was not leading to increased participation of women. Sanders and Schnabel convened a 1-day workshop in 2003, with support from Wardle's budget, of about 30 people from academia, industry, and government to rethink what should be done to broaden participation in computing. As Sanders remembers the findings of that workshop:

> …the recommendations, which basically were…we need a connective tissue. We needed to connect the organizations that cared. We needed to connect organizations to best practices in research. We needed to make sure there were best practices in research. … We needed to make sure we were operating on fact and not reinventing the wheel. (Sanders 2014)

Telle Whitney, the CEO of the Anita Borg Institute, joined forces with Sanders and Schnabel, and they submitted a proposal to NSF. Peter Freeman, the recently hired head of the computing directorate, as well as deputy director Deborah Crawford, were highly supportive; and a large NSF grant started NCWIT on its way. This notion of being the "connective tissue" has been one of the guiding principles of NCWIT throughout its entire history. Another basic principle was added later: "we can also unite [our member organizations] in common action and create platforms for action that they can all plug into and do nationally," (Sanders 2014)

The founders decided early on that they wanted to be more than a grant-funded project at a single university or at a few universities, so they created a nonprofit organization. Sanders identified three reasons for doing so. First, they recognized that to sustain the operation at the level needed to solve the problem, they would need to raise substantially more funding than NSF could offer; and corporations and individuals, they believed, would be less likely to make a donation to NCWIT through a university. Second, they wanted an organization that could control the intellectual property that was created in the process of their work instead of having that intellectual property distributed haphazardly across many universities. The point was not to monetize the intellectual property but instead to consolidate ownership so that action would be easier if someone attempted to misuse the intellectual property. Third, the organizational capabilities of universities make them good at taking in funds but not so good at disbursing them. However, NCWIT wanted to be able to give away funds to people carrying out projects spread across the country[27] (Sanders 2014).

While many of the NCWIT staff conduct their work out of offices on the University of Colorado campus, the organization has staff members scattered around the United States – many of them located at other universities. Sanders, who assumed the responsibility of CEO of NCWIT, was comfortable with a distributed staff because for years she had operated a 650-person engineering operation at Bell

---

[27] NCWIT has no paid employees, though it does have contractors. The NCWIT staff are instead employees of the universities in which they are located: Colorado, Texas, Virginia, Washington, etc.

Labs in which the engineers were scattered around the world.[28] With the right focus and leadership, Sanders believes, a distributed workforce can be effective. Moreover, this distribution has advantages such as diversity of thought that comes from being in different places, and the freedom to choose where to place subcontracts depending on different indirect cost rates.

The support from NSF has been very helpful to NCWIT in its corporate fundraising because NSF has high credibility with corporations. Sanders's first major donation – a million dollars – came from her former employer Avaya.[29] But soon other companies, including Microsoft and Pfizer early on, joined as major donors. Sanders successfully appealed not only to companies in the IT sector but also to companies that "have huge use of tech in the delivery of their business value to their customer base" (Sanders 2014).

NCWIT faced a number of questions about the scope of its mission, especially in its early days: why only women and not all people who are under-represented? Why IT (and what is the scope of IT, anyway)?[30] Why just IT and not all of STEM? Why only the United States and not some international scope? There were two principles behind these delimiting choices. One was the need to stay focused to be effective. This explains why IT and not all of STEM. For example, math is a required course

---

[28] Most of the operational work was led by Sanders, even from the beginning and increasingly as time went on. Whitney and especially Schnabel have continued to be active advisors. ABI had a subcontract on the original NSF grant to conduct research on technical women, but beyond that ABI (or Whitney) has had little involvement in NCWIT's operations. There is good cooperation between NCWIT and ABI, for example related to the Grace Hopper conference. Sanders (2014) describes Schnabel's contributions to NCWIT in this way: "Bobby has been and continues to be, and was then, a leader in the computing community around gender, the broader … computing faculty, computer science administrators and chairs, and deans. He plays that role today. He has always kind of been an ambassador for this. … Of course, he approved the first grant and helped write the first grant, and did some of the things you would think of a PI would do. After the first couple of years… even from the very beginning, all the hands-on stuff, I did. I was like the CEO of the whole thing. Bobby has been always out there, always thinking, always thinking about NCWIT and where we should be going and what we should be doing, and really reflecting on it."

[29] When asked how she settled on the amount to ask from Avaya, Sanders replied: "I wanted to get a million dollars out of the corporation, $250,000 over 4 years commitment. To me, it made sense because it was like … It's funny. I rationalized that it was 1.5-loaded tech-headcount a year. Corporate folks waste that much in a day, probably way more than that in a day … lots of different ways they waste it. So, "Okay. That makes sense. Let's go ask 1.5" (Sanders 2014).

[30] With respect to the scope of IT, Sanders (2014) stated: "We took the name because of the government definition of IT that in fact it includes and they count all of that stuff as part of IT. When they count patents, they count them there. … We look at the broader definition of information technology. We struggled for maybe a year or so when people would go, "What exactly is in our jurisdiction? Is it computing engineering? Is it computer science? Is it electrical engineering? Is it information systems? Is it business systems? Is it informatics?" Finally, we all agreed, we would just call it 'computing' … Once … we [the community] stopped deciding that that was something that needed to be solved, it actually turned out that the broader population who we really need to reach, they don't care about these distinctions. We will sometimes use a specific subcategory … when it makes a difference. For example, in K-12, it makes a difference to call it 'computer science' because the only toehold, we, the computing community, has in the K-12 public education is the AP Computer Science course."

in high school while computer science is not – so they are bound to have different parameters to take into consideration. Focus also explains why addressing only women and not a broader group:

> When you think about what we're doing, our mission, it's hard to do it for a blanket group. When you have a research focus, when you're creating practices, when you're really, really mindful about what the research says about culture; you have to stay focused. Otherwise, you're going to be saying things that are untrue and not very helpful. For example, things about women and the practices for women in technology cultures are different than practices for women in a broader business culture. They're different for African-American men in the culture. In fact, practices for African-American men would be different than practices for Black men from Jamaica. You can't just kind of guess. (Sanders 2014)

The other principle is simply not to take on work until the organization has the infrastructure in place to handle it properly. For example, NCWIT's Aspirations in Computing program (described below) has been successful in the United States and has grown from honoring a few girls to honoring thousands. Sanders deflected efforts to grow this into an international program for now because she does not feel she has the staff and infrastructure to handle a global program effectively.

NCWIT's original organizational structure was based on alliances and hubs. The alliances are groups of like-minded organizations (and in some cases, people). NCWIT's first alliance was its Academic Alliance, to which all of the colleges and universities participating in NCWIT are members. It considers all issues of computing in higher education.[31] Next to be formed was an alliance of social scientists studying women and computing (quickly renamed the Social Science Advisory Board). Later, new alliances were added in the K-12, entrepreneurial, workforce, and affinity group areas.

As Sanders (2014) explains, the hubs were intended to be organizations that were already producing useful results to NCWIT's mission – to give NCWIT a jump start:

> When you're starting out, organizations grow in different ways. Sometimes you start very small and you start you grow, and then all of a sudden you realize you're going national. We all know groups like that, where they start in a few locales. … From our first grant proposal … we had to start big or look big. I said this at the last summit…, "We invented this whole infrastructure, and it wasn't even real. There was no there, there," and the whole audience falls out laughing. But in fact, that was the case. We invented a national infrastructure [which we called] alliances …. We only had two. … The hubs were places where we had imagined they were pockets of existing excellence in the day, like George Tech and Irvine, and that they could help us on some part of the pipeline … when the research started, which is what they did.

The hubs lasted only a few years before being folded into the alliances. The Georgia Tech hub conducted some research on faculty, the Girl Scouts hub worked on K-12 issues, and the UC Irvine hub considered pair programming and other educational practices. These organizations that were once hubs continue to be influential within

---

[31] As of late 2014, the Academic Alliance included over 800 people associated with more than 300 colleges and universities of all types, ranging from majority research universities, to minority-serving institutions, to women's colleges, to community colleges (NCWIT 2014).

NCWIT by contributing significant numbers of people to the NCWIT advisory committee.[32]

When NCWIT was founded, neither the IT industry nor the academic computer scientists were paying much attention to social science research.[33] Nevertheless, from its earliest days, NCWIT strongly supported social science research through an advisory board and through a group of social scientists on the NCWIT staff, led by Lecia Barker and Joanne Cohoon – later adding Catherine Ashcraft and others.[34] The social scientists undertook basic research, identified best and promising practices, and carried out rigorous, scientific evaluations of practices and projects.[35] This work has led to a better understanding of such issues as unconscious bias, stereotype threat, collective intelligence, and changing mindsets – all of which have helped illuminate the causes of under-representation in computing. Sanders believes that corporations are beginning to take an interest in social science research in part because of the presentations they have heard at NCWIT meetings (Sanders 2014).

One snapshot of the place of social science in NCWIT is given in Table 8.1, which presents the authors, titles, and affiliations of the research talks given in a session sponsored by the NCWIT Social Science Advisory Board at the NCWIT Summit in Chicago in 2012.

NCWIT's K-12 Alliance was started in 2005 through the efforts of two major members, the Girl Scouts of America and ACM's Computer Science Teachers

---

[32] ACM was one of the original hubs, although it was not involved in the creation of NCWIT. However, ACM generally and ACM CEO John White personally, have taken an active interest in NCWIT. This is likely to continue in the future since Robert Schnabel, one of the founders of NCWIT, assumed the role of ACM CEO in Fall 2015. ACM was able to contribute to NCWIT in more recent years especially in the K-12 Alliance because of the Computer Science Teachers Association it created in 2005 and its Education Policy Committee, focused on K-12 computer science education policy, created in 2007.

[33] The major exceptions were Xerox PARC and Intel Research, which both employed social scientists to conduct research.

[34] There can be disadvantages to being a social scientist working for NCWIT. The work is of an applied nature, which does not always square well with the criteria by which academic departments evaluate individual scholars for promotion or honors. The NCWIT social scientists do not have as much freedom to select projects to work on, even though the director (Lucy Sanders) is liberal about this issue. On the other hand, NCWIT's Workforce Alliances provide considerable access to corporate America that is not necessarily open to traditional academic researchers, and NCWIT has a powerful set of mechanisms for disseminating results as well as a high reputation for quality, which means that reports from NCWIT get noticed and paid attention to (Ashcraft 2015).

[35] Ashcraft (2015) noted a change in NCWIT research activities over time: "early on I think it was a lot of synthesis and assessing where the state of the field was and translating that into resources, and then I think in the last 5 years or so there's been more of a shift in doing our own research, and that's been important." The NCWIT staff has carried out a number of research projects. For example, the staff did a project together with the firm 1790 Analytics to understand the rates of female patenting over time, which it updated in 2015 and followed with a joint project between NCWIT and the US Patent Trademark Office on a qualitative study of practices that foster or hinder increases in female patenting. Recently, the social science advisory board has become more active in weighing in on NCWIT resource documents and research projects.

Table 8.1 Presentations at the NCWIT Social Science Advisory Board Meeting, Chicago, 2012

Sharla Alegria (U. Massachusetts-Amherst)	Becoming an IT Worker: A Study of Access to Good Jobs in the Knowledge-based Economy
Catherine Ashcraft (NCWIT and U. of Colorado-Boulder)	COMPUGIRLS Intersectionality Study
Catherine Ashcraft and Wendy DuBow (NCWIT and U. of Colorado-Boulder)	Male Influencer Study
Lecia Barker (U. of Texas at Austin and NCWIT)	Faculty Adoption of Practices to Improve Gender Imbalances in Computing
Enobong Hannah Branch (U. of Massachusetts –Amherst)	The Performance vs. Persistence Paradox: Myths About Women in IT
Sapna Cheryan (U. of Washington)	Changing the Image of Computing to Increase Female Participation
Nilanjana (Buju) Dasgupta (U. of Massachusetts-Amherst)	Thriving Despite Negative Stereotypes: How Ingroup Experts and Peers Act as 'Social Vaccines' to Protect the Self
Wendy DuBow (NCWIT and U. of Colorado-Boulder)	Aspirations Program Research
Margaret Eisenhart (U. of Colorado-Boulder)	Female Recruits Explore Engineering (FREE Project) and FREE Pathways
Margaret Eisenhart (U. of Colorado-Boulder)	Urban High School Opportunity Structures, Figured Worlds of STEM, and Choice of Major and College Destination
Nathan Ensmenger (Indiana U.)	Why Guys? How Programming Acquired Its Masculine Identity
Mary Frank Fox (Georgia Tech)	Programs for Undergraduate Women in Science and Engineering: Issues, Problems, and Solutions
Sarah Kuhn (U. Massachusetts-Lowell)	Crocheting the Way to Math Equality: The Effects of Teaching Style on Math Performance
Elsa Macias (Independent Education Consultant)	Preventing Stereotype Threat in Standardized Testing
Rose Marra (U. Missouri)	Leaving Engineering: A Multi-Year Single Institution Study
Jamie McDonald (U. of Colorado-Boulder)	Diversity, Technology, and Occupational Branding: Examining Efforts to Reconstruct the Identity of Computing and IT Work
Irina Nikiforova (Georgia Tech)	Turing Award Scientists: Contribution and Recognition in Computer Science
Maria (Mia) Ong (TERC)	Beyond the Double Bind: Women of Color in Science, Technology, Engineering, and Mathematics
Linda J. Sax (UCLA)	Trends in the Determinants of Gender Segregation Across STEM Majors
Allison Scott (Level Playing Field Institute)	An Examination of Perceived Barriers to Higher Education in STEM Among High-Achieving High School Students from Underrepresented Backgrounds
Gerhard Sonnert (Harvard U.)	Persistence Research in Science and Engineering
Roli Varma (U. of New Mexico)	Gender and Computing: A Case Study of Women in India
Sneha Veeragoudar Harrell (TERC)	The STEM Agency Initiative for STEM Learning Among Marginalized Youth: From Fractal Village to Global Village

Source: NCWIT (2014)

Association (CSTA).[36] The K-12 Alliance has as members formal education organizations, informal education organizations, organizations that serve educators, and organizations that serve children directly. These organizational members together reach practically every girl in the United States.

Some of the K-12 Alliance members are organizations focused generally on girls, not specifically on girls and computing (e.g. Girl Scouts, Boys and Girls Club, Campfire, YWCA). The K-12 Alliance brings to these organizations content-area expertise in computing, e.g. on how to run a successful computing program for girls.

Some of the K-12 Alliance members are organizations that serve influencers (e.g. the Computer Science Teachers Association, the International Society for Technology Education, Guidance Counselors Association, Physics Teachers Association, National Girls Collaborative Project). The K-12 Alliance offers research-based practices and resources for these organizations to do their jobs better in getting girls engaged in computing.

Another group of K-12 Alliance members are organizations that teach coding to students or teachers (Code-org, Bootstrap, Black Girls Code, Girls Who Code). These organizations are interested in being in the K-12 Alliance because it gives them a way to connect with other organizations having a similar mission. It also gives NCWIT a way to ensure that these organizations are teaching coding in an inclusive way.

The number of organizations that might want to become members of the K-12 Alliance is potentially so large – numbering in the tens of thousands[37] – that NCWIT has had to build a new membership model for the K-12 Alliance. To be a full member, an organization has to be national in scope or reach a specific niche of students that NCWIT is interested in. An affiliate membership is open to a much larger set of (often local) organizations. These Affiliates are not invited to the annual NCWIT Summit and they do not help to shape the K-12 Alliance programs, but they do receive information from the K-12 Alliance regularly, participate in its programs, and are eligible to apply to the grant program offered by NCWIT to teach other girls. At the moment, there are approximately 60 full members of the K-12 Alliance and 200 affiliate members (Farmer 2015).

When Catherine Ashcraft arrived as a research scientist at NCWIT in 2006, the Workforce Alliance was just getting started. Ashcraft (2015) notes that the companies that became members of the Workforce Alliance had a tendency to want to focus on outreach, in particular on K-12 activities; and much of the work prior to Ashcraft's arrival had been focused on getting these companies to think about how to reform the situation within their own organization rather than focus on outreach. When the corporate members finally did buy in to the message that this was to be a reform activity, individual member companies wanted to benchmark their own progress against other companies. The only available information was national

---

[36] There is an extended discussion of CSTA in Aspray (2016).

[37] This estimate of how many relevant organizations there are is based on the fact that the National Girls Collaborative Project is working with 18,000 organizations.

Bureau of Labor statistics, and that proved to be not particularly helpful for bench-marking. The alliance spent several years trying to get its members to collect and pool data. This was a largely unsuccessful effort because either the companies did not collect the correct data, or the uniform titles to be used in collecting this group data did not fit the company's occupational titles, or the management was unwilling to release the data. After several years, the Alliance abandoned the hope of providing useful data to its corporate members.

At the same time that the Workforce Alliance was trying to collect data, it was also working with the NCWIT staff to produce best practice sheets to use in the companies. Unfortunately, most of the companies were unwilling to write sheets about their own practices or did not have time to do so, so most (but not all) of the best practice sheets were prepared by the NCWIT staff on the basis of "publicly available stories of companies that had [adopted best practices]" (Ashcraft 2015). It was not until perhaps near the end of NCWIT's first decade that the companies began to seriously discuss best practices for retention and advancement of women within their companies. Before that, the focus had largely been on recruitment. The attitude was:

> [B]ecause if it's recruitment, it's not really a "you" issue, right? … it's just that if they were there, if they'd apply, I'd hire them, right? It's just that they're not applying and so sure, help me find them. I'd be happy, but it's not really anything that I'm doing that's causing this problem. (Ashcraft 2015)

In the past several years, companies had an increased interested in improving their internal operations for the purpose of broadened participation.

> People are starting to pay attention to the fact that we [NCWIT] even have an industry change model. I've been talking about the industry change model here and there for quite some time, but you hear other people mention it now… There's a model we have, it shows the different areas internally that you need to focus on, and so we have been working with companies, sitting down with them and having them actually identify areas of the model that they feel are key areas where they need to focus and then working on practices within that area.
> I think they're all kind of at the beginning stages of that, but running productive team meetings is one … on the … everyday level that I think is strongly felt by most of the companies, and one that we actually try to encourage them to focus on because … it is a huge factor in someone's everyday work life; and so making sure that multiple voices are heard and people aren't getting interrupted or people aren't taking credit for [other] people's ideas. They also … are looking at ways to more actively develop employees and start examining the criteria by which they choose employees to go into talent development programs and then paying attention to who you give what opportunities to. (Ashcraft 2015)

Over the first decade of NCWIT, the Workforce Alliance grew from about 10 companies to more than 50. In the early days, the Alliance meetings involved about a dozen people sitting around a conference table at the annual meeting. Today, most companies send three or more representatives, and a typical annual meeting has more than 200 attendees – a mix of human resources, diversity, and technical managers – seldom senior executives. These Workforce Alliance members come from a number of different industries. One of the interesting dynamics involves the members from the financial or healthcare industries:

They are in with Google and Apple and these other companies that everybody thinks of as tech companies, and so part of their concerns or efforts around recruitment and stuff like that, or image and branding and helping younger people out there – potential candidates – know that we actually are a technology company or that we have a significant technology workforce. That's a huge part of their efforts… (Ashcraft 2015)

Ashcraft (2015) believes that, while the Workforce Alliance was slower to develop and show results than, say, the Academic Alliance, the Workforce Alliance has begun in the last couple of years to take its work more seriously and make increased progress. For example, she believes that NCWIT has raised awareness among its corporate members about unconscious bias and about the roles and values of male allies; and that the companies have started to make more use of NCWIT resources such as the manager-in-a-box toolkit, best practice sheets, and the list identifying the top 10 things to do in order to retain technical women.

I think with Corporate America, it's always hard to tell what's lip service and what's face-saving and real change. I think a lot of people in the room are pretty committed and interested and feel like NCWIT's very helpful in the resources and our guidance, but I think that we feel like there's a shift in the last year or two. The corporate alliance I think has always been lagging behind the other alliances, … Sometimes … they seem to be lagging behind, and then other times they are doing stuff but they don't always want to tell us or it's hard to keep track of what they're doing [more so] than in the Academic Alliance. We know what the Academic Alliance is doing, more upfront, and they have data that they share and show and all of that. None of those conditions are really true in the Workforce Alliance… (Ashcraft 2015)

Part of the reason for recent change in the Workforce Alliance attitudes is a new concern about diversity in Silicon Valley after data about diversity numbers at the major firms had been publicly released in the press.[38]

[S]ince all the firestorm in Silicon Valley with the releasing of the public data, it's definitely been a huge uptick in just the number of companies we go meet with and visit. I would say that the tenor of the conversations feel more real. Sometimes there's actually senior people in the room, so that's always a good sign, and they're actually talking about nitty-gritty details with us, about the questions they have or things they want to start. (Ashcraft 2015)

NCWIT also has an Entrepreneurial Alliance. It has become much more active in the past couple of years. Approximately 150 start-up companies have joined the Entrepreneurial Alliance, though they vary considerably in their involvement in NCWIT. In some ways it has been harder to work with the members of the Entrepreneurial Alliance than with the more established companies in the Workforce Alliance:

---

[38] On the recent interest among high-tech companies in Silicon Valley in diversity, see, for example, Brown (2015), Manjoo (2014), Kang and Frankel (2015), and Guynn et al. (2014). For a more positive view of what Silicon Valley companies are doing to promote diversity, see Silicon Valley Workplace Diversity (http://www.siliconvalley.com/workplace-diversity), which has links to a number of positive stories. For statistics on five companies (Cisco, Dell, eBay, Ingram Micro, and Intel) see CNN Money's interactive website entitled "How Diverse is Silicon Valley", http://money.cnn.com/interactive/technology/tech-diversity-data/ (accessed 21 January 2016). For additional statistics, see Jones and Trop (2015).

Often they, even more so I think than the big companies, are worried about hiring because they're usually growing and trying to just establish [themselves], and so I think that was one thing that was hard initially is, you know, "We just don't have time for this," like everything's a state of panic and urgency even more than normal and so I think they want it fast. They don't have time to read a lot and then not in the study way, but you would, everything's kind of amped up so they have time to read even less. (Ashcraft 2015)

However, in other ways the Entrepreneurial Alliance is easier to work with. They are more willing to share data and are less concerned about their past practices being judged negatively. Their attitude is: "We're new. We're growing. Of course you can't expect us to have done everything right so far" (Ashcraft 2015).

One of NCWIT's earliest hires was Paula Stern, the former chair of the U.S. International Trade Commission, to serve NCWIT's interests in Washington. Although there was pressure from the science policy community to provide a consolidated voice in lobbying the Bush Administration in favor of STEM research and education, NCWIT decided to break ranks and talk about computer science education with federal legislators – particularly K-12 education. None of the computing professional societies was particularly pushing computer science education at the time in Washington. Computing Research Association was advocating for computer science research, but it stayed clear of undergraduate education and to some degree clear of graduate education so as to avoid overlap with ACM. Only in 2005 did ACM take a major step into K-12 computer science education through the formation of the Computer Science Teachers Association (CSTA).

NCWIT connected with Jared Polis, who had a keen interest in computer science education. He had made his fortune in online companies BlueMountain.com and ProFlowers, and his foundation was refurbishing and donating computers to local schools and nonprofits. When he was elected to Congress in 2008 as Boulder's representative, he helped to champion computer science education among his fellow Democrats. NCWIT and ACM's CSTA have worked together as champions of the move to reform K-12 computer science education, change the College Board Advanced Placement exam to make it more about computer science concepts, hire and train 10,000 new computer science teachers for the public schools, and improve articulation agreements between high schools and colleges.[39]

As other strong players have emerged, such as Code.org in 2013, NCWIT has been able to scale back its efforts in the K-12 area. Nevertheless, the NCWIT K-12 Alliance remains committed to these efforts.

More recently, Sanders has been talking to the Office of Science and Technology Policy and various federal agencies about the corporate data they collect. She figures the only way to keep companies accountable and focused on diversity issues is to make sure there are strong metrics by which companies are measured against their competitors. Companies are unlikely to voluntarily provide this data, so the best chance for having a data-driven metric for corporate diversity performance is to have the federal government mandate the reporting of the right kinds of data (Sanders 2014).

---

[39] For a detailed account of K-12 computing education policy, see Aspray (2016).

Sanders has announced on many occasions since the very founding days of NCWIT that her goal is to make enough change that NCWIT can put itself out of business in 20 years. She believes this was an important decision for the organization because it informed many of the other choices NCWIT had to make. It did not have to build an organizational structure, physical plant, and endowment that would last for 50 or 100 years. It could be more agile and move into new programmatic areas as opportunities presented themselves without worrying about long-term commitments it was making for the organization. Perhaps even more importantly, it could readily cede work to others (such as to Code.org in getting girls interested in computing) when those others were successfully covering an area that NCWIT had been working in.

What constitutes enough change to declare a victory? Sanders does not believe it needs to be 50% female participation. However, it does need to involve a high enough percentage of women participating – perhaps a third – that a tipping point is reached in the culture of computing environments, so that bias is mitigated and that it possible for women and girls to have meaningful participation. Anyway, aspiring to a particular number is wrong-headed, she believes. She argues that if one pushed in ways that increased the numbers of women but left the current climate intact, the numbers would eventually tumble again and there would be a renewed need for an organization like NCWIT (Sanders 2014).

Sanders has had moments of doubt,[40] but at the moment she is bullish in particular because of the promise she sees in the Aspirations in Computing program, which she regards as able over time to build a pipeline of 10,000–15,000 women attaining computing degrees. This change would double the number of women in computing.[41] Aspirations, which started out in 2007 as a program to honor the technical accomplishments of a few high school girls in Colorado, has grown into a national award program that spans from middle school through college and into the workforce[42] (Farmer 2013). Companies have formed into 65 clusters. Each cluster raises

---

[40] "It could be easy to get kind of dejected about [what we are doing] because you can't just fix [those big social issues such as the socioeconomic status of minorities or societal attitudes towards women] and it's also easy to feel like, 'Why are you asking me those questions? You know I can't control what society thinks.' I don't have that big a budget, you know?" (Sanders 2014).

[41] The Aspirations program has been built up by Ruthe Farmer, the chief strategy and growth officer at NCWIT. Farmer's first experience with NCWIT came in 2005 as program manager for technology and engineering at the Girl Scouts of the USA, where she was a principal in NCWIT's K-12 hub. After completing an MBA at the University of Oxford, she joined the NCWIT staff.

[42] Farmer (2015) explains how the program has grown. Bank of America supported expanding the program to 10 regions in which they were trying to build a pool of technical talent to hire. She wrote a successful program to the Motorola Foundation to scale the program further. One of the interesting challenges was to collect and review applications: "One key piece of that was that we had to own the data. Because if the girls information, the relationships with the girls are local and national, we have a virtual relationship with the girls through our Facebook community, but the person who's going to see that kid in person is going to be Indiana University or Bank of America in Florida, whomever is hosting the program. That's who's going to have that personal relationship and connection but we need to own all the data of who all these kids are. We spent the money to build an online system that would allow girls to apply, allow their parents to sign a release elec-

funds, organizes an award competition, and recognizes local high school girls not only for their achievements but also for their aspirations. Farmer (2015) contrasts the Aspirations program to the traditional talent search:

> But unlike virtually every other competition in the world, getting the Aspirations award is not the end of something, it's the beginning. And really the award is just our talent search. That's how we find them. It's what we do with them that is the important thing. Like with the science fair, you work for a year, six months, a year on your science fair project. You compete and they winnow it down and there are 10 winners. The science fair throws away all those people that didn't win. Our approach is an abundance approach. Let's find every girl that is exhibiting the potential to be in computing and recognize her and propel her forward. Not: let's find the four that are the best. Because we don't even know if they're the best. We have a system of winners and runners-up in order to increase capacity, and we don't want to discourage girls. I found girls don't apply for things they're not qualified for. It's pretty rare for us to get an application for a girl where we're just like, "What are you thinking?" 95 % of the time, you're like, God I wish I could award all of them. Because girls just don't apply for things that they don't think they have a shot at.

NCWIT provides an infrastructure for girls to apply and technology portals for judging. It also provides toolkits to the clusters to carry out their local programs. Currently, about 1400 girls are recognized as finalists or semi-finalists for their technical abilities, and the numbers are growing rapidly every year. Winners are announced in the schools and in the local press.[43]

---

tronically. This was early in the world of electronic releases and the legality of all that, so we had to go all through the process to make the legal. Allow technology volunteers to read their applications and score and judge them and then allowed our local people to administer their program, send emails connect to the girls – all without anybody having a spreadsheet of girls' personal data because that kind of a breach is dangerous. We built that system and then that enabled to really scale this award because this year, we had well over 5600 applications in the system. Every one of those applications is read by four professionals and scored, and their scores are average. We have pretty high fidelity, confidence in the fidelity of our scoring system. That allows us to rank them and then this year, we're going to award 2500 girls in person, going to an event and having their whole community sit there and honor them for being technical."

[43] Farmer (2015) tells several inspirational stories about girls who have entered computing because of the Aspirations program. One is Grace Gee from Port Lavaca, TX, who graduated from Harvard *magna cum laude*, held the prestigious Thiel fellowship, and co-founded a data science startup company called HoneyInsured that enables rapid enrollment in healthcare.gov. Another example is the story of "this girl Safia Abdalla from Chicago, who is a Somali immigrant, absolutely beautiful; she wore like a full headscarf and she's beaming. She's one of those young women that kind of radiates light. Her teacher in Chicago, she went to the University of Chicago Laboratory School. It's a really good school, smart kid. Freshman year, she's like 'I'm really into computer science.' Her dad was like, 'You're going to be a doctor.' She won our award 3 years straight and she won the national award. It was the second year that her teacher told me … We had a [national] Summit in Chicago and she got the award on stage at the Summit – all these people, huge audience, two female NASA programmers on the stage with her; and her dad took her out and bought her a MacBook Pro. It was like okay and so now, she is the founder of the Society for Data Science and she's graduating from Northwestern in Computer Science. This kid was I mean she was from an immigrant family and an immigrant family understands that doctors get respect regardless of your race. It doesn't matter where you came from. If you're a doctor, that's respected. They just didn't understand what Computer Science could mean for her. Now she's incredibly successful and it's really fun to watch."

Farmer (2015) explains why the Aspirations program is of particular value in getting girls involved in computing:

> The award is important. It's important for kids to have someone tell them they're good at something and validate that. I think what's happening to girls in the school system is that … not only are they not being affirmed in being technical, they're being negatively encouraged. They're getting questions like "are you sure you want to do this? are you sure you belong here? is this something you really want to do?" It's like girls are wading upstream through the river and boys were just kind of going with the flow like all my friends are all taking this class, I guess I'll take this class. There's not a social cost for a boy to sign up to take computer science. Girls, they get the push, I call it push-pull peer pressure so you're pushed out by this male classroom, mostly boys, boy-oriented activities and boy-oriented challenges. You look at some of the curriculum and it's all around like first-person shooter game development and then you look at the posters and the whole environment and often times the teachers are male as well. That's the push, plus the media messages about who belongs, and then the pull is your girlfriends who don't get what you're doing and they're like what are you doing over there when we're over here. We're all doing this and you're doing that. That experience goes well into college.

Farmer (2015) also points out that, by using social networking technologies, the high school Aspiration winners are building virtual communities. These virtual communities offer a means to overcome the isolation that an individual girl might have if there is not a critical mass of girls at a particular high school. The same is true in college. Farmer tells the story of five Aspirations winners who found each other and bonded in a largely male crowd of 700 MIT freshman by the green bag that is awarded to each Aspirations winner. In fact, Farmer asserts, 30 % of the women studying computer science in college today have been Aspiration winners. As the students move in to the workforce, the Aspirations program is becoming a kind of social network to recruit women into technology companies. Apple, for example, enables this social networking by having a special application address only for Aspirations winners. Because these networks are working 24/7 and available virtually, Farmer believes that these may offer the same benefits as more than once-a-year gatherings such as the Hopper conference.

Farmer (2015) also notes that these Aspirations students are cooperative rather than competitive, and that this characteristic is somewhat different from an earlier generation of women in computing:

> They are not competitive with each other. I think this is really important. One of the things that's so neat about this group of young women is that they don't see any difference between real friendship and online friendship. That's one thing that's interesting because they've grown up as sort of a Facebook generation, so they don't differentiate those relationships. If I was applying for a scholarship and there's one scholarship available, would it be smart to share that opportunity with thousands of other girls that might apply for that scholarship? Probably not. But they do it. They share opportunities with each other. Not only do they share, here's a scholarship you could apply for, but then they like get together on a Google Hangout and help each other with their essays. They're very supportive and encouraging. When somebody gets some [award], they were like, "Go you," and when they fail at something, they were like, "Oh that was tough for me too." It's a very warm community. I think there's a group of women a little bit older than me that we often talk about this concept of 'pulling up the ladder'. There's this group of women that came up in the '80s and sort of like scratched and clawed their way through the business world wearing giant shoulder pads

and dressing like a man and acting like a man. Because of their small numbers, they've enjoyed some level of power in their low numbers. It's like I'm the only woman here and so there's sort of this pulling up of the ladder of like 'well, I had to scratch and claw my way here and so do you'.

One of the interesting offerings associated with Aspirations is a grant program that provides $3000 to a girl to partner with another organization, such as the Girls Scouts or the Boys and Girls Club, to teach computing to girls in their local community. NCWIT has not tried to standardize or brand these courses. Instead, it encourages the individual girl from Aspirations to design and teach the program in a way that seems appropriate for that particular environment. Almost 300 girls have taught these courses between 2012 and 2015.

Recently, NCWIT has entered into Project SEED, which is a partnership with the U.S. Departments of Housing and Urban Development, Education, and Energy to offer teaching in broadband-enabled STEM centers associated with the remodeling efforts for five public housing communities. Girls from the Aspirations program provide the computing content in these STEM centers.

The Aspirations grant program to teach computing to other girls is extremely cost-effective. It costs about $100 to educate an individual girl compared to a national average of between $700 and $2000 per student in other informal computing education courses. So far, the program has provided 150,000 h of computing education. Farmer (2015) also points out that the experience is good for the girl from the Aspirations program who is teaching the course:

One of the reasons girls can feel less confident going into computer programs in college is lack of prior experience and exposure. Here's an experience. You've taught this. You have it on your resume. It gives you that boost of like I'm valid, I've done this. It creates a relationship between you and some adult stakeholder who's working with you, whether it's the Boys and Girls Club staff person or a faculty member. You're building that relationship that's going to be someone who's going to do a letter of recommendation or connect you socially to other people who are going to help you. (Farmer 2015)

After the girls go away to college, they remain on a Facebook account with all the other winners – now a community of 4000 young women – and provide peer mentoring to one another. This online contact can mitigate social isolation if they are one of a few young women in their classes or majors, and it gives them a place to discuss social and technical problems they are facing. Members of NCWIT's academic alliance are offering these young women scholarships, and members of NCWIT's workforce alliance are offering them internships and jobs through this network. So far, 63 % of the young women have persisted with computing when they made the transition from high school to college, and 82 % are studying or working in some STEM discipline. Half of the students currently participating are non-white. Data is being collected from the program, with student's permission, to carry out a longitudinal analysis. Aspirations is planning to beef up its college program as well as move into middle schools, where they hope to attract 10,000 girls (Sanders 2014).

Another of NCWIT's innovative programs is its Extension Service program, funded by NSF and modeled after the long-standing agricultural extension program of the federal government. NCWIT's Extension Service aims to make systematic change in undergraduate programs so that they will increase the recruitment and retention of women students. The Extension Service staff, a group of social scientists familiar with the relevant research literature and with assessment methods, provides customized consultation to participating departments. "Consultants support academic departments in identifying opportunities, resources, and peers who are experienced in recruiting and retaining women. They guide departments in developing assessment plans to track progress, suggest resources, and provide ongoing consultation to help clients accomplish their goals" (NCWIT 2014). The Extension Services program is also partnering with ENGAGE: Engaging Students in Engineering (an NSF-sponsored extension service program closely aligned with WEPAN) to provide customized consultation to NCWIT Academic Alliance members in engineering as well as computer science departments, where women's participation is low.

NCWIT's Pacesetter Program, started in 2010, involves partnerships between companies and universities to increase the number of "Net New Women" in computing careers, by motivating technical women who were considering non-computing careers to make a computing career choice instead, or by retaining technical women in computing who were contemplating a departure from the field.[44] NCWIT claims a yield of over 2000 New Net Women in 2014 alone.

Pacesetters matches a top-down approach by executives who "can influence people, policy, and resources within the organization" with a bottom-up approach by change leaders who can build "out an extended team, including people in a variety of key roles across the organization" (Ross et al. 2012). Important to the program has been the requirement that organizations measure progress in numbers of women added in their organization to the computing community by a specific short-term date. Each organization chooses its own methods that are sensitive to its work environment, although the organizations are encouraged to draw upon practices that social science research has shown to be promising. For example, among the university Pacesetter members, the University of Texas at Austin offers a program that reaches out to freshman women who have not yet declared a major; while Virginia Tech sends teams of computer science faculty, staff, and students out to connect with women students in the dormitories and has created a way that students can design their own majors that match computer science with another discipline. On the corporate side, Intel has instituted a workshop in which senior technical women provide career development training to mid-career technical women; while Google has started a program with college women that, for example, gives the students mock interview practice. NCWIT convenes an annual Pacesetters meeting so that

---

[44] According to Ashcraft (2015), one of the reasons for focusing on Net New Women is that the companies do not necessarily want to release before and after data but are more willing to release data about the net change in the number of women over a given period of time. There has also been some opportunity to broaden the metric from Net New Women to metrics that concern number of hires, reductions in attrition, number of women promoted, etc. The companies participating in Pacesetters are also encouraged to have measures that indicate the effectiveness of particular interventions they are trying.

the various organizations can share ideas. NCWIT leadership makes regular site visits to each of the Pacesetter organizations to meet with the leadership and hold them accountable. Ashcraft (2015) argues that one of the greatest benefits of the Pacesetter program has been a deeper level of cooperation between the academic and corporate members of NCWIT.

The most visible part of the program is an advertising campaign called Sit With Me, which NCWIT developed in partnership with the brand marketing firm BBMG. People are invited to sit on a red chair in solidarity with computing through a campaign carried out on web pages, Facebook, Twitter, and at professional meetings, with the theme "Sometimes you have to sit to take a stand" (Ross et al. 2012).

NCWIT sponsors many other programs in addition to those described above. They include programs that: provide resources about computing to high school counselors; engage more community colleges in NCWIT activities; help colleges and universities create and support student women-in-computing organizations; collect and share lecture notes, homework assignments, and projects to enhance broad student engagement (called the Engage CSEdu program); provide a seed fund of up to $10,000 so that individual departments can develop and implement new programs to enhance recruitment and retention of women; conduct research on the organizational culture of technology start-ups, the reasons underlying the low participation of women, and possible interventions; produce audio interviews with women entrepreneurs in IT careers (called the NCWIT Heroes program); and create a partnership between NCWIT and Oxford Economics to study how cultural issues affect women on global tech teams.

8.5   Conclusions

The main purpose of this chapter has been to present four case studies of organizations promoting the participation of advancement of women in computing. We will not provide detailed comparison of these four organizations here. We will simply close with a quotation from Valerie Barr, which reflects on these four organizations and points to their complementary strengths:

> ABI has mastered doing the big event: The Hopper Conference, the Women of Vision awards. They have a level of connections to the corporate side that is pretty phenomenal! NCWIT really brings in the research side and is sort of institutional membership, and the alliance structure, which I think has value in [creating] this collection of best practices and making those available. ... CRA is again institutional membership, and CRA-W is really pushing the research track, really encouraging faculty and grad students and undergrads in the research direction. ACM and ACM-W, we are the only one that is about individual membership, and I think that's important. I think that ... because of the nature of our Celebration events, our scholarships, and our student chapters, reach out to individual students in a way that the other groups don't necessarily. (Barr 2014)

References

Abbate, Janet. 2012. *Recoding gender: Women's changing participation in computing*. Cambridge, MA: MIT Press.

ACM Committee on Women in Computing. 1997. Annual report, July 96–June 97.

ACM Committee on Women in Computing. 1998. Annual report, July 97–June 98.

ACM Committee on Women in Computing. 1999. Annual report, July 98–June 99.

ACM Committee on Women in Computing. 2000. Annual report, July 99–June 00.

ACM Committee on Women in Computing. 2002. Annual report, July 01–June 02.

Ashcraft, Catherine. 2015. *Oral history interview by William Aspray*. Charles Babbage Institute Oral History Collection, December 7.

Aspray, William. 2016. *Paticipation in computing*. London: Springer.

Barr, Valerie. 2014. *Oral history interview by William Aspray*. Charles Babbage Institute Oral History Collection, October 13.

Borg, Anita. 1993. Why systers excludes men. *Computing Research News,* September 3, 5, 13.

Borg, Anita. 1994. Hopper celebration an 'Unqualified success'. *Computing Research News,* September.

Borg, Anita. 2001. Oral history interview by Janet Abbate. *IEEE History Center Oral History Collection,* January 5. . http://www.ieeeghn.org/wiki/index.php/Oral-History:Anita_Borg. Accessed 4 Apr 2016.

Borg, Anita, Annie Warrn, and Mary Jo Doherty. 1994. Conference celebrates women in computing. *Computing Research News,* January 3–5.

Brown, Patricia Leigh. 2015. Silicon Valley, seeking diversity, focuses on blacks. *New York Times,* September 3. http://www.nytimes.com/2015/09/04/technology/silicon-valley-seeking-diversity-focuses-on-blacks.html?_r=0. Accessed 21 Jan 2016.

Camp, Tracy. 1997. The incredible shrinking pipeline. *Communications of the ACM* 40(10): 103–110.

Camp, Tracy. 2002. Women and computing. Special Issue. *Inroads, SIGCSE Bulletin* 34 (2).

Camp, Tracy. 2015. *Oral history interview by William Aspray*. Charles Babbage Institute Oral History Collection, January 9.

Computing Research Association. 2005. CRA-W receives national science board award. *Computing Research News,* September. http://cra.org/crn/2005/09/cra-w-receives-national-science-board-award/. Accessed 4 Aug 2015.

Computing Research Association. 2013a. Center for evaluating the research pipeline (CERP): Director's welcome. *Computing Research News,* April. http://cra.org/crn/2013/04/center_for_evaluating_the_research_pipeline_cerp_directors_welcome/. Accessed 4 Apr 2016.

Computing Research Association. 2013b. Center for evaluating the research pipeline. *Computing Research News,* May. http://cra.org/crn/2013/05/center_for_evaluating_the_research_pipeline/. Accessed 4 Apr 2016.

Danyluk, Andrea. 2013. Collaborative research experiences for undergraduates: The CREU program still going strong at 15. *Computing Research News,* September. http://cra.org/crn/2013/09/collaborative_research_experience_for_undergraduates_the_creu_program_/. Accessed 23 Nov 2015.

Farmer, Ruthe. 2013. Expanding the pipeline – Growing the tech talent pool: NCWIT aspirations in computing program scales up. *Computing Research News,* March. http://cra.org/crn/2013/03/expanding_the_pipeline/. Accessed 5 Jan 2016.

Farmer, Ruthe. 2015. *Oral history interview by William Aspray*. Charles Babbage Institute Oral History Collection, December 7.

Feigenbaum, Joan. 1992. CRA committee is creating database of women scientists. *Computing Research News,* March. http://archive.cra.org/CRN/issues/9202.pdf. Accessed 4 Dec 2014.

Francioni, Joan. 1998. Committee on the status of women in computing research. *Computing Research News,* November 2, 9.

Goral, Cindy, and Dianthe Harris. 2006. From the inside, out. *Computing Research News*, January. http://cra.org/crn/2006/01/from_the_inside_out/. Accessed 23 Nov 2015.

Gurer, Denise. 2015. *Oral history interview by William Aspray*. Charles Babbage Institute Oral History Collection, January 9.

Guynn, Jessica, Paul Overberg, Marco della Cava, and Jon Swartz. 2014. Few minorities in non-tech jobs in Silicon Valley, USA Today finds. *USA Today*, December 30. http://www.usatoday.com/story/tech/2014/12/29/usa-today-analysis-finds-minorities-underrepresented-in-non-tech-tech-jobs/20868353/. Accessed 21 Jan 2016.

Irwin, Mary Jane, and Francine Berman. 1996. CRA-W advancing the status of women in CS&E. *Computing Research News*, September 3–4, 12.

Jamieson, Leah H. 2001. Women, engineering, and community. *Computing Research News*, May. http://archive.cra.org/CRN/issues/0103.pdf. Accessed 23 Nov 2015.

Jamieson, Leah. 2015. *Oral history interview by William Aspray*. Charles Babbage Institute Oral History Collection, February 9.

Jeffries, Robin. 2006. Systers: The electronic community for women in computing. *Computing Research News,* September. http://cra.org/crn/2006/09/systers-the-electronic-community-for--women-in-computing/. Accessed 4 Aug 2015.

Jones, Stacy, and Jaclyn Trop. 2015. See how the big tech companies compare on employee diversity. *Fortune*, July 30, http://fortune.com/2015/07/30/tech-companies-diveristy/. Accessed 21 Jan 2016.

Kang, Cecelia, and Todd C. Frankel. 2015. Silicon Valley struggles to hack its diversity problem. *Washington Post*, July 16. https://www.washingtonpost.com/business/economy/silicon-valley-struggles-to-hack-its-diversity-problem/2015/07/16/0b0144be-2053-11e5-84d5-eb-37ee8eaa61_story.html. Accessed 21 Jan 2016.

Klawe, Maria. 2005. Oral history interview by William Aspray. *Computer Educators Oral History Project, Lisbon, Portugal*, June 27. http://www.cs.southwestern.edu/OHProject/KlaweM/klawe-200506-20100620.pdf. Accessed 11 Dec 2014.

Klawe, Maria, and Nancy Leveson. 1993. CRAW advances status of women in CS&E. *Computing Research News*, November 3–4.

Lopez, Dian Rae, Stephanie Sides, and Ann Redelfs. 1996. New booklet reaches out to young women. *Computing Research News*, January 3.

Manjoo, Farhad. 2014. The business case for diversity in the tech industry. *New York Times*, September 26. http://bits.blogs.nytimes.com/2014/09/26/the-business-case-for-diversity-in-the-tech-industry/. Accessed 21 Jan 2016.

Martin, C. Dianne. 1993. CRA presents windows of opportunity symposium. *Computing Research News,* September 4.

McCorduck, Pamela, and Nancy Ramsey. 1997. *The futures of women: Scenarios for the 21st century*. New York: Grand Central Publishing.

National Center for Women & Information Technology. 2014. *Organization website*. www.ncwit.org. Accessed 20 Dec 2014.

Ross, Jill, Elizabeth Litzler, J. McGrath Cohoon, and Lucy Sanders. 2012. Improving gender composition in computing. *Communications of the ACM* 55(4): 29–31. https://www.ncwit.org/sites/default/files/file_type/improving_gender_composition_in_computing_april_2012_communications_of_the_acm_copy.pdf. Accessed 21 Sept 2015.

Sanders, Lucy. 2014. *Oral history by William Aspray*. Charles Babbage Institute Oral History Collection, December 29.

Soffa, Mary Lou. 2011. CRA-W 20th anniversary celebration. *Computing Research Association Committee on Status of Women in Computing Research*. PowerPoint Presentation. Obtained from the author.

Stout, Jane. 2013. Providing a new way to evaluate diversity initiatives in computing research. *Computing Research Association*, August. http://cra.org/crn/2013/08/cerp_article_august-2013/. Accessed 23 Nov 2015.

Stout, Jane. 2015. Disseminating CERP research findings to promote diversity in computing and other STEM fields. *Computing Research News*, February. http://cra.org/crn/2015/02/disseminating_cerp_research_findings_to_promote_diversity_in_computing/. Accessed 23 Nov 2015.

Weyuker, Elaine. 2014. *Oral history interview by William Aspray*. Charles Babbage Institute Oral History Collection, October 8.

White, John. 2015. Oral history interview by William Aspray, *Charles Babbage Institute Oral History Collection*, June 26.

Whitney, Telle. 2002. Oral history by Janet Abbate. *IEEE History Center, Palo Alto, CA* . July 16. http://www.ieeeghn.org/wiki/index.php/Oral-History:Telle_Whitney. Accessed 5 Apr 2016.

Whitney, Telle. 2014. *Oral history by William Aspray*. Charles Babbage Institute Oral History Collection, September 17.

Chapter 9
Organizations That Help Underrepresented Minorities to Build Computing Careers

Abstract This chapter discusses the creation of organizations focused on opening computing careers to underrepresented minorities. The Black Data Processing Associates (BDPA), formed in 1975, is an organization of individual members with a particular interest in advancing the careers of African-American IT workers, but also concerned with engaging and preparing African-American children for these careers. The Association of Computer and Information Science at Minority Institutions (ADMI) was founded in 1989 to help computing departments at minority institutions to deal with the various issues faced by academic administrators trying to offer an effective educational program. The Coalition to Diversify Computing (CDC) was formed in 1996 primarily to prepare undergraduate and graduate students in computer science for research careers in universities or industry. The chapter closes with a discussion of the Center for Minorities and People with Disabilities in Information Technology (CMD-IT), formed in 2007, which serves a similar role for underrepresented minorities and people with disabilities as NCWIT does for women.

This chapter discusses the creation of organizations focused on opening computing careers to underrepresented minorities. Most of the organizations that had broadening participation in computing as a specific mission were formed in the late 1980s or the 1990s. One of the four organizations discussed in this chapter, however, was created much earlier and one much later. The Black Data Processing Associates (BDPA) was formed in 1975, more in line with the founding of the broadening STEM organizations that came about as a direct response to the civil unrest in the 1960s and early 1970s (as discussed in Chap. 7). BDPA is an organization of individual members with a particular interest in advancing the careers of African-American IT workers, but also with a mission of engaging and preparing African-American children for these careers. The Association of Computer and Information Science at Minority Institutions (ADMI) was founded in 1989 to help computing departments at HBCUs, HSIs, and TCUs deal with the various issues faced by academic administrators when trying to offer an effective educational program. The Coalition to Diversify Computing (CDC) was formed in 1996. It is primarily focused on programs to prepare undergraduate and graduate students in computer science for research careers in universities or industry, and to a lesser extent for other IT careers. We close with a discussion of the Center for Minorities

© Springer International Publishing Switzerland 2016
W. Aspray, *Women and Underrepresented Minorities in Computing*,
History of Computing, DOI 10.1007/978-3-319-24811-0_9

and People with Disabilities in Information Technology (CMD-IT), which grew out of conversations begun in 2007. In some ways, it serves a similar role for underrepresented minorities and people with disabilities that NCWIT does for women. Together with the four organizations discussed in Chap. 8 (ABI, CRA-W, ACM-W, and NCWIT), the four organizations discussed here represent the primary organizations focused on broadening participation in computing.

9.1   Black Data Processing Associates (BDPA)

Black Data Processing Associates is the oldest and largest organization focused on minorities in the IT professions.[1] The organization was founded by Earl Pace and David Wimberly.[2] Pace had studied business with an emphasis on personnel issues and sought a job in labor relations to support his family. He applied to the Pennsylvania Railroad for a position as a labor relations specialist. The company had already filled this position but asked Pace if he wished to be considered for a computer programmer position in the newly created information systems department. He knew nothing about programming but agreed to take the aptitude test and was offered a position. He took advantage of training provided by the company and became a competent programmer. After 2 years, he left the railroad to join a start-up financial services firm.

In his creation story about BDPA, Pace discusses two shaping incidents. One occurred after he had been promoted to a management position with the financial services firm. One of the company directors asked him if he was trying to recreate the United Nations, referring to the large number of people of color he had hired. He replied that he hired on the basis of applicant talent, and the board member went away unhappy. The lesson that Pace drew from this episode was that that people who are in positions of hiring have significant power to shape the workplace and in particular to give opportunities to qualified minority professionals. The other episode concerned a professional IT conference in Arizona that he attended; of the 200

---

[1] The organization expanded its name to BDPA—Information Technology Thought Leaders in 1996 in hopes of becoming better known.

This section is based on the organization's website (bdpa.org); BDPA Education and Technology Foundation (2006), Norton (2013), Anon. (2009, 2010, 2012), Gibson (1997), Summers (2013), Warfield (1991), Hicks (2011), Fairley (1995), Bates (1993), Davis (1997), Watson (2014), Berkley (1999), Muhammad (1996, 1998), Sherman (2011), Larson (2011a, b), Roach (2006), Burgess (2004), Carter (2014), Haskell (2010), Tennant (2009), and Pace (2014).

[2] Wimberly had an office on the same floor as Pace, and they discussed the needs for an organization like BDPA. Pace and Wimberly shared the cost of a hotel meeting room to convene a group of African American IT professionals to discuss the need for an organization like BDPA. Wimberly died soon thereafter, even before the BDPA bylaws were written. Wimberly met few of the early volunteers working to build BDPA, but he did have an influence on Pace, who has been a driving force in BDPA throughout its entire history (BETF 2014).

people there, he was the only African American. He recalled the isolation that he had felt as a result.

In order to build up a community of African American IT professionals, Pace first created BDPA in 1975 and then his own company, Pace Data Systems, the following year. Stand-alone BDPA chapters were founded in Philadelphia, Washington DC, and Cleveland between 1977 and 1979, and Pace then worked to reorganize BDPA as a national organization with local chapters.

The organization currently has approximately 2000 professional members. The majority of them hold positions as system administrators, system analysts, or IT managers. The almost 50 local chapters, which are located primarily east of the Rocky Mountains, are the locus of most BDPA activities. The organization has a greater percentage of female members (42 %) than most IT organizations.[3]

BDPA's motto is "Advancing Careers From the Classroom to the Boardroom." As one might expect, much of the focus is on advancing members through their careers, from beginning programming positions, through various levels of management, to executive positions. There is a strong culture of expectation that BDPA members will give back to the community – to help children as well as adults who have not progressed as far in their career. So, in addition to professional development activities, BDPA has an active program to interest children in IT careers and help them gain an appropriate education.[4]

BDPA professional activities include networking events, technology certification programs, webinars on best practices and technology trends, an online job board for posting both jobs and resumes, a career fair at the annual meeting, an IT Senior Management Forum for mentoring Black professionals who want to advance to executive-level positions, and a professional magazine. BDPA also partners with its corporate members to help them recruit, retain, and advance minority professionals in their organizations and to help these companies to reach out to the communities in which they operate. For example, BDPA runs IT training workshops and career development seminars in IT in the local communities of their corporate members, teaching adults about the opportunities for careers (as a means to inform both the adults and their children).

BDPA's flagship program for students is the Student Information Technology & Education Scholarship (SITES). Local chapters provide training programs for children in their community. In a typical year, BDPA provides training to between 800 and 1200 high school students in computer programming and web development.[5] Its

---

[3] Almost half of the leadership positions at both the national and local level have been held by women since the founding of the organization, and this pattern continues today (Anon. 2012).

[4] Larson (2011b) describes the poster child of BDPA's student and professional activities: Stephanie Brown, who had participated in one of the organization's Saturday programming classes in Washington, DC when she was in ninth grade; was inspired at age 14 by the BDPA annual conference; completed internships with Northrop Grumman, Lockheed Martin, and Deloitte by age 21; graduated with a major in management science and engineering from Stanford; was recruited out of college by Microsoft without applying; and quickly moved up the management ranks.

[5] Watson (2014) describes an example of a web programming, web development, and database design course offered at Bloomfield College in New Jersey, in collaboration with BDPA.

computer literacy programs at elementary and middle schools reach thousands of children.[6] These educational activities are intended to promote general academic achievement, interest students in IT careers, and teach skills about the business world such as teamwork and professional business behavior. Each chapter sends a student team of 3–5 students to the national conference each year to compete in the National High School Computing Competition against the teams from other chapters.[7] The winning teams receive scholarship awards for each of their team members. BDPA also supports a Youth Technology Camp for high school students, providing them with hands-on experience and chances to network with young entrepreneurs.

One of the key events is the annual BDPA Technical Conference. The first one was held in 1979.[8] This 4-day conference moves around the country from year to year. It includes a technical forum for IT professionals, entrepreneur events, a career fair, and a technology expo.

In 1992 BDPA created the BDPA Education and Technology Foundation to raise funds from corporations for student scholarships and various BDPA programs. In a typical year, BDPA provides more than $100,000 in scholarships to high school and college students.[9] Students have used these scholarships, for example to attend HBCUs such as Bowie State and Hampton as well as PWIs such as Rice, Illinois, Michigan, and Washington.

9.2   Association of Computer/Information Science and Engineering Departments at Minority Institutions (ADMI)

In 1989 the Association of Computer and Information Science at Minority Institutions (ADMI) was founded as a national organization to identify and address the problems experienced by computing-related departments at minority institutions. During the 1980s, a number of the computer science faculty and chairs from minority-serving institutions attended ACM's computer science education conference (SIGCSE). Larry Oliver, a program officer at NSF, would often organize a session at SIGCSE concerning the interests of minority institutions, e.g. NSF programs that might be of interest to the minority institutions. At one of these sessions, the audience began to discuss the formation of a new organization for

---

Larson (2011a) describes the Charlotte (NC) High School Computer Academy.

[6] On computer literacy camp for younger students, see Warfield (1991).

[7] For an account of the national competition, see Gibson (1997), Summers (2013), and Carter (2014).

[8] For a list of the annual conferences and their themes, see http://www.bdpa.org/?page=PastBDPA Conferences.

[9] For more information about BETF and its scholarship programs, see BETF (2006), Muhammad (1998), and Hicks (2011).

minority-serving institutions in computing. Two organizational models were considered: one was an individual membership model similar to that of the National Association of Mathematics, an African-American organization; the other model was a departmental membership model similar to that used by Computing Research Association for its biennial department chairs meeting at Snowbird, Utah. The computer science model was selected, and the chairs began meeting with support from NSF that Oliver arranged (ADMI 2014; Lawrence 2014).

The meetings included discussions of challenges that the computer science departments at MSIs were facing, but there were also visitors who came to present information about new developments in the field or provide professional development training. Challenges commonly faced by these departments included recruitment of qualified faculty in sufficient numbers, as well as obtaining sufficient resources. In the beginning years of ADMI, resource issues primarily revolved around building an adequate Internet infrastructure for departmental needs; and NASA helped some of these schools to obtain that infrastructure. There was also considerable discussion about keeping the curriculum up to date. With heavy teaching loads and typically lack of a track record in obtaining external research grants, it was difficult for these departments to be competitive for NSF and other external research funding. One of the other NSF program officers, Harriet Taylor, recorded the names of people attending the ADMI meetings and invited them to NSF as reviewers, so that they could learn first-hand what made for a successful proposal. Over time, the Internet infrastructure issue was resolved but new equipment needs continued to crop up. Faculty retention remained a major issue, and finding scholarship funds for students so that they did not drop out of college became increasingly important (Lawrence 2014).

Beginning in 1996, ADMI moved from a department-chair meeting model to a symposium model. Under the new format, which continues to this day, the organization's major event is an annual symposium. Faculty from both member and non-member departments attend to listen to research lectures. Students are also invited and given the opportunity to report on their own research efforts through posters or talks (Lawrence 2014).

At the 1997 ADMI symposium, for example, CRA organized an Opportunities in Computing Research Fair. The fair enabled minority graduate students to explore computing research jobs in academia and industry and hear discussions about how to have a successful research career. Minority undergraduates had the chance to meet with representatives of graduate departments and also talk with minority graduate students to learn what it is like to attend graduate school. Minority high school students had a chance to meet with representatives of undergraduate programs in computer science (Cartwright 1997).

ADMI has not changed its mission statement since 2003. It states that ADMI offers a forum for faculty and students to exchange information and ideas about how to make computing education more effective at minority institutions. This involves enhancing curriculum, research, and equipment, as well as promoting the professional advancement of faculty. ADMI identifies problems and seeks out their solutions; communicates these issues to various interested parties in the government, industrial, and academic sectors; and raises funds to solve these problems.

The annual conference is ADMI's main activity.[10] Consider the 2003 annual conference, when the mission statement was first proclaimed by ADMI president Andrea Lawrence of Spelman College. This was the 16th annual conference and the 8th to include presentations by both faculty and students. It included "[s]tudent research presentations, reports by faculty on successful experiences, reports on ways to search for success, a proposal logistics workshop, poster presentations, a graduate opportunities workshop, and an opportunities fair" (Lawrence 2003). In a typical year, about 50 students attend the ADMI Symposium, but there have been years when funding was more readily available in which more than 100 students attend[11] (Willis 2015).

Ten years later, in 2013, the annual conference remained the biggest event of the year – that year with a theme of cloud computing (ADMI 2013a, b). However, because of additional funding ADMI was able to undertake other projects. A group of faculty, students, and staff from ADMI member institutions attended a summer school on use of the cloud for computational science and engineering organized by Geoffrey Fox, a professor at Indiana University, together with one of his doctoral students, Jerome Mitchell, who himself was a former ADMI student member. ADMI was also participating in NSF-funded projects sponsored by the Center for Remote Sensing of Ice Sheets, with the University of Kansas as the center's home – and with Elizabeth City State University (North Carolina) leading ADMI's involvement.[12] Related activities included an 8-week research-training program including ADMI students, support for students to travel to Germany to report their research at the IEEE Geoscience and Remote Sensing Conference, and participation of several ADMI institutions in an NSF-funded project on Cyberinfrastructure for Remote Sensing of Ice Sheets (ADMI 2014).

Another benefit of ADMI is that it serves as a support network that informs the faculty about opportunities to share with their students.

---

[10] Links to the ADMI symposia flyers, which include some history of the past year, from 2002 through 2014 are available at ADMI (2014).

[11] The highlight of the Symposium for the students is the banquet, where the awards are presented to students on the basis of their research talks. However, the tutorials or skill-building exercises are also important. The symposium is of particular importance to the faculty, as Willis (2015) explained: "Kind of like a birds of a feather [sessions] where you get together, you talk. You could discuss your problems. You can discuss different types of solutions and of course the research that's presented. You learn some other things. Of course, is the ability to join forces as far as maybe some grantsmanship or research projects concerned, collaboration."

[12] Under the direction of Linda Hayden at Elizabeth City State University, students are given the opportunity to participate in a Research Experiences for Undergraduates associated with NSF's CReSIS program. The students who participate at Elizabeth City are typically younger students who have no previous research experience. Based on what they learn there, these students are able the following year to engage in a more advanced research experience at a remote site such as the University of Kansas or Indiana University. Many of the students involved with this program pursue masters programs in computing or computational science, and this engagement has also led several students to undertake doctoral study (Lawrence 2014).

Hampton University's involvement gives a good picture of ADMI's impact. Hampton had created its computing programs out of the math department in the early 1980s. At first it had a math-centered curriculum that involved taking a number of advance math courses as well as typical computer science courses, but eventually it adopted ACM's model curriculum. In addition to an undergraduate computer science degree, Hampton offered an information systems degree. Between these two programs, some years Hampton enrolled as many as 300 majors – making it one of the largest producers of computing degrees among the HBCUs. These numbers dropped off to about 150 students after the dot-com bust. Since 2015 Hampton has been offering a third undergraduate degree in cybersecurity, which complements a long-standing a masters degree in the same field.

Hampton is not only one of the largest computer science programs at an HBCU, it is also one of the academically strongest. The department receives significant external funding from both NSF and the Department of Defense. Its students are drawn from across the entire United States. Graduates are commonly hired by defense companies such as Lockheed Martin, computing companies such as Google and Microsoft, and financial companies such as Merrill Lynch. Only about 2 % of the students go immediately to graduate school, but over time almost half of them do so:

> ...to me it's astonishing that so many of our students, of our general graduates from the department, within four or five years they enroll in a graduate program somewhere. All of that preaching that we do about the graduate school may not sink in quite as [quickly as we might like.] ... When they graduate they're looking at okay, I can get this high salary. There are things that I want to buy, etc. After they get out and they're in the job, a lot of the things that we advise them about, they see that some of the things we talk about are actually true, and then they realize that in order for them to move up or to gain better professional standards or to improve themselves, they need to go to graduate school. (Willis 2015)

Hampton had an active ACM chapter on campus. It tried to institute a BDPA chapter, but it never took hold. Hampton was a member of ADMI from the beginning, and once the symposium model of operation was established in 1996, Hampton students attended the symposium every year.[13] Each year, the Hampton faculty would select at least six students to attend. "They were selected on the basis of research of a quality that [they] can make a presentation and also have [their] paper published. I think that's very important. The second thing is interest in graduate school because one of the core elements is to get the students to go to graduate school" (Willis 2015). The students consider it a great honor to be invited to the ADMI Symposium, and Willis regards it as important to their academic and professional development:

> The first thing to me is that they come to the symposium, they're with 50 other students from minority institutions, and they see the quality of work that's being done. It's like, I'm

---

[13] Hampton, Spelman, University of the District of Columbia, Florida A&M, Winston Salem State University, and Elizabeth City State University students regularly attended. Other schools, such as Fisk or Xavier in St. Louis, would send students when there was external funding available (Willis 2015).

a Hampton student and in this department we have maybe 150 other students. We do good work. I think we're doing great work, etc. Then they go and they see all these other students that are like them that are doing great work as well. Now it's an enhancement of their self-image, if you will, and you can almost always notice an improvement in their attitudes when they return to the home school. Not only in Hampton, but I talked to the other faculty and they say the same thing. They become more serious. They become more professional. They are now thinking about how well prepared are they going to be for whatever comes next. (Willis 2015)

9.3   Coalition to Diversify Computing (CDC)

The Coalition to Diversify Computing (CDC) was formed in 1996.[14] It is jointly operated by three computing professional societies: the Association for Computing Machinery (ACM), Computing Research Association (CRA), and the Computer Society of the Institute of Electrical and Electronics Engineers (IEEE CS). The ACM has been the largest and most steady funder of CDC, while CRA has provided the greatest degree of program development, management, and assistance. CDC is primarily focused on programs for undergraduate and graduate students in computer science – to prepare them for research careers in universities or industry, and to a lesser extent for other IT careers.

In 1995, Bryant York, a computer scientist from Northeastern University, convened a workshop with NSF funding at the Airlie House in Warrenton, Virginia to discuss ways to increase diversity in computing. The meeting was stimulated by the Taulbee Survey results from 1993, which had shown the serious underrepresentation of minorities in computer science: less than 1 % of the Ph.D.s in computer science were earned by African Americans and less than 2 % by Hispanics, even though a much larger percentage (21 %) of the U.S. population was African American or Hispanic. The intention of the workshop was to identify root causes of this underrepresentation and make suggestions on how to address it.[15] Apparently the importance that mentoring contributed to the success of minority students in computing had only become widely recognized within the 5 years before the workshop; and mentoring was the most widely discussed solution, including how to scale it. Other topics discussed at the workshop included giving greater emphasis to teaching (and metrics to enhance effective teaching) in the colleges and universities, whether MSIs should create their own doctoral programs or schools should focus instead on

---

[14] This material about the Coalition to Diversify Computing is based primarily on the twice-a-year reports that CDC makes to the Computing Research Association board of directors (CRA 2000 – 2013) and CDC (n.d.). This section also draws on interviews that the author conducted with Andrew Bernat (2015), Valerie Taylor (2015), Elaine Weyuker (2014), and John White (2015).

[15] On why American Indians were not one of the underrepresented groups discussed at the meeting, the final report states: "…because the number of Native Americans enrolled in CS programs is so small, their academic experience is significantly different from that of other minorities. We acceded to the suggestion that we focus the workshop on issues relating only to African Americans and Hispanics." (York et al. 1995)

being the best at their primary mission, and various improvements in computing instruction at the K-12 levels.[16] There was considerable hand wringing about how difficult it would be to effect real change in an era of federal budget slashing, and this was accompanied by a stern criticism of the affirmative action programs of the 1960s, which "missed the broader issue of human resource development" (York et al. 1995).

Perhaps the most important outcome of the workshop was the formation of the Coalition to Diversify Computing. The three computing societies – ACM, the IEEE CS, and CRA – had each formed its own committee to look into issues concerning the situation for women in computing. However, they decided that – especially since they wanted only PhD-holding faculty members to serve on this new committee on diversity in computing – there were too few qualified minorities for each of the three computing organizations to populate its own committee. So a single committee, with the three computing societies as parent organizations, was formed in 1996. Sandra Johnson from IBM and Andrew Bernat from the University of Texas at El Paso were selected as the first co-chairs (Taylor 2015; White 2015; Baylor and Redelfs 2000).

CDC was modeled after the highly successful CRA-W organization (described in Chap. 8). However, the fact that the three parent organizations had slightly different missions raised some early questions within CDC about its mission. CRA-W, like its parent organization CRA, was focused on computing research; whereas ACM and the IEEE Computer Society had a broader scope, involving computing practice as well as research; and both ACM and the Computer Society gave more attention to undergraduate education than CRA did. The situation was eventually resolved by creating a broad range of programs that most closely mapped on to ACM's purview (including in the first few years some K-12 and undergraduate programs). Some of CDC's most successful program were its mentoring programs, which were modeled after those of CRA-W (and after 2006 the CDC programs were merged and run in collaboration with those of CRA-W). CDC also followed CRA-W's model that every member of the committee must be actively engaged in a project. According to one discussion by the chair of the Coalition:

> Some strategies employed by CDC projects that have been effective in achieving these objectives are: 1) accessibility to role models, 2) enhancement of student confidence, 3) availability of financial and emotional support, and 4) awareness of education and career possibilities. (Teller 2004)

The three-parent organizational scheme proved somewhat burdensome to CDC volunteer leaders over the years. It required the CDC chair to become familiar with all three organizations and make separate formal reports to each of them.[17] There was also confusion about each parent organization's responsibility to

---

[16]York et al. (1995) provides demographics of the workshop participants: 26 academic (mostly professors, but 13 had also been deans or department chairs), 4 government, and only 2 industrial. Of these, only 5 were women, 20 were African American, 7 were Hispanic, 7 were White, and 1 was Middle Eastern. The group came from 9 minority institutions and 8 majority institutions.

[17]CDC had particular difficulty in its communications with IEEE Computer Society (Taylor 2015).

CDC. Eventually, in 2003, all parties signed a memorandum of understanding, which spelled out that ACM would handle financial matters, IEEE Computer Society would handle the website and listservs, and CRA would handle administrative matters.

One major programmatic area for CDC involves four activities concerning research experience and mentoring for undergraduates: Collaborative Research Experiences for Undergraduates (CREU), Distributed Research Experiences for Undergraduates (DREU), Discipline-Specific Mentoring Workshops (DSW), and the Distinguished Lecture Series (DSL) supported through a grant from NSF's Broadening Participation in Computing program.[18] Three of these activities were started by CRA-W and CDC became a joint partner around 2009, when the programs were expanded to include underrepresented minorities as well as women. The Discipline-Specific Mentoring Workshops were joint activities with CRA-W from the beginning. CDC had some antecedent activities in these areas, but they had not been as successful as those run by CRA-W. These four activities are described in Chap. 8, in the section about CRA-W (CDC n.d.).

The other major activity of CDC has been the Richard Tapia Celebration of Diversity in Computing Conference. In 2000, a small meeting was convened at the Chicago airport by the new CDC chair, Valerie Taylor. The group included Taylor, Johnson, Bernat, and York. They discussed the limited effectiveness of CDC so far and the fact that it was not well known in the minority computing communities. Taylor had attended the first Grace Hopper conference in 1994, was on the program committee for the 1997 Grace Hopper conference and was chair of the 2000 program committee. She was impressed by the publicity and recognition among computing women that came through that conference. So this small group meeting in Chicago decided to sponsor a similar conference for minorities.[19] They decided that, like the Grace Hopper conference, it should be named after someone. The obvious choice was Richard Tapia, the computational scientist at Rice University who had a national reputation for his work with minority students in STEM and who already had named in his honor the Blackwell-Tapia conference, which recognizes

---

[18] CDC had some additional early activities. These included a Traveling Graduate School Workshop in which graduate students and faculty members from research universities would visit HBCUs and other Minority-Serving Institutions "to provide students with an honest picture of the value and the downsides of enrolling in Master's and Ph.D. programs. Presentations by visiting graduate students and faculty detail what graduate school is like, how to apply, how to get financial aid, and the benefits of attending. Further, the workshop plays a role in connecting students with possible institutions for further study." Another early example was a Pre-Conference Minority Networking Event before a Usenix conference so that the students would not feel so isolated and so that they would learn how to get the most out of a conference. CDC also organized a best practices workshop for recruiting and retaining minority graduate students (Baylor and Redelfs 2000).

[19] "[Valerie Taylor a]nd I were at Grace Hopper several times over the years, and at one of the meetings she invited me to participate in this organizational meeting for Tapia. At that point I don't think it was named, they just wanted to have something to reach minorities in computing ... in the same way that Grace Hopper was reaching women; and there was a meeting in Atlanta and we got it organized and actually it became something that was basically run by the CDC..." (Lawrence 2014)

mathematical excellence by minority researchers. When CDC asked Tapia if he would be willing to have the new conference named in his honor, he demurred twice, the second time replying to the effect "do you know something I don't know? Am I dying? Is that why you want to name this conference after me?" But an impassioned letter from York explained to Tapia why it would serve the needs of the community to name it after him, and Tapia then agreed (Taylor 2015).

The first Tapia Conference was held in 2001, only a month after the 9/11 terrorist attacks.[20] The organizers expected 70 or 80 people, but 164 people showed up. The conference was subsequently held every other year through 2013 – with attendance growing to over 500.[21] Richard Tapia gave the keynote address at the first conference. There were also technical sessions on topics ranging from geographic information systems to "culture-specific approaches to e-learning." (Taylor 2002) Over the years, the conference has been helpful in fighting isolation often experienced by minority students and in enabling people to find mentors and role models. Speaking of the 2005 Tapia conference, program co-chair Elaine Weyuker (2014) said:

> [A] young man stood up and he said, "I just have to tell everybody that this is the first time in my life that I have been at a professional, technical thing where I look around the room and I see other short brown men who look like me; and you have no idea what does this for me." And you know you could feel everybody's eyes welling up.

For the tenth anniversary conference – in 2011 – the organizers chose David Patterson, the well-known computer scientists from Berkeley who had served as both the president of ACM and the chairman of the board of CRA, as the conference chair in order to give the conference a makeover. Patterson held discussions with past organizers about what the conference should be about and also surveyed students. The program committee decided, after significant discussion and some disagreement, that:

> The goal of the Tapia Conferences is to bring together undergraduate and graduate students, professionals, and faculty in CS&E from all backgrounds and ethnicities to:
>
> - Celebrate the diversity that currently exists in CS&E;

---

[20] CRA was too small an organization to take the financial risk of being an official sponsor for the Tapia Conference, e.g. signing the hotel contract for the meeting, but it did provide $10,000 for travel scholarships. Moreover, Jim Foley, then the CRA board chair and acting dean of the College of Computing at Georgia Tech, donated some scholarship funding from his college – and several other CRA board members did the same from their departments. ACM took on the entire financial risk for the first Tapia conference; for some of the later meetings, the IEEE Computer Society shared this risk. But ACM remained the most active partner of the three. In recent year, IEEE Computer Society has dropped its support for CDC (Taylor 2015).

[21] While there has been some concern that the Tapia Conference is getting too large and impersonal, Juan Gilbert is all in favor of the growth: "They definitely increased the size. I think it's a good thing. Now, more people are aware. I think that's very important. More and more students are coming – all types of students from all over the country. It has that appeal. The only problem we've had is that students lose attention. We need to make sure that they're aware that, "Okay, we have a keynote. You should go to the keynote, these are very high-profile people." Perhaps they don't know them yet. That doesn't mean they won't be interested. ... It's been very good. I think connecting entering students [to one another]" (Gilbert 2015).

- Connect with others with common backgrounds, ethnicities, and gender so as to create communities that extend beyond the conference;
- Receive advice from and make useful contacts with CS&E leaders in academia and industry; and
- Be inspired by great presentations and conversations with successful people in CS&E who have similar backgrounds, ethnicities, and gender to the attendee. (Patterson 2011)

The new format retained some of the successful activities from the previous Tapia conferences: a doctoral consortium, resume preparation workshop, graduate school workshop, career advice workshop, student poster session, town hall, banquet, and dance. But it added some new activities, including a luncheon with successful local people, a meet-up to explore professional opportunities, and a tour of the city. Whereas there had always been plenary sessions, the number was increased to eight and higher-profile members of the academic community were enlisted to make the presentations.

The eighth Tapia Conference, the most recent at the time of this writing, was held in 2014 as the organizers moved to an annual schedule in response to the strong interest. The 2014 conference continued many of the activities that had been established for the 2011 conference. The 2014 conference, for example, included a technical program, a poster session, a doctoral consortium, and a robotics competition. Many students were able to attend because of scholarships provided through CDC (CDC n.d.).[22] Today, according to the conference website (http://tapiaconference. org), the conference is organized by CDC, sponsored by the computing professional society ACM, and presented by CMD-IT (which is discussed below).

Over the years, CDC has sponsored some additional programs. For example, early in its history it created a brochure, entitled *Faces of Computer Science*, to educate and interest high-school students in computing education and careers. CDC also sponsored Distributed Rap Sessions, which involved getting undergraduate minority students involved in research projects to use the Access Grid, a web-based grid portal funded by NSF to provide researchers who were spread across the nation with access to advanced computing systems. CDC also built a database of underrepresented minorities in computer science as part of its effort to build community. CDC has a long history of sponsoring students to attend research conferences. It also organizes a traveling graduate school forum – held at conferences as well as at individual colleges and universities – with the aim of informing undergraduate minority students about graduate school and encouraging them to continue their education.

Today, in addition to the Tapia Conference and the joint projects with CRA-W, CDC organizes a workshop for underrepresented junior faculty covering topics such as grant proposal writing and preparing for tenure. This workshop is a place where these junior faculty members can network with nationally prominent minority leaders. CDC is also operating two other community-building activities: a Diversity in Privacy and Security Seminar and a Women of Color in Computing affinity group (CDC n.d.).

---

[22] For a detailed descriptions of the 2007 Tapia conferences, as an example, see Martinez-Canales (2007) and Redelfs (2008).

9.4   Center for Minorities and People with Disabilities in IT (CMD-IT)

In 2007 a group of researchers who had been involved in broadening participation in computing activities for many years began to have conversations about forming a new alliance under NSF's Broadening Participation in Computing program.[23] These individuals included Valerie Taylor (Texas A&M), Bryant York (Portland State), Ann Gates (UTEP), Richard Ladner (University of Washington), and Ron Eglash (RPI). They noted that, while there were organizations focused on one or another underrepresented group in computing (e.g. CAHSI for Hispanics or AccessComputing for people with disabilities), there was no group that covered all the underrepresented groups. They acknowledged that, while there are some issues that are particular to any given underrepresented group, there are many issues that are common across the groups.

The group considered whether the Coalition to Diversify Computing could take on this role. Taylor, a former chair of CDC, and Eglash, then the current chair of CDC, both argued that CDC was not suitable for this purpose. It had certain responsibilities – and limitations – created by its relationships with its three parent societies (ACM, CRA, and the IEEE Computer Society). Because of these ties, the group believed, CDC's ability to act agilely and independently was limited; and a change in CDC's charter might have a negative impact on the diversity efforts of the three parent societies.

NSF provided funds to convene a workshop at Texas A&M at which 35 people discussed whether there might be value in creating a new umbrella organization for all underrepresented populations in computing. The workshop included representatives from a number of the existing Broadening Participation in Computing alliances, other organizations working in this space, and some other organizations with an interest such as the National Federation of the Blind and the American Association for the Advancement of Science. The mission of CMD-IT was established at this workshop, and further feedback was solicited at a session at the 2009 Tapia conference.[24] As one can see from the official mission statement, the mission for CMD-IT has two parts, both a traditional goal of increasing the participation of underrepresented groups and a less common goal of enabling these groups to give back to society.

> Our mission is to ensure that under-represented groups are fully engaged in computing and information technologies, and to promote innovation that enriches, enhances, and enables these communities, such that more equitable and sustainable contributions are possible by all communities. (CMD-IT 2015)

---

[23] For a history of the NSF Broadening Participation in Computing program, see Aspray (2016).

[24] The name of CMD-IT (pronounced like "command it") was thought up by Ann Gates during one of the planning calls. CMD was taken from the command line in a programming language, and IT referred of course to information technology. However, the acronym also stands for the organization's full name, Center for Minorities and People with Disabilities in Information Technology.

In 2010, CMD-IT formed a high-powered advisory group including Dan Reed (now vice president of research at the University of Iowa but formerly in leadership positions at NCSA, RENCI, and Microsoft Research), Telle Whitney (ABI), Lucy Sanders (NCWIT), Tony Smith (Schlumberger), Stu Feldman (Google), and Miriam Briggs (Briggs and Briggs). The following year the organization incorporated as a 501(c)(3) nonprofit. Taylor assumed the role of executive director, and a small office was opened at Texas A&M. Currently, CMD-IT funding comes primarily from NSF grants as well as surpluses from the Tapia Conference, for which CMD-IT has taken operational responsibility.[25]

One of CMD-IT's first tasks was to use federal data to report out useful statistics about underrepresented groups in all of computing – going beyond the material reported by CRA's Taulbee Survey, which covers only the PhD-granting institutions. The goal is to grow the CMD-IT staff so that it can conduct some of its own research.

Professional development workshops have been among CMD-IT's major activities, in addition to the Tapia Conference and data collection. Because of the low numbers of minorities in the computing fields, many minority computer scientists feel isolated and do not receive strong mentoring. CMD-IT is trying to address this issue through a series of ongoing mentoring activities. Since 2009 it has organized an annual Academic Career Workshop for Underrepresented Junior Faculty and Senior Graduates Students. Since 2012 it has organized an annual National Laboratories Professional Development Workshop for Underrepresented Participants focused on mentoring senior graduate students, postdocs, assistant and associate level faculty, and laboratory professional junior staff. Annual Student Professional Development Workshops started that same year. At these workshops, industry professionals undergraduates and masters students. In 2011 CMD-IT began a student competition in which students design projects for an introductory CS course intended to be attractive to underrepresented groups. CDM-IT also supports a series of research workshops on Diversity as an Innovation.

> The goal of this workshop series is to bring together leaders and sociologists focused on different underrepresented groups in computing to collectively identify how projects, which seek to increase the number of students/professionals from underrepresented groups in computing, can use that opportunity to foster innovative projects. Each workshop will focus on a particular topic to address a key problem. The workshop series will provide a mechanism whereby the underrepresented communities are engaged in developing innovations that address some specific problems initially, and broader problems in the long term. (CMD-IT 2015)

The first of these workshops, held in 2012, focused on mathematics education for grades 6 through 8 in under-served communities. Another project, which ran from 2009 to 2012 with NSF funding, involved Incorporating Cultural Tools for Math and Computing Concepts into Boys and Girls Clubs. This project included some of

---

[25] CMD-IT also runs the Grace Hopper Underrepresented Women in Computing Committee, which takes responsibility for content related to underrepresented minorities and people with disabilities at the Grace Hopper conferences.

Eglash's research on Culturally Situated Design Tools, such as teaching kids about geometry through cornrow hair design or learning about angles through breakdancing (See Taylor 2015 or http://homepages.rpi.edu/~eglash/eglash.htm for more information.)

CMD-IT also works to build community across these various underrepresented groups by maintaining a Community Calendar of events sponsored by all relevant organizations, and by holding Community of Practice teleconferences every 2 months with all the major stakeholders.

Until recently, the population of women in computing has been much larger than the number of underrepresented minorities in computing. However, it is expected that, by the year 2050, the United States will be a majority minority nation, with 60 % of the total population made up of a collection of minority populations. For this reason, there will be at least as many underrepresented minorities and people with disabilities to draw upon as there are women. This means that there will be as much of a need for a center like CMD-IT as there is for a center like NCWIT. At this time, NCWIT has 20 times as many staff members as CMD-IT and has built up well-established programs, networks of partnering organizations, Washington connections, a track record with major corporate fundraising, and an extraordinary leadership team. CMD-IT has been in business only about a third as long as NCWIT. Despite its highly promising start, CMD-IT faces traditional startup problems. As superb as Taylor is, it is likely to be impossible for her to serve full-time as executive director of CMD-IT and also as a full-time academic with an active research career. CMD-IT has to confront the challenging task of embracing all the differences among the various underserved groups it represents and not just their points of commonality. It is too soon to know how CMD-IT will grow in the future or whether it will be the organization that represents the underrepresented populations in computing over the long term.

References

Anonymous. 2009. Pilgrim software receives 2009 BDPA award. *Quality,* November: 14–15.

Anonymous. 2010. Making the list: Best employers for blacks in technology. *Chicago Citizen,* May 26: 6.

Anonymous. 2012. African American women and girls take center stage at national technology conference, *Chicago Weekend,* August 1: 3.

Aspray, William. 2016. *Participation in computing: The national science foundation's expansionary programs.* London: Springer.

Association of Computer/Information Science and Engineering Departments at Minority Institutions. 2013a. *Cultivating innovation: Riding the new waves in computing.* http://www. admiusa.org/admi2013/2013ADMIflyer.pdf. Accessed 28 Oct 2014.

Association of Computer/Information Science and Engineering Departments at Minority Institutions. 2013b. *Surfing through a sea of data.* https://www.admiusa.org/admi2014/. Accessed 11 Jan 2016.

Association of Computer/Information Science and Engineering Departments at Minority Institutions. 2014. *The symposium on computing at minority institutions.* http://www.admiusa. org. Accessed 28 Oct 2014.

Bates, Ivory. 1993. Black EXPO accents outreach to African-American community. *Richmond Afro-American*, October 16: A12.

Baylor, S.andra Johnson, and Ann. Redelfs. 2000. Coalition to diversify computing addresses minority issues in CSE. *Computing Research News*, May. http://archive.cra.org/CRN/ issues/0003.pdf. Accessed 23 Nov 2015.

BDPA Education and Technology Foundation. (BETF). 2006. Annual Report.

BDPA Education & Technology Foundation (BETF). 2014. *BETF scholarships.* http://www.betf. org/scholarships/scholarships.shtml. Accessed 11 Jan 2016.

Berkley, Thomas L. 1999. Coalition tackles 'digital divide': Publishers join to create 'soul of technology' to highlight black talent. *The Sacramento Observer*, October 20: G5.

Bernat, Andrew. 2015. *Oral history interview by William Aspray.* Charles Babbage Oral History Collection, June 23.

Burgess, Michael C. 2004. National black data processing associates. *Congressional Record*, 150 Cong Rec E 1418, July 19.

Carter, Perry. 2014. Top coders from the national capitol region capture high school computer competition. *The Washington Informer*, August 21: BS4.

Cartwright, Robert Corky. 1997. Fair for minority students at ADMI 97. *Computing Research News*, March. http://archive.cra.org/CRN/issues/9702.pdf. Accessed 23 Nov 2015.

Center for Minorities and People with Disabilities in IT (CMD-IT). 2015. *Organization website.* http://www.cmd-it.org. Accessed 11 Jan 2016.

Coalition to Diversify Computing (CDC). n.d. *Programs.* http://www.cdc-computing.org/pro-grams/. Accessed 13 Dec 2014.

Computing Research Association. 2000–2013. *Board Briefing Books.* Computing Research Association Board of Directors. Twice a year meetings. Made available by the executive director.

Davis, Anthony. 1997. Black professional organizations celebrate their diversity in unity. *The Philadelphia Tribune*, January 3: 2-C.

Fairley, Juliette. 1995. Black data processing associates. *Black Enterprise*, June: 44.

Gibson, Stan. 1997. BDPA: Helping teens who will shape the future. *PC Week*, August 25: 79.

Gilbert, Juan. 2015. *Oral history interview by William Aspray.* Charles Babbage Institute Oral History Collection, January 29.

Haskell, Angela. 2010. From counter to boardroom, pioneer still blazing trail. *The Philadelphia Tribune*, February 26: 2B.

Hicks, Wayne. 2011. Creating IT futures foundation provides funding towards college scholarships for 25 students at the 2011 National BDPA High School Computer Competition Championship. *BETF Media Advisory, BDPA Education Technology Foundation*, July 28.

Larson, Eric. 2011a. It's time to show IT off. *Membership News Blog, CompTIA*, October 3. http:// www.comptia.org/about-us/newsroom/blog/11-10-03/it_s_time_to_show_it_off.aspx. Accessed 15 Apr 2015.

Larson, Eric. 2011b. Two foundations team up to provide scholarships for talented minority students. *Membership News blog, CompTIA*, August 31.

Lawrence, Andrea. 2003. Message from the ADMI president. *ADMI 2003 Our Cyber Future.* http://nia.ecsu.edu/ureoms2003/admi03.html. Accessed 28 Oct 2014.

Lawrence, Andrea. 2014. *Oral history by William Aspray.* Charles Babbage Institute Oral History Collection, December 19.

Martinez-Canales, Monica. 2007. Tapia conference to focus on passion, diversity, and innovation. *Computing Research News,* May. http://cra.org/crn/2007/05/tapia-conference-to-focus-on--passion-diversity-and-innovation/. Accessed 5 Aug 2015.

Muhammad, Tariq K. 1996. New name, same game. *Black Enterprise*, August: 38.

Muhammad, Tariq K. 1998. Creating a human IT network. *Black Enterprise*, March: 36.

Norton, Eleanor Holmes. 2013. Congratulating the national black data processing associates. *Congressional Record*, 159 Cong Rec E 1202, August 1.

Pace, Earl. 2014. *Is BDPA a promise unfulfilled?* Keynote lecture to the Philadelphia Chapter of BDPA. https://www.youtube.com/watch?v=dgD2ShxlEPc&feature=youtu.be. Accessed 15 Apr 2015.

Patterson, David. 2011. Tapia conference 2011: Reshaped by feedback. *Computing Research News*, January. http://cra.org/crn/2011/01/tapia_conference_2011_reshaped_by_feedback/. Accessed 23 Nov 2015.

Redelfs, Ann. 2008. Tapia celebration of diversity in computing: 2007 event strongest ever; Next event planned. *Computing Research News*, March. http://cra.org/crn/2008/03/tapia-celebration-of-diversity-in-computing-2007-event-strongest-ever-next-event-planned/. Accessed 5 Aug 2015.

Roach, Ronald. 2006. Minority computing group establishes IT Institute at Auburn university. *Diverse Issues in Higher Education*, September 7: 15.

Sherman, Brian. 2011. IT hall of fame honors innovators at CompTIA annual member meeting. *Membership News Blog, CompTIA*, April 7.

Summers, Margaret. 2013. BDPA helps bridge 'digital divide'. *The Washington Informer*, August 22–28: 22.

Taylor, Valerie. 2002. Coalition to diversify computing launches Tapia Celebration of Diversity in Computing series. *Computing Research News*, January. http://archive.cra.org/CRN/issues/0201. pdf. Accessed 23 Nov 2015.

Taylor, Valerie. 2015. Oral history interview by William Aspray. Charles Babbage Oral History Collection, January 7, 8, and 15.

Teller, Patricia. 2004. Coalition to Diversify Computing (CDC). *Computing Research News*, November. http://archive.cra.org/CRN/issues/0405.pdf. Accessed 23 Nov 2015.

Tennant, Don. 2009. Earl A. Pace, Jr. *Computerworld*, February 16: 12, 14.

Warfield, Carolyn. 1991. BDPA's Computer Camp Gives Students Right Start. *The Michigan Citizen*, May 18: 9.

Watson, Jamal Eric. 2014. Bloomfield college partners with nonprofit to help minority and computer science students. *Diverse*, July 31: 6, 8.

Weyuker, Elaine. 2014. *Oral history interview by William Aspray*. Charles Babbage Oral History Collection, October 8.

White, John. 2015. *Oral history interview by William Aspray*. Charles Babbage Oral History Collection, June 26.

Willis, Bob. 2015. *Oral history interview by William Aspray*. Charles Babbage Oral History Collection, November 24.

York, Bryant W, Andrew Bernat, Robert Cartwright, Don Coleman, Roscoe Giles, Valerie Taylor, and Ramon Vasquez-Espinosa. 1995. *Final Report of the Workshop on Increasing Participation of Minorities in the Computing Disciplines (CDA-9401736)*. May 4–7, Airlie: Virginia. Personal property of Bryant York.

Chapter 10
Building Educational Infrastructures for Broadening Participation in Computing

Abstract This chapter addresses the issue of what makes for a successful higher education program to attract women into a science or technology department. The first section describes the characteristics of a successful program for attracting women into its STEM education programs, based on the research of social scientists Mary Frank Fox and Gerhard Sonnert. The next section, based on the work of education professor Frances Stage and her colleague Steven Hubbard, identify colleges and universities that have done the best job of attracting women into science and technology education. The final section, based on oral histories with participants as well as a wide public literature, provides case studies of departments that have been successful at recruiting and retaining women. These include the general engineering major at Olin College, both the engineering and computer science department at Smith College, the combined department of electrical engineering and computer science at the University of California at Berkeley, and the computer science departments at Carnegie Mellon University and Harvey Mudd College. More abbreviated accounts are given of computer science at Bryn Mawr, Mills, and Wellesley Colleges, the University of Colorado Boulder, and Georgia Tech.

This chapter addresses the issue of what makes for a successful higher education program to attract women into a STEM department, especially one that has historically had low participation of women such as a computer science or engineering department. The first section describes the characteristics of a successful program for attracting women into its STEM education program. This section is based on the work of two highly experienced social scientists, Mary Frank Fox at Georgia Tech and Gerhard Sonnert at Harvard, and their colleagues. The next section, based on the work of the well-known education professor Frances Stage and her NYU colleague Steven Hubbard, identify colleges and universities that have done the best job of attracting women into STEM education. The final section, based on oral histories with participants as well as a wide public literature, provides case studies of departments that have been successful at recruiting and retaining women. These include general engineering at Olin College, both engineering and computer science at Smith College, the combined department of electrical engineering and computer science at the University of California at Berkeley, and computer science at Carnegie Mellon University and Harvey Mudd College. In addition, more abbreviated

© Springer International Publishing Switzerland 2016
W. Aspray, *Women and Underrepresented Minorities in Computing*,
History of Computing, DOI 10.1007/978-3-319-24811-0_10

accounts are given of computer science at Bryn Mawr, Mills, and Wellesley Colleges, the University of Colorado at Boulder, and Georgia Tech.

10.1   The Characteristics of a Successful Program for Attracting Women into STEM Education

In two thoughtful and well-researched papers, the social scientists Mary Frank Fox, Gerhard Sonnert, and Irina Nikiforova have examined why programs intended to recruit, retain, and advance undergraduate women in the STEM disciplines are more successful in some departments than in others.[1] The most important difference they find concerns whether the interventions that the departments make are focused on fixing the system or fixing the individual. Fixing the system is harder to implement but more successful.[2] As the authors explain the distinction:

> From the individual perspective, the status of women in science and engineering is attributed to, or thought to correspond to, women's individual characteristics. These individual characteristics include attitudes, behaviors, aptitudes, skills, and experience of women that may affect their participation and performance in science… For example, women's lower level of self-confidence in mathematics and lower internal sense of ability or potential for scientific achievement can be seen as barriers to pursuing scientific careers in these fields… Likewise, levels of motivation to perform in scientific areas may be regarded as supports or barriers to pursuing scientific careers. From the institutional or structural perspective, the status of women in science and engineering is more strongly attributed to factors beyond individual characteristics, that is, to features of the settings in which women are educated and in which they work. These factors may include, for example, patterns of inclusion or exclusion in research groups, selective access to human and material resources, and different practices and standards of evaluation that may operate for women compared to men… From this structural perspective, factors also include science and engineering teaching environments that may isolate students from social concerns, portray science and engineering as highly competitive, masculine domains …, and intend to 'weed out' students in the curricular process… (Fox et al. 2009)

In the first paper (Fox et al. 2009), the authors compare the five most successful to the five least successful (out of 49 undergraduate STEM programs they have

---

[1] Fox had written an earlier paper (Fox 2000) about doctoral education for women in the STEM disciplines. It presaged but did not have as advanced a level of analysis as Fox et al. (2009, 2011). On the complexities of theory and practice related to programs for women graduate students in science and technology, see Fox (1998).

[2] Fox et al. (2009) make this astute remark: "The distinction between individualistic and institutional/structural perspectives or explanations for the status of women in science and engineering is important. This is because a long-standing and controversial debate exists about the extent to which it is the women or the social systems of education and work that need to be 'fixed' to improve the participation and performance of women in these fields… Further, although social scientists who have studied the phenomenon have tended increasingly to lean to the structural perspective, natural scientists and engineers themselves typically leave unexamined the structures in which students are educated (and work), and expect students to 'shape themselves' to prevailing environments…"

Table 10.1 Comparison of most and least successful departments in graduating undergraduate women in science and engineering

Issue	Most successful programs	Least successful programs
Definition of the problem	Multiplicity of issues, including structural issues such as faculty and classroom bias against underrepresented groups, weed-out grading systems, and support structures that work better for men than women	Recruitment and retention numbers, without clear focus on the underlying structural issues, or that any of these issues were within their control
Current and desired future initiatives to solve the problem	A broad range of activities including bridge programs from high school to college, living/learning residence halls, mentoring, faculty buy-in through workshops, undergraduate research projects with faculty, hands on experiences for students	Peer mentoring and more peer mentoring
Desirable characteristics for the program leader	Communicate well with the students but also be resourceful in reaching out to administrators and others within the environment beyond the students; having an advanced science degree so as to have respect within the university system to get things done	Approachable, empathetic, encouraging so as to effectively operate a student counseling service; having an advance science degree so as to understand the students
Linkages to institutional context	Report to someone for the purposes of impact, visibility, and academic connection; deep concern in cases where there was not faculty buy-in to the program	Report to someone who has access to material resources; little concern about lack of faculty buy-in because the programs were run largely independently of the faculty
Aspirations for the program's future	More visibility, greater scope and range of activities especially in partnership with faculty	More of the same (scholarships, recruiting, etc.)

Source: Fox et al. (2009)

studied) in terms of the percentage of degree recipients who are women – as of the 2001–2002 academic year.[3] Their findings are summarized in Table 10.1.

As the authors explain, in the successful schools, the programs were intended to be integrated with the faculty and the rest of the university; they were also intended to change the university environment in ways that were good for undergraduate women students studying in the STEM disciplines. In other words, they were intended to fix the system. By contrast, the least successful programs were operated independently, largely outside of the existing university system. Peer mentoring, for

---

[3] This material is based on interviews with program directors and others from the five programs with the most successful outcomes and the five programs with the least successful outcomes in undergraduate degrees awarded to women in science and engineering, out of the 49 programs that the authors studied. The 49 programs were all of the ones that the authors could identify for under-graduate programs in science and engineering in the United States.

Table 10.2 Perceived importance to women in science program directors about individual and institutional obstacles to undergraduate women in science and engineering

Individual obstacles	Institutional obstacles
Women's self-confidence (3.2)	Classroom climate (3.1)
Knowledge about careers in STEM (3.0)	Peer relationships (2.9)
Career commitment (2.7)	Faculty advisors (2.9)
Motivation to succeed (2.4)	Faculty commitment to the success of undergraduate women (2.8)
Commitment to academic work (1.7)	Campus climate of gender equity (2.7)
Independence of students (1.7)	Administrative commitment to the success of undergraduate women (2.4).
Academic ability (1.3)	
Overall mean (2.3)	Overall mean (2.8)

Source: Fox et al. (2011)
Scale: very important (4), moderately important (3), slightly important (2), and not at all important (1)

example, was favored by these less successful departments – the authors argue – because the onus of the work is thrust upon the students themselves, not on the faculty, administrators, or other centers on campus. The existing system for these less successful departments was not one to be changed, but instead one to be adopted; the system was viewed as an immutable, exogenous force around which the woman's program must operate. Thus the most successful programs were centered upon institutional or structural efforts, while the least successful ones were focused on the individual students. As the authors report, some of the more successful programs had started with an approach similar to those of the least successful programs but changed their approach over time. As one of the program directors they interviewed told them:

> At first the focus was on helping the women cope. Then [we] realized the women were not the problem, the system was. This was an evolution in thought. We were scared, it felt overwhelming [to have] to challenge [our] home base. (Fox et al., 2009)

The second paper by these same authors (Fox et al., 2011) is based on survey data from the 49 programs with undergraduate women in science and technology programs. However, this paper provides more detail about the attitudes of the people running all 49 of the programs, their beliefs about the importance of various specific individual and structural obstacles to the participation of women in their undergraduate programs, goals of these programs, and matches and mismatches between obstacles and goals. The authors measure the perceived importance to program directors of individual and structural obstacles on a four-point scale. Table 10.2 presents the perceptions of the perceived importance of various individual and institutional obstacles to providing a successful environment for recruiting, retaining, and advanced female students in science departments.

The authors point out that the overall mean on importance of individual factors is 2.3, while the overall mean on importance of structural factors is 2.8; thus indicating the program directors on average have a structural definition of the problem, i.e. that institutional rather than individual factors are more important.

The authors also report on the mean prevalence of various activities in these 49 programs. Presenting these activities in order, from major activity, to minor activity, to no activity, provides the following list: career seminars, peer mentoring, social activities, linkages with other programs, academic tutoring, scholarships, living-learning programs, study/social lounges, diversity training for faculty, faculty mentoring, research on women in science, curriculum courses, and graduate student mentoring. As the authors observe, there is a mismatch between the perceived problems and the perceived solutions. This list of activities emphasizes individual issues, even though the structural issues were perceived by the program directors as more important. In particular, these activities leave mostly unaddressed issues that were reported by these program directors as particularly important: classroom climate and faculty commitment to female undergraduate education. Moreover, the activities that require the engagement of faculty appear toward the bottom of the list.

The authors draw some additional conclusions from their data. The typical program reports to a mid-to-low level person in the university administration, has relatively high funding from multiple sources, has a full-time director who is generally not a faculty member, and has weak buy-in from the faculty. Those programs that have highest level of administrative reporting, highest level of funding, have been founded more recently, have full-time instead of part-time directors, and have high faculty buy-in are correlated with the programs focused on structural rather than individual obstacles.

10.2   Which Colleges and Universities Attract Women to STEM Education?

Stage and Hubbard (2008) analyzes which types of universities have produced the most women for mathematics and science careers in the period 1995–2004. More specifically, they track baccalaureate origins of women PhDs in the STEM disciplines. As they expected to find, women's colleges, Ivy League universities, and technical institutes have been and continue to be large producers of female baccalaureate degrees in the STEM disciplines. Other studies have identified some additional types of schools that have a history of producing high percentages of women STEM baccalaureates: highly selective, small coeducational institutions that have high per-student academic expenditures as well as schools with large numbers of women on their STEM faculties. Historically Black women's colleges, historically Black coeducational colleges and universities, and women's colleges have produced higher percentages of Black women scientists than other types of institutions. Although the women's colleges produce scientists at higher rates than the research

Table 10.3 Top producers of baccalaureate degrees to women who go on to receive a Ph.D. in a STEM discipline, 1995–2004

Carnegie classification	Unexpected top producers
Research I	Cornell, MIT, Harvard
Research II	UC Santa Cruz, Rice, RPI, SUNY Albany, Vermont
Doctoral I	William & Mary, SUNY Binghamton, Loyola of Chicago, Marquette
Doctoral II	Dartmouth, Maine
Comprehensive I	Trinity U.
Comprehensive II	LIU Southampton, Tampa, Chestnut Hill
Liberal Arts I	Swarthmore
Liberal Arts II	Evergreen State, Lebanon Valley, Texas Lutheran

Source: Stage and Hubbard (2008). Extracted from Table 5.5. Within a Carnegie category, the schools are listed in decreasing order of the number of STEM doctorates awarded

universities, their overall capacity is much smaller. There are some other schools that, for individual reasons, produce high percentages of women scientists, mathematicians, and engineers. The unexpected top producers – i.e. those not predicted by the general trends just mentioned, as organized by Carnegie classification, are given in Table 10.3.

10.3   Success Stories in the Education of Women in Engineering and Computing

This section provides case studies of individual colleges and universities that have had recent successes at educating women in the engineering or computing disciplines. We cover these institutions in the chronological order in which they began these broadening participation programs: Berkeley and Mills (1977), Carnegie Mellon (1995), Olin (1997), Smith (1999), and Harvey Mudd (2000). The chapter ends with brief mention of some other schools that have had success in attracting women: Bryn Mawr, Georgia Tech, Spelman, University of Colorado Boulder, and Wellesley.

10.3.1   University of California Berkeley and Mills College

The first two institutional programs intended to increase the number of women in computing were reentry programs established in the early 1980s at UC Berkeley, a large research university with one of the top computer science programs, and Mills College, a nearby, small women's college focused on undergraduate liberal arts education.

Sheila Humphreys was the common denominator between the Berkeley and Mills programs. In the 1960s, Humphrey had relocated to the west coast and accepted a job in the admissions office at Mills College. The sociologist Lucy Sells had written a thesis arguing that mathematics was the critical filter that prevented many women from majoring in any field except for those in the humanities. Working under Professor Neil Smelser, the mathematician Lenore Blum had based her fast-track precalculus course on Sell's research. Blum familiarized Humphreys with Sells' results and, for several years in the second half of the 1970s, Humphreys worked to recruit high school students with a possible interest in mathematics and science to Mills and get them enrolled in the pre-calculus program that Blum had developed. After several years of this work in the admissions office, Humphreys was recruited away to Berkeley by Margaret Wilkerson, the head of Berkeley's women's center. Humphreys was appointed as associate director, and one of her responsibilities was to start a women-in-science program on campus. (Private communication from Sheila Humphreys, 2 April 2016)

Humphreys explains how she became involved with computer science at Berkeley:

> At the Women's Center, I was very fortunate to meet and begin collaborating with one recent PhD graduate, Dr. Paula Hawthorn,[4] and one student who was still there, Dr. Barbara Simons. Both of these women were themselves single mothers [and] reentry women who had not majored in computer science because it was such a new field before they got into the doctoral program. They were absolutely passionate – passionately committed to the idea of opening more access to undergraduates and graduates to the field of computer science. That's how I got started working with them out of the UC Berkeley Women's Center. (Humphreys 2015)

Paula Hawthorn had been a straight-A math major as an undergraduate and had planned to become a high school math teacher. However, she was arrested during sit-ins at the University of Houston, which made her ineligible to participate in the student teaching program; and thus she was barred from being a public school math teacher. Someone suggested that she take a course in computer science, and she was quickly caught up in the field. Because of a messy divorce, she decided to leave Houston. Her major professor in the University of Houston computer science department convinced her that she would do well in a doctoral program in computer science. She accepted an offer from Berkeley to enroll in its doctoral program not only because of its excellence, but also because her parents had relocated to northern California and she figured it might be easier to negotiate graduate school as a single mother with her parents nearby as a support network. (Hawthorn 2015)

The environment at Berkeley for graduate students who were mothers was not welcoming. As Hawthorn explains her initiation at Berkeley:

> [M]y first day at University of California at Berkeley, when I was on campus, I left my kids with my parents and went to meet with my major field advisor, who was not my research

---

[4] Actually, at the time that Humphreys met Hawthorn, the latter was still a doctoral student. Hawthorn was one year ahead of Simons in the doctoral program. (Private communication, Barbara Simons, 3 April 2016)

advisor. ...[A]t UC, you have a major field advisor who is just supposed to make sure you're taking the right courses and so forth. I went to meet with him, and when he discovered that I had children, he told me that I should immediately leave and go back home because no one could have children and be serious about pursuing a PhD at UC Berkeley. ... If for some reason I didn't want to do that, then I should make sure that my husband ... take the kids. Because there's no way that I could be a student at Cal with two children. And the next year, the question 'Do you have children?' did appear on the application.[5]

Barbara Simons was an older graduate student – older than some of the faculty members, which was unusual at the time at Berkeley. She was separated from her husband – and later divorced – but at the time she entered Berkeley her three children were living with their father in Switzerland, which made it easier for Simons to begin her course of study. She was also in much better financial shape than Hawthorn, which made it easier to be a graduate student. Both students had an interest in politics, though at the time Hawkins had more of a track record as an activist. So it was not surprising that, with these similarities, second-year graduate student Simons bonded with the slightly more advanced graduate student Hawthorn. Simons was appreciative of the help that Dana Angluin, one of the few other, older female graduate students offered her (in her case helping her prepare for her oral exams) and the other female grad students; and Simons was interested in supporting the female students who came after her in a similar way. Simons, today a confident and outspoken member of the computer science community, describes herself as insecure and afraid to speak up in class as a graduate student. (Simons 2015)

As Simons (2015) remembers:

Paula and I became a team, which we've been for many years. We decided that we wanted to try to get more women into computer science at Berkeley. We went to the Women's Center and we said, "We'd like to run a seminar series through the Women's Center which would reach out to women already on campus and encourage them to take computer science classes and maybe consider becoming computer science majors". ... We decided to do the seminars, so we went to the Women's Center and they were particularly ineffectual. We just didn't get much out of it. We left thinking that this is a real joke. However, shortly after we first reached out to them, Sheila Humphreys went to work at the Women's Center. ... [W]e ended up working with Sheila in her role at the Women's Center. We did actually put on a seminar series.

Humphreys, Hawthorn, and Simons began to invite women to speak in computer science and they held a regular informal lunch that evolved into being a formal women's group. As Hawthorn (2015) remembers the purpose of this group:

[S]o the first thing I did when I got on campus was to start a women's program, a women's group.... [I]t became the Women in Computer Science and Engineering ... We needed a women's support group. The professors ... were... one set of issues. But the other set of issues was, being such a tiny minority in a very, very male environment. ... remember,

---

[5] Hawthorn (2015) also talked about what she called the "extreme academic elitism" at Berkeley: "the atmosphere was one of extreme academic elitism. It didn't matter that I was a woman, what mattered was that I had those two kids and needed to get rid of them. I never felt that I was discounted because of my sex. [However,] I felt that I was discounted because I talked like a southerner. And because I didn't participate in a lot of extracurricular activities. I just couldn't. Because I don't have the east coast pedigree that a lot of people do that came in to Cal."

we're talking a very long time ago; nobody worked from home. And so you were working late at night in the basement of Evans Hall ... Big room, filled with terminals and with students. Mostly males, and a few females and young male graduate students who, you know, often were approaching women in ways that women didn't want to be approached... So there was always a need to support one another and to figure out how to basically navigate through that.

Various female computer scientists, including Marie-Anne Neimat, Susan Eggers, and Linda Lawson, as well as the sole undergraduate computer science major engaged in these activities, Deborah Estrin, would sometimes join them.[6] The formal women's group in EECS, known as Women in Computer Science and Engineering (WICSE), was created in 1977.[7] When Barbara Simons learned from Mike Ubell that funds were available from the university for student organization's activities, they decided to sign up the group as an official organization with the university's student affairs office.

After the first conference, WICSE would occasionally do projects with other established national organizations that were on campus, such as the Society of Women Engineers. But for the most part they worked on their own. They would also collaborate with other organizations on their own campus, including the Career Center, the Women's Center, and the math department's graduate women's group (the Noetherian Ring named after the famous algebraist, Emmy Noether). Part of the reason was that some WICSE members felt that SWE was addressing a different audience – students interested in a professional engineering career rather than a path to doctoral study and a computing research career – and did not particularly reach the needs of the WICSE members[8]:

SWE, of course, was on campus, and I didn't especially like SWE. I called them the 'My daddy was an engineer women'. ... I found them to be ... always so positive and never really wanting to deal with any of the issues that people had ... I just wanted to have our little group that talked about stuff we were going through. 'My Daddy was an Engineer', so

---

[6] Diane McEntyre, who was an advanced graduate student in the school of education studying computer science education, was a supporter of the early WICSE group. (Private communication from Sheila Humphreys, 2 April 2016)

[7] In later years, after the Hawthorn and Simons era, some of the people who had taken an active role in WICSE include Faith Fich of Toronto, Dana Randall of Georgia Tech, and Valerie Taylor of Texas A&M and the founder of CMD-IT (see Chapter 9), Amy Wendt who is a professor of electrical and computer engineering at the University of Wisconsin-Madison, Dawn Tibury who is a professor of mechanical engineering at the University of Michigan, and Daisy Wang who is a professor in the CISE department at the University of Florida.

[8] Similarly, WICSE did not have much to do with the general women's movement, which was heating up during this same time period. As Simons (2015) noted: "Like with the women's movement, I think that there was some hostility towards the whole, how can I put it, the stereotypical engineer. They had hostility because they didn't understand it. It goes both ways. I think that there's a problem with people who go into engineering science in that many of them are not well balanced educationally. They tend to have a very narrow, and sometimes [have a] quite warped view of the world because they [are] focused so much on the particular discipline that they're in. It goes both ways. But, to me I was coming at it as someone coming out of computer science and math and interacting with women, many of whom were math-phobic and therefore were hostile. We had nothing to do with the national women's movement; nothing, absolutely nothing."

what that is about, the young women whose fathers were engineers and who encouraged them to become engineers, simply do not understand that anyone could feel, in any way, discriminated against, or, in any way even slightly not having the same equal treatment as other people. It just isn't in their vocabulary. And that's great. You know, I just wish everybody could be like that. Because when they have … this, such a well-grounded kind of aura … I don't think that they're targets of as much discrimination. (Hawthorn 2015)

WICSE's first major activity was to organize a Women in Engineering conference on the Berkeley campus in 1978, with funding from corporate sponsors of the department as well as the College of Engineering.[9] More than 800 people attended. Every woman on the engineering faculty participated, including the sole female computer science faculty member (Susan Graham). One notable speaker was Professor Elizabeth Scott, an astronomer, chair of the statistics department, and a well-known feminist who wrote an influential report for the American Association of University Women about the discrepancies in salaries between male and female scientists at every level. The conference "created a community of people across the campus who wanted to take some action." (Humphreys 2015)

Several students who were in attendance challenged the deans who were in the room about why there were so few women students admitted into the engineering school; and one dean in particular was particularly defensive, making the argument that standards had to be maintained in admissions. Scott was indignant about his comments, and she followed the conference with a memo to the associate dean responsible for admissions in engineering, asking that "the door be opened a little bit wider, and pointing out that the under-preparation of women at the high school level was a broader problem and it didn't have to do with their talent." (Humphreys 2015) The admission criteria were studied and changes were made a couple of years later with the intention of broadening the pool of students admitted in engineering.

In computer science, enrollment pressures were building at both the undergraduate and graduate levels.[10] Simons and Hawthorn noticed the declining numbers of women entering Berkeley's graduate programs in the early 1980s. As Hawthorn observed:

We learned that as the enrollment pressure on the department built up, the prerequisites for acceptance into the EECS graduate program became more stringent, so that people who had not followed a standard engineering program of study practically from junior high school

---

[9] Over the years, WICSE has organized six major conferences concerning opportunities or barriers associated with women in computer science and electrical engineering and has held WICSE reunion symposia for the 10th, 20th, 25th, and 30th anniversary's of the group's founding. As a result of the first conference, attendance and activity within WICSE grew rapidly, even though that was not the intended purpose of the conference. (Simons 2015)

[10] In the period 1970–2000, there were fewer women by percentage in engineering than there were in computer science. In one way, it makes sense to lump all of engineering together because they were typically lumped together organizationally and financially within the university. However, there are significant variations by percentage in female enrollments by particular engineering discipline. Typically, biomedical, chemical, and civil engineering have had greater percentages of female students at the undergraduate level than computer science, while mechanical and electrical engineering have been lower. See the National Center for Education Statistics for details.

had a hard time being accepted into the EECS graduate program.[11] We wanted to give "non-standard" students, like ourselves, a better chance. (Paula Hawthorn, as quoted in Humphreys and Spertus 2002)[12]

Simons and Hawthorn, working with Humphreys and several others, developed a plan for a reentry program in computer science, which would open up graduate education in computer science to these non-standard students. In 1983 they convinced the Berkeley computer science department to establish a reentry program for students with high academic promise, with particular attention to women and underrepresented minorities. The goal was to prepare these students to enter a computer science graduate program.[13] Simons (2015) remembers she and Hawthorn making a

---

[11] Hawthorn and Simons called this phenomenon the "Engineeringification of Computer Science" and they attributed it to the move of the CS program into the engineering school and a move to add more admission requirements that looked like those in other engineering departments. (Hawthorn 2015)

[12] Hawthorn (2015) amplifies on this issue: "[S]o I started asking questions, and asked to do a study. And I actually got the cooperation of the Computer Science Department to try to find out why women who had been accepted were not coming. You know what I mean? They'd been accepted but then they decided to go to other schools. ...

I did a survey. ... I sent out letters to people and got their responses and then presented it to the faculty. And to the women in WICSE. We found that they were not being offered financial aid in the same proportion that men were. ... And so, I said, 'Clearly there's bias going on here', which caused ... a lot of men in the department [to say], 'Oh you're being silly,' and so forth. But the next year they managed to up the number of women who got financial aid. ...

And then later on – Barbara and I spent a long time in graduate school; it took me almost six years to get my PhD and her as well – and we found that while we were in school, ... the number of women kept dropping. So we started looking for what was going on with that. And that is when we found that the criteria for admission, as the enrollment pressure went higher, the enrollment criteria for admission was more and more skewed toward lots of math, lots of science, and many women just didn't have the prerequisites to get in. And in graduate school, a lot of women had not taken Math and Science kinds of undergraduate degrees and couldn't get into graduate programs. So that's when we started the re-entry program..."

Hawthorn remembers that Martin Graham and Eugene Lawler on the faculty were particularly supportive of Hawthorn's concerns; however, most of the faculty were neutral on these issues of broadening participation. As Hawthorn (2015) remembers: "most of them just wanted to get the research done and just wanted to get the work done and didn't care what sex you were. What your orientation was, what color you were. The only thing was, did you get results, are you working hard on my research project? And so I know that my own advisor, Mike Stonebraker was lightly discouraging about participating in things like that because I already had these two kids and I already had enough to do, and don't you have enough to do that you don't need to be taking on that kind of stuff. And I know that another, I think the department head, even, at the time, said, 'This is not anything that you'll ever be able to put on your CV. Why are you wasting your time on it?'. You know, those kinds of things. Which honestly, I think is an understandable reaction. When you're talking about this selective group of people who are really just trying to get world-class research done, and that is their focus."

[13] The Berkeley re-entry program was not unlike an NSF program funded several years earlier, but lasting only a few years, called the Career Facilitation Program. The plan offered promising women students who had interrupted their education a chance to catch up on their education at 17 universities and then enter the labor force. However, one major difference was that the Berkeley program was more focused on women obtaining advanced degrees, while the NSF program was

pitch to the faculty to secure their buy-in since the faculty were the ones who were going to have to teach the courses and administer the program:

> We decided to have a meeting with the faculty, we were still graduate students, to discuss the program. So we called for them to meet, and we provided lunch and wine. We paid for it ourselves. And, I'll tell you, it doesn't take much to get people to come to your meetings. If you say there's going to be free food and wine, they come. It was a really cheap way of getting all the faculty to turn up. Not only that, because we were giving them free food and wine, it's a pleasant thing. It's a positive thing. I think it makes people more positively inclined to something. Here we're being really nice, doing outreach. … Basically the faculty bought into it. It didn't cost them any money. We said we would raise all of the money to cover the costs, and we did. Again, it was fortuitous that this was during a period when there was a lot of money, when companies were making money hand over fist. They wanted to be able to show that they were being good citizens.

The faculty champion for the program was Eugene Lawler, but it was also supported by department chairs, Manuel Blum, Richard Karp, and Domenico Ferrari. The engineering dean, Karl Pister, gave the initial approval to establish the program and fund the first coordinator (Sheila Humphreys).[14] (Humphreys 2015; private communication from Humphreys, 2 April 2016)

The re-entry program offered students a concentrated set of upper-level undergraduate courses as a transition into graduate study.[15] Students were required to have already taken calculus, discrete mathematics, and computer programming prior to entering the reentry program and were expected to have high scores on the College Board's Graduate Record Exam.[16] Successful completion of the program did not confer either an undergraduate or a graduate computer science degree, nor

---

more focused on women entering the technical workforce. (Humphreys 2015; Simons, private communication 3 April 2016)

[14] Others who were supportive of the computer science re-entry program included Professor Thelma Estrin at UCLA and Fran Allen at IBM. Dean Pister, who was a member of the National Academy of Engineering, brought various female members of the National Academy to the Berkeley campus to lecture and serve as visiting professors; and this helped to enhance the climate for women in engineering at Berkeley. (Humphreys 2015)

[15] Two examples of students who were served by the re-entry program are Nina Amenta, who had received an undergraduate degree in classics at Yale and is now the chair of computer science at University of California, Davis; and Diane Greene, who had studied naval architecture and became the founder and CEO of VMware.

[16] There had been an unfortunate case of a female student in the Berkeley CS program in Hawthorn's first year (thus occurring before Simons arrived at Berkeley) who was unable to handle the academic pressure and dropped out. This had a bearing on the design of the re-entry program. As Simons (2015) remembers: "[T]hat event had a real impact on Paula, … We wanted to make damned sure that people were qualified. … [T]he whole reason … for doing a re-entry program was to make sure that women were qualified to go to graduate school, because we didn't want a repeat of somebody being admitted because they felt they should get some women [but] who wasn't qualified. That's bad on all counts. It's not fair to the person. It reinforces negative stereotypes of women's capabilities and so on. We definitely wanted to make sure that … [the re-entry program] would be preparing students to be successful in graduate school."

was it a certificate program.[17] However, it did add to the students' record and knowledge the core upper-level undergraduate major courses such as data structures and operating systems, which enabled them to be regarded by computer science faculties as prepared for admission into a graduate computer science program.[18] The students received credit from the University of California Extension, but they were taught in regular Berkeley undergraduate courses. Humphreys moved into a new staff position in the EECS department to manage the reentry program but also to work on recruitment and retention of women students. She raised the initial funds from Chevron to pay the tuition, so the classes were free to the reentry students. The companies often took a strong interest in the students.[19] (Humphreys 2015; Hawthorn 2015)

While some of the graduates of the reentry program attended Berkeley's graduate degree programs, others left for graduate study at MIT, Stanford, UCLA, Washington, Texas, or other strong programs. More than 150 students went through the reentry program, resulting in program graduates earning more than ten doctorates and 40 master's degrees. (Humphreys and Spertus 2002; private communication, Sheila Humphreys, 2 April 2016)

> One of the interesting side effects of the re-entry program was that faculty loved it once it was in place. Because the re-entry students were older, more mature and they were there because they really wanted to be there. Unlike some of the undergraduates, who didn't quite fit that bill. They really liked the re-entry program." (Simons 2015)

---

[17] Because the program did not lead to a degree or a certificate, it was regarded as unusual and it took more than a year to get official approval for the re-entry program. It would have been easier to have established this as a regular set of special courses in the UC Extension program, but that would have had certain disadvantages: "[W]e wanted to get these women into classes with regular students, rubbing shoulders with regular students so that the professors could see them, could experience them, could see whether these were good enough to go into graduate schools and become part of their research teams. But also so that the women themselves could see what that is like. You know? And so it was important to us not to have special programs that they were special students. But they were in regular classes." (Hawthorn 2015)

[18] Humphreys (2015) elaborated on the design of the program: "in this period, computer science classes were highly sought after. It was hard to get into the computer science classes. Berkeley is a public university. Obtaining a second bachelor's degree was not and is not something, I think, the legislature approved [of]. Everybody is entitled to one bachelor's degree. No more. We had to set it up either as a graduate program or as an undergraduate program, and we really couldn't do that. We set it up as an in-between program and that is rather unique. Students did not get a degree; they got an education. They got an accelerated education that replicated a set of upper division computer science courses plus discrete math if they needed it, and advising from CS faculty. Thus, they got letters of recommendation for graduate school. They would get access into the classes, which was very difficult at the time. The faculty agreed to that in the spirit of increasing diversity. They really changed the culture of the department by exposing faculty to a non-standard student, which was unusual at the time."

[19] This continued industrial support to Berkeley for women in the STEM disciplines. Earlier, under Dean Ernest Kuh, Ford and General Motors had provided funding to initiate programs to recruit women students into engineering. (Private communication, Sheila Humphreys, 2 April 2016)

Humphreys disagrees with Hawthorn and Simons about the degree of success of the re-entry program. Humphreys has always thought it was extremely successful, whereas Hawthorn and Simons were less satisfied. As Hawthorn (2015) argues:

> We didn't produce more than about four or five PhDs. [Actually, the program produced 10 PhDs.] We had a lot of women that got Master's degrees. We got a lot more women who got good jobs. You know, that wasn't what I wanted. I wanted women at the top. Frankly, I still do. I think that having more women, as you know, as professors, as more role models who can encourage other women, to say 'I can do that'. And so, yeah, in most cases, people did stop with their masters.

The program, which Humphreys (2015) believed worked "very well", was closed down in 1998 by the introduction of Propositions ST1 and ST2 by Ward Connerly of the University of California Board of Regents, the passage of California Proposition 209 by California voters, and an edict from California governor Pete Wilson halting use of educational preferences based on gender and ethnicity.

Both EECS and the college of engineering at Berkeley had to rethink its programs to broaden participation in the face of Proposition 209[20]:

> [A]fter Proposition 209 was passed in 1996, not only this program but many programs on the campus suddenly needed to be reexamined. There was a great swerve and a great worry, and concern about what was legal and what was illegal. We had a very, very good leader as dean of our graduate division at the time, a chemistry professor, Joseph Cerny. We spent a lot of time figuring out how we could work within the law and not be sued. I think there were faculty at the beginning who were worried about being sued. In fact, NSF sent out a lot of things about what you could and couldn't do. I think NSF was sued. ... People were at best cautious. (Humphreys 2015, slightly modified by Humphreys 2 April 2016)

In her staff position in EECS, Humphreys created "an array of programs under the general rubric of excellence and diversity." (Humphreys 2015) Recruitment of women and minority students became more active. The department produced publications intended to attract a diverse community of students. A number of faculty members (including Michael Lieberman, Paul Gray, David Hodges, Andrew Neureuther, and others) would regularly meet with prospective students when these professors attended conferences or went to other research universities to lecture; and they would also make special trips to minority-serving institutions such as Tuskegee, Howard, and North Carolina A&T to lecture and meet with students and faculty.

---

[20] Humphreys (2015, slightly revised by Humphreys 2 April 2016) did note that: "Proposition 209 allowed for consideration of socioeconomic disadvantage and disability and other criteria, which are often overlapped with some of these candidates who increase diversity." There was also a way for the department to get around Proposition 209 through their industrial partners, as Humphreys noted: "In the '90s, for example, we had a lot of support from Intel for programs ... but direct support to the department for programs to help women and programs to help minorities. And Cisco in particular started a scholarship program, which was very useful after Proposition 209 ... was that they gave us scholarships to award to incoming freshman minorities and women, which we kind of directly do based on gender and ethnicity .... We made this information about the scholarship available to admitted applicants and they applied directly to Cisco, which chose the students; which is something that we couldn't do, but which Stanford, Harvard, Princeton, everybody else who's private can engage in that kind of financial aid.

By the end of the 1980s, EECS had built up a sizable cohort of minority graduate students. In 1986 the department established its EECS Excellence and Diversity Student Programs. The department began to pay the minoriy graduate students to go back to their undergraduate institutions to talk to undergraduate students, and this proved to be highly effective. The official admission criteria in Berkeley's doctoral program were not revised, but at the encouragement of department chair Eugene Wong, the admissions committee decided to take some risks on promising applicants from schools other than MIT, Carnegie Mellon, Stanford, and the University of California campuses. The department provided financial support and space to the minority student and women student organizations, and these groups were given the chance to address the entire faculty at the annual faculty retreat.

Senior women computer scientists and electrical engineers, who could serve as role models, (including Fran Allen, Thelma Estrin, Millie Dresselhaus, Sheila Widnall, Ruth Davis, Jeanne Ferrante, Mary Lou Soffa, Maria Klawe, Barbara Liskov, and Barbara Grosz) were invited to campus as visiting faculty. Students were funded to attend the Grace Hopper conference and the annual National Society of Black Engineers conference. WICSE held weekly lunches, supported by the department and took responsibility for making women visitors feel welcome in the department.[21] (Humphreys 2015)

In 1984 Professor Lenore Blum established a New Horizons certificate program in computer science at Mills College for both male and female students who already held an undergraduate degree in a field other than computer science. An underlying reason for the program was that many students, especially women, did not identify their interest in computer science early enough to select it as their college major. In the New Horizons program, students enrolled in two undergraduate courses per term for four semesters, typically while working full time.

Later, Mills established a master of arts in interdisciplinary computer science, which was attractive to the New Horizons talent pool, given their interest in computer science and their undergraduate degree in another field. Indeed, a number of the certificate students entered and completed this master's degree. Three of the four computer science faculty members at Mills at the time were women, and this seemed to be particularly attractive to female students. The culture was welcoming, with small class sizes, freedom to ask questions in class, and the chance for children to accompany their parents to the labs or classrooms. By 2002, seven students had been awarded the New Horizons certificate, while many others had left the New Horizons program before earning their certificate in order enroll in the

---

[21] Dan Garcia, a Hispanic who studied at Berkeley and MIT, is a member of the faculty as a long-term lecturer. His project (Beauty and Joy of Computing) in reforming the large, introductory computer science courses has received national attention for attracting a broad collection of students to computer science at the undergraduate level. (See Aspray 2016 for further discussion of this program.) Another program, which started in 1991 and continues today that is intended to interest undergraduates in computing, is called Summer Undergraduate Program in Engineering Research at Berkeley (SUPERB), funded by NSF's Research Experiences for Undergraduates program. Gary May, a Berkeley graduate who is now the dean of engineering at Georgia Tech, has replicated the SUPERB program there (called SUPREME).

interdisciplinary masters program or a graduate degree program elsewhere. By 2002, 58 students (12 men and 46 women) had earned the interdisciplinary masters degree. In more recent years, Berkeley and Mills, which are located nearby one another, have found small ways to cooperate with one another with respect to reentry students. One of the WICSE alumna, Almudena Ordonez, is a faculty member at Mills. Both programs received corporate support, which was useful not only to fund the program, but also to validate it.

10.3.2   Carnegie Mellon University

While the programs at Berkeley and Mills started much earlier, the most famous and most influential institutional effort to increase women in computer science was carried out at another of the top-ranked computer science programs, Carnegie Mellon.[22] In 1995 the Sloan Foundation funded Allan Fisher, a computer scientist, and Jane Margolis, an education scholar specializing in gender issues, to study why there was such a low participation rate (under 10%) of women in Carnegie Mellon's computer science program and to craft interventions to increase female participation. The percentage of women among freshman majors in computer science increased from 7% in 1995 to 42% in 2000; and the national computer science community took notice.

Fisher and Margolis's research reached some findings that were widely circulated in the computer science education community:

1. males and females often had different motivations for participating in computing, with men often interested in hacking for its own sake, and women often interested in it for its connections to other topics such as medicine or the arts;
2. the male students tend to have more pre-college computing experience than the females, and the males become interested in computing earlier in their lives;
3. while many young people do not relate well to the geek stereotype of hackers, the stereotype is more damaging in general to women; and
4. women transfer out of computing in their undergraduate years twice as often as men, and this loss of interest is often preceded not by lower grades but instead by a loss of confidence in their abilities to do computer science.

The researchers proposed and the school implemented a number of interventions: admissions requirements were changed to deemphasize the importance of prior computing experience and take a broader look at an applicant's background; diversity issues were discussed with the faculty; students were told repeatedly that prior experience with computing was not a factor of success in the undergraduate major and that computer science was much broader than hacking; undergraduate courses were added that offered a human as well as a technical dimension, such as

---

[22] This account of Carnegie Mellon is written primarily from Fisher and Margolis (2002), Margolis and Fisher (2002), Frieze and Blum (2002), and Margolis et al. (2000).

an undergraduate concentration in human-computer interaction and courses that used multi-disciplinary approaches to study actual social problems; the entry courses into the major were re-designed, segregating students by prior computing experience but nevertheless ramping them up to the same point in the undergraduate major by their sophomore year; a women in computing group was started to lessen isolation for women, and that group was given access to decision-makers within the school[23]; the curriculum was monitored for trouble spots, and interventions such as peer mentoring programs were added as needed, e.g. for courses in which students had difficulties.

Margolis and Fisher's 2003 book, *Unlocking the Clubhouse*, was widely lauded by the computer science community as a (the?) solution to the problem of representation of women in computer science. Indeed, the changes that had occurred at Carnegie Mellon in the number of female students was indeed impressive.

However, there are reasons that the computer scientists should have been more questioning about whether the Carnegie Mellon intervention would work at their own academic institution. Carnegie Mellon is an unusual place, and what happened there might not translate well to other universities and colleges. The undergraduates entering the computer science major were elite students, with an average score on 1450 on the 1600-point College Board exam. The computer science department was the jewel program of this small research university, so it was much easier to get the attention and resources of the higher administration when changes needed to be made by its most honored department. From 1997 to 1999, the NSF funded summer schools at Carnegie Mellon for high-school computer science teachers who taught the AP courses on computer science, to prepare them to teach the C++ programming language that was to be the language tested in the revised College Board Computer Science Advanced Placement exam. Almost one-fifth of the CS AP teachers in the United States attended one of these workshops, so one might believe that these teachers had an affinity to Carnegie Mellon when they were advising their brightest female high school students on where to attend college. Thus one might interpret Carnegie Mellon's results as their gaining a larger piece of the same-sized pie, rather than increasing the size of the pie. If pie eating is a zero-sum game, one might even argue that Carnegie Mellon's recruitment efforts might make it harder for computer science departments at other universities to attract female students.

Margolis and Fisher's work was an ethnographic study, and a primary value of ethnography is to build theory that can be tested. This was exactly the way in which their book was received by thoughtful education specialists and social scientists. The Carnegie Mellon study stimulated a wave of new scholarship on the causes of underrepresentation of women in computing. This stimulus to research was arguably the greatest success of the Carnegie Mellon program.

---

[23] On Carnegie Mellon's women in computing group, see Frieze and Blum (2002). Blum, who organized the women's group in computer science at Carnegie Mellon, is the same person who had earlier initiated the program at Mills College, as described above. The typical story told about Carnegie Mellon's success in increasing the number of women students perhaps underplays the importance of the knowledge that Blum brought with her when she relocated at Carnegie Mellon.

10.3.3   Olin College

There was a period of substantial engineering education reform in the 1990s. It led, for example, to the highly influential EC2000 model curriculum of the nonprofit Accreditation Board for Engineering and Technology (ABET) and to a number of engineering curricular reforms driven by the National Science Foundation.[24] One outcome of all this reform effort was the creation of Olin College of Engineering in Massachusetts by the F. W. Olin Foundation in 1997. The college matriculated its first class of 75 students in 2002.[25] Among institutions focused on engineering, Olin has among the highest percentages of women students and women faculty. Currently, the 343 students are about evenly divided between men and women; and throughout its short history, the college has been approximately evenly gender-balanced for both students and faculty. The plan was to build a new kind of engineering school – one that gives students training in entrepreneurship as well as technology, and that pays close attention to the social, political, and economic contexts of engineering. The school opened on a campus in the Boston suburbs, adjacent to Babson College, which has a highly regarded entrepreneurship program that Olin planned to draw upon. Olin graduated its first class in 2006 and received ABET accreditation the following year.

The school's founding precepts are found in Table 10.4. It is clear from them that the school intends to focus on a single kind of education – undergraduate engineering – and to provide that education in a way that attends to the business and societal, not just the technical dimensions of engineering, while at the same time providing the most up-to-date technical education possible.

The curricular vision operationalized the founding precepts in a straightforward way. It called for a curriculum based on engineering education equivalent to that offered at the best engineering schools in the country, with a focus on entrepreneurial thinking, and containing elements that teach students creativity, innovation, and design.[26] The first two years of the curriculum primarily emphasize general educational study, while the following two years are more focused on specialized study in a major together with a real-world-based capstone project. Each year of the degree program a student engages in at least one hands-on design project. There is also

---

[24] In the 1980s and 1990s, the NSF called for changes in engineering education that included "a shift from disciplinary thinking to interdisciplinary approaches, increased development of communication and teaming skills, and emphasis on engineering practice and design throughout the curriculum." (Stolk and Spence 2003) For a review of NSF programs to reform undergraduate engineering education, see Shipp et al. (2009).

[25] This account of Olin College is based on Thys (2014); Rubin (2006); Anonymous (1997); Stolk and Spence (2003); Sanoff (2000); and the Olin College website (www.olin.edu).

Franklin Olin was a civil engineer and professional baseball player who amassed a fortune through his eponymous corporation. From the formation of his foundation in 1938 until 1997, the foundation built buildings on more than 50 university campuses. The trustees then decided to put all their funding into the founding of a new college, which they supported with a grant of $460 million.

[26] See (http://www.olin.edu/sites/default/files/curricular_vision.pdf).

Table 10.4 The founding precepts of Olin College

1.	Focus on undergraduate engineering education;
2.	Recruit students with high academic achievement, but with an eye towards diverse student and faculty bodies;
3.	Reform engineering education in a way that incorporates "interdisciplinary and integrated teaching, hands-on learning and research opportunities for students, improved communication skills, students working as members of teams (the way that engineers in industry work), exposure to other cultures or an international experience, and a better understanding of business and management practices";
4.	Be student centered and teach the student body about the role of philanthropy in American life;
5.	Offer full tuition scholarships to all students (provided that finances permit it);
6.	Collaborate with Babson College on management and entrepreneurship education, and with other nearby colleges and universities, such as Brandeis and Wellesley;
7.	Provide long-term contracts but no tenure to faculty so as to better assure a faculty that is up to date with the latest advances in science and technology;
8.	Stay independent from government funding, other than for competitive, peer review funding programs; and
9.	Adhere to free enterprise, capitalistic economy, and democratic nation ideals.

Source: (http://www.olin.edu/sites/default/files/olin_founding-precepts.pdf)

continuing work throughout the course of study on oral and written communication skills and on driving home the philanthropic value of giving back to society. In an effort to promote interdisciplinary education, there are no academic departments.

Three majors are offered: mechanical engineering, electrical and computer engineering, and engineering (the latter being a major in which each student has an individual plan of study, some of which have become regularized once enough students have walked a particular path). The choice of major determines only 20% of the student's coursework. The most computing-intensive coursework is typically part of the engineering rather than the electrical and computer engineering major; so the engineering major is the one taken most commonly by students who later enroll in graduate study in computer science[27]; however, it is possible to get significant computer science coursework as part of any of the three majors. (Stein 2015)

One of the principal goals of Olin is to train the students to be good learners, so that they continue to learn when there is no faculty member or formal course structure present. For example, in one course professor Lynn Andrea Stein teaches, entitled Foundations of Computer Science, the material covered includes a smattering of topics drawn from traditional courses in computational complexity, automata

---

[27] The Olin faculty intentionally use the term "computing" rather than "computer science" when talking about their curriculum. While their degree program is ABET-accredited, it is accredited as an engineering degree program, not as a computer science and engineering program. (Stein 2015) They differentiate the course of instruction they offer from a traditional computer science undergraduate major. The goal of the Olin faculty is to provide a compressed version of all the fundamental ideas of computer science so that their students have enough foundational knowledge to learn the remainder of the standard computer science core curriculum. (See Downey and Stein 2006.)

theory, and programming languages. The real goal of the course is not to cover any of this material in depth, but instead to teach about a few basic structures such as trees and give the students a visceral understanding of the fundamental importance of the logarithmic and exponential processes that govern many results across many different subfields of computer science.

Olin has struggled to find an appropriate way to fit entrepreneurship into its mission. There are some students who are interested in entrepreneurship in a traditional business sense. A number of students have taken traditional entrepreneurship courses at Babson, and several Olin graduates have created successful start-up companies. However, there is another sense of entrepreneurship that has gained a foothold at Olin, tied closely to the tasks that an engineer may confront in her everyday work and built into a recently introduced course on Products and Markets that is required of every first-year student:

> The question the people behind that course really wrestled with was: what is it about entrepreneurial thinking? What is it about making something maybe in the face of inadequate resources? About figuring out what it takes to actualize and not just [build] the prototype but [build something that is] really adaptable, adoptable, sustainable? What aspect of that does every engineer need to know? (Stein 2015)

The theme of social responsibility of the engineer is strongly tied into the curriculum. There is a first-year course entitled Engineering for Humanity, jointly designed by an anthropologist and computer scientist on the faculty. In this course, the students do a semester-long project with a particular client, and this project generally puts the students in contact with older adults in the local community. The course is intended not only to teach the students the engineering process of going from need-finding to delivery, but also to help them understand "the situation of the individual community member they were working with and more broadly some of the issues of aging and changing demographics, and how the particular concrete thing they are doing fits into this larger issue." (Stein 2015) One type of senior capstone that is chosen by some students, for example, concerns affordable design and entrepreneurship. These projects are carried out for communities with limited resources. Projects have been carried out, for example, in Morocco, Ghana, and India outside the United States; but also in impoverished communities in Mississippi and Boston.

Olin has no tenure system; instead, faculty members receive multi-year contracts (currently, for six years). Nevertheless, Olin has been able to attract top faculty talent – in particular people who are enthusiastic about participating in a grand educational experiment.[28]

---

[28] For example, Lynn Andrea Stein, a member of the original faulty and the first computer scientist hired, was encouraged by her department head at MIT to accept the Olin offer, saying "It's not a once-in-a-lifetime opportunity. Opportunities like this don't come along nearly that often." (Stein 2015) Stein, who had already been very active in computer education, including some of the earliest use of robotics for educational purposes, was attracted by the opportunity of a "clean slate" to rethink undergraduate education. At MIT, she had been especially excited by her research on the philosophy of computing, to try to find the central models of the computing field. This research did not particularly interest her MIT colleagues; they were instead much more interested in narrowly

The academic excellence, sense of community, and free tuition help Olin to attract top-level students who might have gone to MIT, Cal Tech, or Stanford. It is still too soon to know the outcome of this educational experiment, although there are some promising early signs. The Olin faculty learned by trial and error that it is very effective to use students to recruit new students of exceptional quality. They found that their original student body tended to include many people who were not averse to risk if they thought the rewards might be high. This is understandable, given that the first several student classes were recruited before there was a well-established curriculum, before the school had been accredited (which only can happen after the first class of students graduate), and without many physical buildings.

Over time, Olin has honed a two-stage admission process, which it has used since its initial round of admissions. The process first narrows the applicant pool to those students who are academically gifted based on traditional criteria such as grade-point averages, College Board test scores, and outside interests. All of these students are then invited to a weekend on campus, during which they are placed in small teams and given a fixed time to carry out some engineering-related task, e.g. building the tallest structure possible using a hot wire cutter and an eight-foot piece of insulating Styrofoam.

> It was very much [about] being thrown into a crazy environment, and given something ridiculous to do, and seeing for whom that was an intriguing opportunity and for whom that was terrifying. We didn't actually care whether people could create stable structures. We know how to teach people [that]… What we didn't know how to teach people was how to be excited about that kind of experience… It's designed so that it's impossible… they do all sorts of things to make sure this cannot be misunderstood to be about your skill as an engineer. It's about your comfort level and interest in being asked to do things that are a little bit ambiguous and uncertain, and combining with people around you to make progress. (Stein 2015)

This weekend is a make-or-break experience for many of the students. Either it cements their enthusiasm for Olin as their first choice of college to attend, or it helps them decide that Olin is not the place for them.[29]

---

focused areas of artificial intelligence research. She felt she might have a better chance to engage these broader, more fundamental questions in a setting such as Olin. Stein's only concern was that Olin would turn out to be mediocre, which is no longer a concern ten years later.

[29] Stein (2015) notes that Olin pays attention to diversity in its admission. It has done well in having the student body about equally male and female, which is a higher percentage of female engineering students at any program except Smith, which is all women. Olin's racial-ethnic diversity numbers are about average for undergraduate engineering programs in the United States overall – although there is continuing interest in improving this diversity.

The Olin faculty has learned that having equal numbers of men and women in the classroom has in itself not solved all the gender issues: "we have many of the same issues of gender dynamics that you find elsewhere, including female students who believe, and sometimes are told by students or outsiders, that they were only admitted because they were female, which by the way is definitely not true." The faculty began to talk explicitly with the students about these issues, and they found that this explicit talk can be "very empowering for the students." The faculty discuss with the students the impact it has when a female student receives the remark "Really, you're an engineer?" This leads them to talk about biases: "not biases that come out of wanting to be unjust but biases in the sense of tilt. Things that aren't level. We do think computer scientists look like Mark

The students who were in Olin's first ten graduating classes have done well for themselves. Employers praise the Olin graduates for communication, teamwork, and self-direction. The profile of these students looks different from those of other engineering students through the lens of the National Survey of Student Engagement. As a recruiter from Microsoft said: "If we want people who are technically skilled, there are a lot of schools we can go to. If we want somebody who can walk out of [their] undergraduate experience ready to be a project manager, Olin is the place we need to go." (Stein 2015) The Olin students may not have been exposed to as much content in a particular engineering area as someone who studied in a traditional degree program in that particular engineering field, but the Olin students are quick learners, having been taught how to learn for themselves and acquire knowledge in an ambiguous situation. The academically strongest Olin students have been admitted and have done well in top-ranked graduate engineering programs. For example, Olin students have attended doctoral programs in computer science at MIT, Carnegie Mellon, and Cornell.[30]

10.3.4 Smith College

Smith College, the largest women's college in the United States, was the first women's college to offer an engineering degree.[31] Women's colleges have historically been a strong feeder of women into graduate STEM programs; however, over the past half century the number of women's colleges in the United States has dropped from 230 to 46 through closure or coeducation. With strong support from the math, computer science, and physics faculties, Smith College decided in 1999 to establish the first engineering program on a woman's college campus in the United States.[32]

Zuckerberg, not like me [the female student]. What are the implications of that? Part of the implication is that the experience of being a female computer scientist is quite different from the experience of being a male computer scientist. Just knowing that fact is a first step towards mitigating the potentially negative impact..." (Stein 2015)

[30] It is hard to generalize from the small number of Olin students, but within this small population there have been growing numbers over time of students interested in computing; and while the early student populations of students interested in computing were primarily male, they are now more than half female. (Stein 2015)

[31] The account of engineering at Smith College is based on Pfabe and Easwar (1999); Ellis et al. (2010); Farrell (2004); Deitz (2001); Anonymous (2004); Loftus (2003); Voss and Ellis (2002); Grasso (2004); Voss (2015); and the department's website (http://www.smith.edu/engin/).

[32] Only a dozen years before Smith College opened its engineering program, the civil engineer Samuel Florman, well known for his writings about engineering and culture, had visited Smith to participate in a seminar on the role of technology in modern society. Reflecting on that visit, he wrote an essay for *Harper's Magazine* that gave a class argument about why there were so few women in engineering and why Smith would never have an engineering program: engineering has attracted bright boys from blue collar families; girls from blue-collar families have not been given the educational opportunities as boys; upper-class families "do not esteem a career in engineer-

Its first class of 20 engineering majors graduated in 2004 – with these graduates admitted into leading graduate engineering programs at Harvard-MIT, Michigan, Dartmouth, Cornell, Princeton, Berkeley, and Notre Dame.[33] (Anonymous 2004) As of 2004, five of the nine faculty members in the engineering program were women and 5 % of Smith students were engineering majors. In its early years, the program received substantial financial support from industry, notably from Ford Motor Company and Hewlett Packard. (Deitz 2001)

Smith defined its engineering program as "the application of mathematics and science *to serve humanity*," (Grasso 2004, emphasis in original) and thus the education of the Smith engineering students included significant course work in the humanities and social sciences as well as in mathematics and science.[34] As Domenico Grasso (2004), the founding director of Smith's Picker Engineering Program,[35] argued:

> Engineering decisions rarely hinge entirely on science. Rather, engineers must also consider many other factors such as economics, safety, accessibility, manufacturability, reliability, the environment and sustainability, to name a few. Engineers must learn to manage and integrate a wide variety of information and knowledge to make sound decisions.

For example, in a study of the 1985 Mexico City earthquake in the continuum mechanics course, the students considered not only the statistics and dynamics, but also "the social implications of the quake, the city's preparedness, and its response in the aftermath." (Loftus 2003) The education was intended not only to be

---

ing"; hence there will not be girls in engineering. (Florman 1978) There are of course many problems with this argument, as the creation a decade later of a successful engineering program at Smith implies.

Another predecessor was a proposal in 1992 by the anthropologist Barbara Lazarus, the associate provost for academics at Carnegie Mellon and former dean at Wellesley, to create a women's engineering institute. The idea got some attention at the time in lectures at a WEPAN meeting, several National Academy of Sciences talks, and several other professional meetings; but it was not implemented. Some people were opposed because they believed it would further isolate women in the engineering community. (Emmett 1992)

[33] Ruth Simmons was president of Smith College at the time the engineering program was established. She was the first African American to serve as president of any of the elite Seven Sisters colleges. She took an interest in diversity in the STEM disciplines, not only the engineering program, for example supporting a campus organization called Union of Underrepresented Science Students, which focused on retaining students in the STEM disciplines. (Dietz 2001)

[34] As the engineering program because established, students used internships, research experiences for undergraduates, and capstone programs to pursue engineering in service to humanity. Smith has a program called Praxis, which guarantees every Smith student a stipend of a few thousand dollars if they want to pursue an unpaid summer internship. The majority of students fulfill their year-long capstone requirement by working through the program's design clinic, in small teams of three or four students, to work with an industry, government, or nonprofit client on a real-world problem. (Voss 2015)

[35] Grasso had a deep influence on the direction of the program. He hired all of the original engineering faculty members, choosing people who could integrate undergraduate students into research and who appreciated the strengths offered by having the engineering program set in a liberal arts environment. Grasso was an influential voice in discussions of what this program should be. It was not going to be a program that produces "the stereotypical engineer" and teaches its graduates "only to solve the math of [engineering]." (Voss 2015)

interdisciplinary, it was intended to train students to use their education in service to society.[36] Perhaps because of this aspect, the engineering program has attracted students to the major who had never considered majoring in engineering when they were in high school.[37]

It was important to the Smith faculty that the program be accredited by the principal engineering education standards organization, the Accreditation Board for Engineering and Technology (ABET), to ensure the academic integrity of the program and to satisfy industry employers of Smith graduates. The learner-centered educational philosophy adopted at Smith followed the Engineering Criteria 2000 (EC 2000) established by ABET.[38] Instruction in the Smith program draws upon many different methods, including "critical thinking using techniques usually associated with study in the liberal arts and through structured problem solving, which is typically associated with an engineering education." (Grasso 2004) Rather than offering different degrees in different engineering specialties such as mechanical or electrical engineering, there is a single engineering science degree focused on the fundamentals of engineering. In the first two years, the students take the same courses in fundamental topics such as circuit theory, thermodynamics, statics, and fluid mechanics. As each individual student comes to understand what area of engineering interests her most, she has significant choice over courses during her final two years – nevertheless, this course of study does not offer the same amount of specialization that would be offered in, say, an electrical engineering degree program elsewhere.[39]

---

[36] Smith's engineering program was selected in its first year as one of the initial sites for a Virtual Development Center for Anita Borg's Institute of Women and Technology (now called the Anita Borg Institute). See Chapter 8 for further information about the Virtual Development Centers and the Anita Borg Institute.

[37] The facilities of the engineering program were unprepossessing: a temporary building "where [the faculty] had one large open space and... ran all of [their] labs out of it. [They] were constantly putting equipment away and setting equipment back up." Faculty offices were located in an old house three blocks away. As the faculty expanded, offices were found here and there for them across campus. In 2010 the engineering program became an occupant in a state-of-the-art building (Ford Hall). This move has helped somewhat with the rapid growth in the number of students majoring in engineering over the past 5 years. (Voss 2015)

[38] Much in line with ABET's learner-centered standards was an influential report from the National Research Council on how people learn (NRC 2000). Learning-centered pedagogy starts with "the knowledge, skills, attitudes, and beliefs that learners bring to the educational setting." (from NRC 2000, as quoted in Voss and Ellis 2002). For a discussion of how women-friendly ABET's EC 2000 education standards are, see Loftus (2003) and Rosser (2001). Voss and Ellis (2002) provides an early example of an attempt to teach an engineering course (circuit theory) using a learner-centered approach. Also see Glazer-Raymo (2008).

[39] The current department head, Susan Voss, saw no down side to this curriculum. 80 to 90 % of the engineering majors pursued careers in engineering. The top students were admitted and did well in the top engineering departments; and employers were generally eager to hire the Smith graduates because they had stronger training in communication skills, working with people, thinking about the big picture, and managing projects. These remarks are based on the experience of the almost 300 students who had graduated from Smith with engineering degrees by 2015. (Voss 2015)

While the faculty members implement their courses in many different ways, considerable attention is given to building a community of learners among the engineering students. For example, in 2009, Smith introduced a knowledge-building pilot study in its engineering program, in an elective course entitled Techniques for Modeling Engineering Processes. As the instructors of this course explain:

> Knowledge building pedagogies place great emphasis on community rather than individual knowledge creation, on the crucial role of discourse, and on the shared goal of idea improvement rather than seeking a final answer. Students are cast as knowledge workers, engaged in the same social, intellectual, and discourse practices as those found in all knowledge producing organizations. (Ellis et al. 2010)

Faculty enabled this knowledge-building environment by assigning problems that students care about and by creating an environment that is friendly to collaboration among students. In an assessment of the class, the students stated that it prepared them well for their future work in the information age:

> Most students (84%) felt that knowledge building helped prepare them for working in the knowledge age. They cited that it helped them express ideas and see other viewpoints; broadened their perspective, knowledge, and awareness of outside resources; improved their ability to think creatively and critically; and improved their ability to interact electronically. Only one student (5%) disagreed. (Ellis et al. 2010)

Smith already had a computer science major at the time the engineering program was established.[40] In 1988 Joseph O'Rourke, a professor at Johns Hopkins University, was hired to build a computer science department; and he remains a member of the department today. There had already been a major in computer science, but no department. Courses had been taught by Bert Mendelson, a professor in the mathematics department. Smith was an example of a larger trend in liberal arts colleges in the late 1980s; with a dawning awareness of the importance of computing in society, a number of computer science departments were forming – many of them as spinoffs from mathematics departments. The Smith computer science department has stabilized with 6.25 positions, about equally divided between male and female faculty. Department enrollments have tracked with national trends, so there have been periods of feast and famine in enrollments; and at times the faculty have been stretched thin to meet demand.

After following a traditional curriculum for its first few years, Smith liberalized its computer science course requirements around 2000. Instead of requiring specific courses, the faculty implemented distribution requirements that required each student to take some intermediate or advanced level course in each of the systems, programming, and theory areas. There was latitude in what courses would satisfy each of these requirements. The department was an early adopter of the programming language Python and a heavy user of applications in its introductory computer science course. Today, the department has a grant from Project Kaleidoscope to encompass diversity to a greater extent in the introductory course. The department has also flattened the hierarchy in course offerings, so as to reduce the number of

---

[40] The discussion here of the Smith computer science department is based on O'Rourke (2015).

prerequisites to enroll in some of the more interesting applications-oriented courses such as artificial intelligence.

In recent years, the students have shown strong interest in web design, mobile computing, and software engineering, but not in most of the hardware areas of computer science. Unlike the engineering program, there is no formal commitment in the computer science department to a curriculum that emphasizes the social responsibility aspects of computing, although individual faculty members and students sometimes pursue this theme.

The students who major in computer science have high variation in their computing, math, and science backgrounds when they arrive on campus. The department gives students many opportunities to participate in research projects or do independent studies with professors. Students who want access to a larger number of advanced computing courses are able to take courses for no extra charge at the University of Massachusetts in nearby Amherst. Once students decide to major in computer science, there is practically 100% retention; the big decision for many students is whether to elect a second major. Economics and computer science is a common double major. There are also strong ties of computer science majors to math and art, and courses in human-computer interaction and computer graphics are popular.

About a quarter of the computer science students go on to graduate school. But the students are much in demand in the marketplace when they graduate, and this is a disincentive to further education.

Engineering and computer science is less common as a double major because of the high number of required courses to obtain the engineering degree. Especially in the past few years, a few engineering students every year have wanted to do a combined degree in computer science and engineering, and many of the engineering students have started to take extra classes in computer science. In 2014, the large number of incoming engineering majors swamped the program, and the engineering faculty had to send some of the new students over to instead take computer science in their first semester while they waited for there to be room in the introductory engineering course. At least a few of these students found that they were strongly attracted to computer science and decided to major in it rather than returning to begin their study of engineering. (Voss 2015)

Several schools have contacted Smith about building similar programs at their institutions. One is Sweet Briar College in Virginia. Sweet Briar opened its engineering program in 2004 and received accreditation from ABET in 2011. It offers a Bachelor of Science degree in engineering science and a Bachelor of Arts degree in engineering management. At the time of accreditation, it had 30 engineering majors, with plans to triple the size of the program. Important to recruitment into the program were a series of Explore Engineering events held on campus for high school girls. (House 2011) Unfortunately, the long-term viability of the college is open to question, having one decision in 2015 to close the college to be subsequently overturned.

10.3.5   Harvey Mudd College

Harvey Mudd College is an unusual place because, while it is a liberal arts college, it is also one of the leading colleges for the study of engineering, science, and mathematics. The college has no non-STEM majors, and throughout most of the twentieth century the student population was approximately 90% male. It was founded in 1955 in Claremont, California and is part of the Claremont Colleges consortium.

Harvey Mudd hired its first computer scientist to the faculty in the 1970s, but he stayed only one year before leaving. The computer science program began in earnest in 1981, when Mike Erlinger joined the faculty. He had worked at Hughes Aircraft while completing his PhD in computer science at UCLA. Computer science was established as a department in 1985 or 1986, and a major in computer science was established in 1992.[41] While the major originally enrolled only a few students, it grew quickly by drawing not only Mudd students who might otherwise have majored in math or engineering, but also by attracting students from the other Claremont colleges. There were four faculty members when the major was established, and the number has grown to 13 today.

Erlinger characterized the students in the 1980s as "strongly nerds." (Erlinger 2015) They were of high academic caliber and had a tradition of working very hard. It was, Erlinger noted, "a very male-dominated environment" where the students tended to be narrowly focused on scientific topics. For example, the faculty had difficulty introducing the pair programming educational method into the computer science curriculum because the students were so used to and so comfortable in working alone. However, during the current century, the student demographics have been consciously made more diverse. Today, there are approximately equal numbers of male and female students, and the student body has much more widely varied interests than before.

Since 1981, all Harvey Mudd students have been required to take a programming course in the computer science department in the first semester of their freshman year as part of the core science curriculum. For the most part, the computer science curriculum follows the ACM/IEEE model curricula. However, the curriculum emphasizes systems, theory, and algorithms and largely ignores course offerings in computer engineering.[42]

---

[41] At the time Erlinger was hired, Harvey Mudd also did not have a biology department. Biology and computer science developed according to the same approximate schedule in the 1980s and early 1990s. The faculty in the other science and engineering disciplines were nervous about the additions of these two majors because they meant the same budget and same number of students had to be divided up into more parts. (Erlinger 2015) Much of the material for this section on Harvey Mudd is based on Erlinger (2015) and Alvarado (2016), and secondarily on Shellenbarger (2013). However, also see a number of other papers cited below on particular aspects of the departmental change that led to more women enrolling in the major.

[42] As Erlinger (2015) explains the absence of computer engineering in the computer science curriculum: "We paid very little lip service to that, and the reason was that the college's engineering program is a general engineering program, which had … some computer engineering. They were always the strongest department in terms of faculty and students and the least willing to cooper-

The change in the department that led to broadened participation began as a result of a visit by Jane Margolis – by this time in 2000 at UCLA – to give a colloquium about her previous research on attracting women into computer science at Carnegie Mellon. As Erlinger remembers the response to this colloquium:

> This was not meant in a derogatory way, but a couple of students asked, why do we care? We have plenty of CS majors in the department, it's a great department, why do we care? I thought that was very interesting because we started to ask ourselves why do we care for the department and its faculty members. It wasn't that we were against women; there weren't any women at Harvey Mudd College, in a way. The numbers were fairly small. The fact that we didn't get women was just … the way it is. We started to ask: "Is this really what we should be doing, and is this what we want to be?" We started questioning it. It isn't clear that we knew what to do; first of all, the admissions [process] had to change in order that there were more women. We were never against women … There just weren't a lot of women and we didn't really know what to do. We were … at the end of the pipeline in terms of who came to Mudd.

Under Jon Strauss, who was the president of Harvey Mudd from 1997 to 2006, a major overhaul was made to the admission process in order to recruit more women and minorities. As a result, the percentage of women students began to rise, starting about 2000.

> [T]he department started to ask: "All right, there [are] women here now. Why aren't they in CS?" It is not clear to me that we had a real plan; I'm not sure that we ever had a plan. It sounds crazy, but I think we just said: "We need to do something and we're not sure exactly what. We need to start thinking about this".

The change began to happen in earnest in 2005, when Christine Alvarado joined the faculty.[43] Erlinger was then the department chair, and he fully supported Alvarado's efforts. One of her principal changes was to rethink the introductory course. It had been a Java programming course; but now it was to be turned into a breadth-first course – what some would now call a "principles-based course"[44] – that had "five or six different modules that were not … There's programming in

---

ate… There were always some students that found a way to cross over and do a CSE degree by specialization, et cetera. It's never been real cooperation in that aspect between the two departments. That's slowly changing now, partly because a lot of historical faculty in engineering are gone and new faculty are asking, why, why is it this way? … Because these students are always self motivated in Mudd a lot of them found a way to do something by working through the two departments. In that sense it's a little bit different than other places, so we don't have any hardware labs, for instance, in CS at all. We have a robotics lab, but then that was really it."

[43] Alvarado had studied computer science at Dartmouth and MIT before joining the Harvey Mudd faculty. Her involvement in a newly formed Women's Technology program at MIT had spurred her interest in both teaching and underrepresentation of women in computing. She was attracted to Harvey Mudd because it had a large computer faculty (10) for a school with a small number of students (700). Moreover, she described the atmosphere as "vibrant", with "everybody…doing really exciting things." (Alvarado 2016)

[44] This is in reference to the CS Principles Course that has been created with support from the National Science Foundation and the College Board by a committee led by Owen Astrachan of Duke University and Amy Briggs of Middlebury College. See Aspray (2016) for a history of the development of this course. Alvarado (2016) indicates that the Harvey Mudd introductory course included significantly more programming while still having the same philosophical goals of the CS Principles course.

there all along, but it's not the focus. The focus was really to understand the feel of computer science in a different way." (Erlinger 2015)[45] As Alvarado described it, the course was intended to be "not only more appealing to women but basically more appealing to any student who wasn't well versed in computer science." (Alvarado 2016)

The students were divided into separate sections: some sections for those students with a year or more of high school computer science, and other sections for the larger group of students that had less high school experience.[46] (Dodds et al. 2007, 2008; Klawe as interviewed in Hoffmann 2012) Alvarado (2016) described this sectioning practice as "significant" and explained its purpose:

> They were in theory the same course, but the track for students with experience went deeper. It was like an honors track, just to have a little more enrichment, but the goal was not to get those students further ahead but rather to stall them working on interesting problems to give the students with no experience a little time to catch up to get into the field.

Maria Klawe arrived as the new president of Harvey Mudd in the Spring of 2006, just as the new introductory computer science course was being taught for the first time. She had a long track record in supporting women in computer science, and she expanded the efforts initiated by President Strauss to attract scientifically qualified women to attend Mudd. It was under President Klawe's administration that females rose to become half of each entering Mudd class.[47]

Computer science at Mudd had always attracted many of its majors on the basis of student experience in the introductory course, because many of the students had not had the chance to experience real computer science in high school. With the

---

[45] As Alvarado (2016) amplified on the description of the new introductory course: it was a "broad introduction to computer science where we touched on different styles of programming, a little bit of hardware, a little bit of computer science theory. All along the way there were applications to other scientific disciplines that would appeal to … the broad science and engineering base that we have at Harvey Mudd."

[46] Based on preliminary empirical findings, "the data suggest that CS for Scientists succeeds in three ways:

students 'get' the importance of CS to their future scientific endeavors and its importance as a stand-alone discipline; students succeed in future CS courses to a greater extent; and they continue studying CS as often – and perhaps more – than from traditional alternatives. In each measure women benefit at least as much as men." (Dodds et al. 2008)

[47] The college is making efforts, but it is not as far along in attracting minority students as it is in attracting women students. "We decided to go to eight hundred from roughly seven hundred students, and part of the discussion of that was to try and make 40 %, roughly, of that eight hundred students of color. That's starting to happen, the infrastructure's being created better to support these students … think we've expanded our schools that we visit and people that we talk to and we're starting to see more students of color. That started in roughly 2000, so this is a continuing effort." (Erlinger 2015)

increase in the number of women and the change in the introductory course, the number of women majors in computer science began to rise.[48]

To complement the change in the introductory course, the department began to send students to the Grace Hopper conference – at first about ten students, and today about 40.[49] At Mudd, students do not declare a major until the end of their sophomore year. During the summer between freshman and sophomore years, the department asked all the female members of the incoming sophomore class, who had by now had all taken their introductory programming class, if they wanted to attend the Hopper conference.

> [I]t didn't necessarily give us majors out of that group of women that went [to the Hopper conference], but it gave us people who understood computer science and, no matter what their major was, started to see computation as an important aspect of their program, whatever it was. That turns out to us to be really important because we see computer science becoming more and more recognized as a required science within any major, much like math is ... We were really happy with that result too, in the sense that it just made the whole campus much more cognizant of computer science and how it might fit into the program, whatever it might be. (Erlinger 2015)

Another change introduced at the same time in 2006 was that the department began to place women in summer research programs, even though the students had typically taken only one or at most two computer science courses by this time.[50] There was a long tradition at Harvey Mudd to engage students in research, but in computer science in the past, the students had typically not become involved in research until their junior year. The faculty found that engaging the younger students in research worked:

> surprisingly well. I think the key to making it work was finding projects that weren't necessarily research *per se* but were either contributions to existing research projects ... Maybe

---

[48] Reflecting on the new introductory course, Erlinger (2015) noted: "The funny thing is that in hindsight when you really look at it, there's more programming in that course then there was in the original Java course. But, it's spread out among all these other topics. There's a lot of short Python programs, some of them are rather large, but they're much more interesting programs. There's a lot more interesting material for students. Java class had been ranked among the freshmen as the lowest class in terms of like. This new Python-based class is ranked the highest."

[49] The rationale for sending the students to the Hopper conference were given in the GHC Evaluation and Impact Study that attendance at the conference leads students to feel less isolated, be more committed to CS, and be more inspired. The Mudd faculty informally polled their students who attended the Hopper conference in 2007 and 2008, and in 2009 and 2010 they more formally assessed their students who attended. They found in the more formal analysis that the Hopper conference gave most students a better understanding of both CS and its culture; it increased their desire to take another CS course and become a major; Hopper attendance was positively correlated with taking another CS course (52% who attended the Hopper conference did take another CS course, compared to 31% of female students who did not attend the Hopper conference took another CS course. (Alvarado and Judson 2014) Also see DuBow et al. (2012) and Alvarado (2016).

[50] Alvarado et al. (2012) summarized the lessons they had learned in providing research experiences to Mudd undergraduate women as a means to increase their participation in computing: (1) provide open-ended projects with sufficient scaffolding; (2) have the students work in teams, with a mix of experienced and inexperienced students; (3) daily communication and mentoring is important; and (4) broadening participation benefits all students.

the faculty member and their senior students have a project, but there are some tasks that need doing that these first year students could come in and support. These are just straight-forward like maybe data analysis or building an interface or something that's relatively basic in terms of programming skills but that still contributes very strongly to the actual project as a whole. Then the students get a chance to contribute, and they get a chance to get involved with those students who are more advanced than they are. ...

I think that the two keys of the summer research experience for the students were, one was a sense of accomplishment. When we did our follow-up survey for that, one of the most common comments we heard was, "I didn't know I could do so much with my skills. I'm shocked at how good I am now, even though I don't know very much." It gave them a sense of confidence. Then the other thing that we heard a lot was that they got more involved in this community, that they felt really a part of the CS community. (Alvarado 2016)

However, these changes were costly. It cost an average of $1000 to send a student to the Hopper conference and $6000 to enter a student in a summer research program. These funds were secured through the NSF REU program, faculty grants, and sources that President Klawe helped the faculty to find.

The number of women on the faculty increased over time. The faculty took great effort to mentor all the students, not only the female students. A culture was developed in which the students also mentor other students.[51] Today the introductory class typically enrolls 200–300 students. Approximately 150–200 "grutors" (a neologism created at Mudd by merging the words 'grader' and 'tutor') are assigned to both work in the labs and grade in the courses.

The whole idea of Grutoring and mentoring other students has become a way of life at Mudd. When I first got there, there were two people who helped somebody grade a math class or whatever. In CS now, everybody does this. This is part of their education; this is part of what they want to do. (Erlinger 2015)

Today, there are three majors offered by the computer science department: computer science; a combined major between math and computer science, focused on algorithms; and a combined major between biology, math, and computer science, focused on biocomputation.[52] Of the entering Harvey Mudd class of approximately 200 students, some 60 students – or 30 % of the entering class – end up majoring in

---

[51] Alvarado (2016) spoke of the change of culture in the department: "It was very much almost an overnight cultural shift. As you probably know, the numbers went from about thirteen percent women to about forty percent women in the span of about two years. It felt like it went from CS being this place that's kind of weird for women, a lot of stereotypical geeky guys. Nothing against geeky guys. I love them, but [only as] one kind of student in CS. Then suddenly there were so many different kinds of students in CS, and that was just sort of how it was. Nobody thought that was weird. It just happened. Then when the new students came in, it felt like the norm. ... It was almost surprising that the culture shift was such a non-issue. It just sort of happened magically. I reel from that."

[52] Alvarado left Harvey Mudd to take a faculty position at the University of California at San Diego, so she has not witnessed the growth of the various majors; but from while she was there, having three majors, including one with a biological orientation that might be more attractive to women, she stated: "I'm not sure about whether that made a difference. Certainly I think the Bioinformatics or Computational Biology Major, there were more women percentage-wise in that major, but when I left Harvey Mudd, it was still a very small major. There were not many students in that major at all. Then in terms of Math-CS versus the CS, I think it was about the same in terms of the repre-

one of these programs offered by computer science. In fact, in some years that percentage has been as high as 45%. They also get students from Pitzer, Claremont McKenna, and Scripps – three of the other Clarement colleges – who take courses at Mudd for a computer science major or minor since there is not a computer science major at those colleges. The number of females in the computer science majors is 40% or higher, with the percentage of women in the computer science and biocomputation programs actually tracking around 50%.[53]

The culture in the department is also quite important. In the old days, people tended to "do their own thing" and there was not a lot of communication. With recent hires, the faculty is much younger and about 40% female (whereas there were zero women faculty members in the science departments and only one in the humanities when Erlinger arrived at Mudd in 1981). Now,

> the department is really active, a very communicative department, a very cooperative department.[54] Everybody's door is open all the time and faculty come in and out of offices all the time.
>
> The other thing is it's also been very open for the students in the sense that Mudd's always been a place where faculty are available. We as a department, right or wrong, faculty are available 24 hours a day, almost. There are some [who] come in late, work very late at night, and students will e-mail them at 2:00 in the morning and they'll respond. There are others [who] come in early in the morning and deal with the students who get up early in the morning.
>
> I know there is a hierarchy of faculty and students, but in some ways it doesn't exist. It's more like a group that's trying to develop and learn from each other. (Erlinger 2015)

Today, about 20% of the Mudd computer science majors attend graduate school immediately upon graduation, although a larger number of the students eventually attend graduate school after a few years in the workforce. As Erlinger (2015) explains:

> Harvey Mudd is a very expensive school. People leave with well above the average in terms of loans, et cetera. And so, the market is where they'll end up. They're going to have to go to Google or Apple or somebody else. I don't see that changing. We're really pushing a number of people to go to graduate school and the ones that do, do well at it. It's difficult (given that the demand of the marketplace [is so great]) for students to make that choice.

---

sentation of women in both of those majors. I don't know that the number of majors made that much of a difference." (Alvarado 2016)

[53] Erlinger (2015) indicates that the male and female students generally tend to get along well: "We don't have a lot of all-male tables at lunch or projects that are only male. It's a very mix. There are obviously problems at times among relationships as there is anywhere, but overall, that has not turned out to be an issue. There have been only one or two situations in terms of language and other things over the years that I've been involved between a male and a female. Not actually any action, just unhappy with a choice of words."

[54] Erlinger (2015) that the average course load in computer science in terms of number of student contacts is about triple the average for the rest of the college, and that this has caused some stress. However, the good communication has enabled the department to get through this stressful situation without any meltdowns.

Traditionally, computer science had placed many of its students in programming jobs in the aerospace industry, which has a concentration in the Los Angeles area. However, over time, increasing numbers of their graduates have gone to Apple, Google, and other companies in the computing industry.

> Also, we've seen a tremendous number of our students go on to small companies. Not necessarily start-ups, but small companies where they feel they really can do something. Even though the money's there, our students are also looking for something that would be exciting and fun for them, whatever their capabilities are. They're more interested in the project than sitting down and coding. They like building things, they like designing things. (Erlinger 2015)

Many outsiders, knowing of Maria Klawe's success in promoting women in computing at the University of British Columbia, Princeton University, and the ACM, have naturally attributed the success that Harvey Mudd to her efforts. The story is more complicated than that.[55]

> [This giving the credit to Klawe has] been a real stickler in the department, because in the time I was chair – and I was chair most of the time Maria was there – Maria came the department only once. It was not to discuss broadening participation in general – and I can't even remember what she wanted, but she came for some one reason. That's it. Harvey Mudd started to change [in broadening participation] long before Maria got there. What Maria's really provided, and I don't think she denies this – I just don't think people listen to what she's saying – she really has provided the top-level support, the attitude that really allows everybody else to change and do this. And she's done that unbelievably well. (Erlinger 2015)[56]

---

[55] On his personal experiences as department chair with President Klawe, Erlinger states: "I love Maria, I think she's fantastic, [but] she's a complete pain to me because when I go to her and tell her about things I want to do, and can we really do this, her answer always is: 'Do it. You guys get down there and do it'. … She's never asked me to do something that's not right, but she's really been very supportive [but also] very demanding that we really do do it.

"I think you could have somebody different at top and we would have done some of this, but we would not be where we are. When I wanted to pay for Grace Hopper or other things, I could go and Maria would give me pointers and help me try figure out how to do it. She didn't always have the money, but we would figure out together how to do it. She's always been that supportive … The school started [to change] before she got there; I think the department and the school recognized it needed to change. It's just that she's kind of turned up the heat and turned up the fire."

[56] Alvarado (2016) has a somewhat different view from Erlinger: "right around the same time in 2006, Maria Klawe joined as President, and she was very excited to hear about things that we had already gotten started on and just threw her support in one hundred percent, with funding, with her connections, with just talking to people, playing it up. That's why I say it just wouldn't have been possible without her support. Even though most of the stuff was already underway, she acted as a giant catalyst that got everything moving so quickly."

10.3.6   Other Computing Departments Attracting Women

It is beyond the scope of this study to review all of the major academic programs that have been successful in attracting females to study computing.[57] There have been several leaders among co-educational institutions. The University of Washington has revised its introductory computer science course to emphasize creativity and real-world applications; ran summer schools for high-school teachers and summer camps for high school students; and worked closely with WEPAN to increase participation of women, and with the DO-IT Center and the Access Computing Alliance to increase participation of people with disabilities.[58] Georgia Tech has built a wide interest in computing through its university-wide required computing course and its undergraduate computational media program. The University of Colorado at Boulder has found success in attracting female students to its minor and certificate programs in Technology, Arts, and Media, as well as in a technology for community course.[59] Among the women's colleges, there are strong stand-alone computer science programs at Wellesley[60] and Bryn Mawr,[61] and Spelman College has a successful five-year degree program in computer science with Georgia Tech.

References

Alvarado, Christine. 2016. *Oral history interview by William Aspray*. Charles Babbage Institute Oral History Collection, January 7.

Alvarado, Christine, and Eugene Judson. 2014. Using targeted conferences to recruit women into computer science. *Communications of the ACM* 57(3): 70–77.

Alvarado, Christine, Zachary Dodds, and Ran Libeskind-Hadas. 2012. Increasing women's participation in computing at Harvey Mudd College. *ACM Inroads* 3(4): 55–64.

Anonymous. 1997. Olin College joins list of engineering schools. *Civil Engineering*, August 23.

Anonymous. 2004. Smith college graduates country's first all-female class of engineers. *Black Issues in Higher Education* June 3: 12–13.

Aspray, William. 2016. *Participation in computing: The National Science Foundation's expansionary programs*. London: Springer.

---

[57] Miller (2015) discusses some of the schools that have succeeded in attracting women into their computer science program. Propsner (2013) provides links to a number of women's colleges that have programs to attract women to STEM disciplines and careers.

[58] See Aspray (2016) for a discussion of the work of Sheryl Burstahler and Richard Ladner at the University of Washington concerning people with disabilities.

[59] On TAM versus computer science, see Barker and Garvin-Doxis (2004), Barker et al. (2002), and Barker et al. (2005). On the technology for community course, see Jessup et al. (2005); Barker and Jessup (2006); and Jessup and Sumner (2005)

[60] On computing at Wellesley, see Rowell (2001) and Shaer (2014).

[61] On computing at Bryn Mawr, particularly its use of robots for instructional purposes, see Blank (2006), Kumar and Meeden (1998a, b), Blank et al. (2003), Summet et al. (2009), Blank et al. (2006), and Blank et al. (2004).

Barker, Lecia J., and Kathy Garvin-Doxas. 2004. Making visible the behaviors that influence learning environment: A qualitative exploration of computer science classrooms. *Computer Science Education* 14(2): 119–145.

Barker, Lecia, Kathy Garvin-Doxis, and Michelle Jackson. 2002. Defensive climate in the computer science classroom. *ACM SIGCSE Bulletin* 34(1): 43–47.

Barker, Lecia J., Kathy Garvin-Doxas, and Eric Roberts. 2005. What can computer science learn from a fine arts approach to teaching? *ACM SIGCSE Bulletin* 37(1): 421–425.

Barker, Lecia J., and Elizabeth R. Jessup. 2006. *Student and faculty choices that widen the experience gap.* https://www.cs.colorado.edu/~jessup/SUBPAGES/PS/gender.pdf. Accessed 8 Dec 2015.

Blank, Douglas. 2006. Robots make computer science personal. *Communications of the ACM* 49(12): 25–27.

Blank, Douglas, Lisa Meeden, and Deepak Kumar. 2003. Python robotics: An environment for exploring robotics beyond LEGOs. *ACM SIGCSE Bulletin* 35(1): 317–321.

Blank, Douglas, Deepak Kumar, Lisa Meeden, and Holly Yanco. 2004. Pyro: A python-based versatile programming environment for teaching robotics. *Journal on Educational Resources in Computing* 3(4): 1–15.

Blank, Douglas, Deepak Kumar, Lisa Meeden, and Holly Yanco. 2006. The pyro toolkit for AI and robotics. *AI Magazine* 27(1): 39–50.

Deitz, Roger A. 2001. Smith College: New engineering program the first at a women's college. *The Hispanic Outlook in Higher Education*, February 26: 18–20.

Dodds, Zachary, Christine Alarado, Geoff Kuenning, and Ran Libeskind-Hadas. 2007. Breadth-first CS1 for scientists. *ACM SIGCSE Bulletin* 39(3): 23–27.

Dodds, Zachary, Ran Libeskind-Hadas, Christine Alarado, and Geoff Kuenning. 2008. Evaluating a breadth-first CS1 for scientists. *ACM SIGCSE Bulletin* 40(1): 266–270.

Downey, Allen B., and Lynn Andrea Stein. 2006. Designing a small-footprint curriculum in computer science. *36th Annual Frontiers in Education Conference. October 28–31.* San Diego, 21–26.

DuBow Wendy, Elizabeth Litzler, Maureen Biggers, and Mike Erlinger. 2012. Implementing evidence-based practices makes a difference in female undergraduate enrollments. *ACM SIGCSE'12*, 479–480.

Ellis, Glenn W., Alan N. Rudnitsky, and Mary A. Moriarty. 2010. Using knowledge building to support deep learning, collaboration and innovation in engineering education. *40th ASEE/IEEE Frontiers in Education Conference, October 27–30.* Washington, DC, T2J1–T2J5.

Emmett, A. 1992. A women's institute of technology? *Technology Review* 95(3): 3.

Erlinger, Michael. 2015. Oral history interview conducted by William Aspray. Charles Babbage Institute Oral History Collection, October 7.

Farrell, Elizabeth F. 2004. Smith College's first engineers feel like rock stars. *The Chronicle of Higher Education*, May 28, 3 pp.

Fisher, Allan, and Jane Margolis. 2002. Unlocking the clubhouse: The Carnegie Mellon experience. *ACM Inroads* 34(2): 79–83.

Florman, Samuel C. 1978. Engineering and the female mind. *Harper's*, February 1: 57–63.

Fox, Mary Frank. 2000. Organizational environments and doctoral degrees awarded to women in science and engineering departments. *Women's Studies Quarterly* 28(1–2): 47–61.

Fox, Mary Frank. 1998. Women in science and engineering: Theory, practice, and policy in programs. *Signs* 24(Autumn, 1998, 1): 201–223.

Fox, Mary Frank, Gerhard Sonnert, and Irina Nikiforova. 2009. Successful programs for undergraduate women in science and engineering: Adapting versus adopting the institutional environment. *Research in Higher Education* 50(4): 333–353.

Fox, Mary Frank, Gerhard Sonnert, and Irina Nikiforova. 2011. Programs for undergraduate women in science and engineering: Issues, problems, and solutions. *Gender and Society* 25(5): 589–615.

Frieze, Carol, and Lenore Blum. 2002. Building an effective computer science student organization: The Carnegie Mellon Women@SCS action plan. *ACM SIGCSE Bulletin* 34(2): 74–78.

Glazer-Raymo, Judith (ed.). 2008. *Unfinished agendas: New and continuing gender challenges in higher education*. Baltimore: Johns Hopkins University Press.

Grasso, Domenico. 2004. Engineering and the human spirit. *American Scientist* 92(3): 206–208.

Hawthorn, Paula. 2015. Oral history interview by William Aspray. Charles Babbage Institute Oral History Collection, October 1.

Hoffmann, Leah. 2012. What women want. An interview with Maria Klawe. *Communications of the ACM* 55(9): 120-ff.

House, Megan L. 2011. Engineering program moves into accreditation – Sweet Briar College celebrates recently acquired accreditation status. *Lynchburg Business*, November 6.

Humphreys, Sheila. 2015. Oral history interview by William Aspray. Charles Babbage Institute Oral History Collection, September 28.

Humphreys, Sheila, and Ellen Spertus. 2002. Leveraging an alternative source of computer scientists: Reentry programs. *ACM Inroads* 34(2): 53–56.

Jessup, Elizabeth, and Tamara Sumner. 2005. Design-based learning and the participation of women in IT. *Frontiers: A Journal of Women's Studies* 26(1): 141–147.

Jessup, Elizabeth, Tamara Sumner, and Lecia Barker. 2005. Report from the trenches: Implementing curriculum to promote the participation of women in computer science. *Journal of Women and Minorities in Science and Engineering* 11(3): 273–294.

Kumar, Deepak, and Lisa Meeden. 1998a. A robot laboratory for teaching artificial intelligence. *ACM SIGCSE Bulletin* 20(1): 341–344.

Kumar, Depak, and Lisa Meeden. 1998b. Robots in the undergraduate curriculum. *The Journal of Computing in Small Colleges* 13(5): 105–112.

Loftus, Margaret. 2003. A new era. *ASEE Prism* 12(8): 26–29.

Margolis, Jane, and Allan Fisher. 2002. *Unlocking the clubhouse: Women in computing*. Cambridge, MA: MIT Press.

Margolis, Jane, Allan Fisher, and Faye Miller. 2000. The anatomy of interest: Women in undergraduate computer science. *Women's Studies Quarterly* 28(1–2): 104–127.

Miller, Claire Cain. 2015. Making computer science more inviting: A look at what works. *New York Times*, May 21. http://www.nytimes.com/2015/05/22/upshot/making-computer-science-more-inviting-a-look-at-what-works.html. Accessed 11 Jan 2016.

National Research Council. 2000. *How People Learn: Brain, Mind, Experience and School*. Washington, DC: National Academies Press.

O'Rourke, Joseph. 2015. Oral history interview by William Aspray. Charles Babbage Institute Oral History Collection, July 6.

Pfabe, M., and N. Easwar. 1999. The Picker Engineering Program at Smith College: Building a new educational paradigm and bridging the gender gap. *American Journal of Physics* 67: 849.

Propsner, Diane. 2013. *Why first-year STEM girls attend women's colleges*, Huffington Post, August 29, http://www.huffingtonpost.com/dianepropsner/girls-in-stem_b_3810383.html.

Rosser, Sue V. 2001. Will EC 2000 make engineering more female friendly? *Women's Studies Quarterly* XXIX (3/4): Fall/Winter, 164–186.

Rowell, Marjorie. 2001. Women & technology: How Wellesley College recruits, trains and retains student staff. *ACM SIGUCCS'01*, 169–171.

Rubin, Debra K. 2006. Olin College trains new engineers by going its own way. *Engineering News-Record*, October 30.

Sanoff, Alvin P. 2000. Creating a masterpiece at Olin College. *ASEE Prism* 10(1): 20.

Shaer, Orit. 2014. WHCI Lab, Wellesley college. *ACM Interactions* May–June: 14–17.

Shellenbarger, Sue. 2013. How one college boosted female STEM graduates – Fast. *Wall Street Journal*, November 26. http://blogs.wsj.com/atwork/2013/11/26/how-one-college-boosted-female-stem-graduates-fast/tab/print/. Accessed 27 Apr 2015.

Shipp, Stephanie, Nyema Mitchell, and Bhavya Lal. 2009. *Portfolio evaluation of the national science foundation's grants program for the Development-level Reform (DLR) of undergraduate engineering education*. Science and Technology Policy Institute, Document D-3724, Washington, DC: Institute for Defense Analysis.

Simons, Barbara. 2015. Oral history interview by William Aspray. Charles Babbage Institute Oral History Collection, September 29.

Stage, Frances K., and Steven Hubbard. 2008. Developing women scientists: Baccalaureate origins of recent mathematics and science doctorates. In *Unfinished agendas: New and continuing gender challenges in higher education*, ed. Judith Glazer-Raymo, 112–141. Baltimore: Johns Hopkins University Press.

Stein, Lynn Andrea. 2015. Oral history interview by William Aspray. Charles Babbage Institute Oral History Collection, May 22.

Stolk, Jonathan, and Sarah Spence. 2003. Olin College: It's alive. *33rd ASEE/IEEE Frontiers in Education Conference. November 5–8,* F3H-1.

Summet, Jay, Deepak Kumar, Keith O'Hara, Daniel Walker, Lijun Ni, Doug Blank, and Tucker Balch. 2009. Personalizing CS1 with robots. *ACM SIGCSE Bulletin* 41(1): 433–437.

Thys, Fred. 2014. Why women don't study engineering – And what 1 mass. College is doing about it. *WBUR*, May 27.

Voss, Susan. 2015. Oral history interview by William Aspray. Charles Babbage Institute Oral History Collection, May 28.

Voss, Susan E., and Glenn W. Ellis. 2002. Applying learning-centered pedagogy to an engineering circuit-theory class at Smith college. *32nd ASEE/IEEE Frontiers in Education Conference. November 6–9,* F2F-1-F2F-6.

Index

© Springer International Publishing Switzerland 2016
W. Aspray, *Women and Underrepresented Minorities in Computing*,
History of Computing, DOI 10.1007/978-3-319-24811-0

Printed in the United States
By Bookmasters